D0960938

THE HUMAN COSMOS

ALSO BY JO MARCHANT

Decoding the Heavens: Solving the Mystery of the World's First Computer

The Shadow King: The Bizarre Afterlife of King Tut's Mummy

Cure: A Journey into the Science of Mind over Body

THE HUMAN

HUMAN COSMOS

CIVILIZATION AND THE STARS

JO MARCHANT

DUTTON

DUTTON

An imprint of Penguin Random House LLC
penguinrandomhouse.com

Library of Congress Cataloging-in-Publication Data

Names: Marchant, Jo, author.
Title: The human cosmos : civilization and the stars / Jo Marchant.
Description: New York : Dutton, Penguin Random House LLC, [2020] |
Includes bibliographical references and index.
Identifiers: LCCN 2020004943 (print) | LCCN 2020004944 (ebook) |
ISBN 9780593183014 (hardcover) | ISBN 9780593183021 (ebook)
Subjects: LCSH: Cosmology--History.
Classification: LCC QB981 .M218 2020 (print) |
LCC QB981 (ebook) | DDC 523.1--dc23
LC record available at https://lccn.loc.gov/2020004943
LC ebook record available at https://lccn.loc.gov/2020004944

Printed in the United States of America
10 9 8 7 6 5 4 3 2 1

Book design by Lorie Pagnozzi

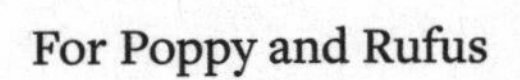

For Poppy and Rufus

CONTENTS

Prologue

ALMOST FOURTEEN BILLION YEARS AGO, EVERYTHING BURST out of nothing. Our universe pricked into being as an unimaginably hot, dense point, then almost instantaneously exploded outward, the very fabric of space expanding faster than the speed of light, until all of existence was roughly the size of a grapefruit. After that, the universe continued to expand and cool, and the first matter formed. Within the first second, a dense soup of particles—neutrons, protons, electrons, photons, neutrinos—jostled in a smashing, searing heat that scattered light like fog.

By the time it was about 380,000 years old, this cosmic bubble had expanded to tens of millions of light-years across and cooled to a few thousand degrees, mild enough for atoms to hold together, and for the first time the universe became transparent to light. There was an initial flash of illumination, then darkness fell. It took several hundred million years for the attractive force of gravity to work on subtle density variations, inexorably collapsing clumps of gas to form the first stars and galaxies, and one by one, the celestial lights switched on.

Most guides to cosmology tell some version of this sequence of events. Mysteries remain: Was this Big Bang really the start of everything, or is our universe just one inflating bubble in a much larger multiverse? What is the epic force that still pushes space apart? Will it keep expanding forever, or eventually collapse again in a Big Crunch? But the general nature and story of the universe is agreed. Reality has been revealed as a vast and sophisticated machine, composed of physical particles and forces governed by mathematical equations and laws.

This book tells a different story. The scientific account of the universe is a pinnacle of our modern civilization, a vision so powerful that its rivals have been all but obliterated. Cosmology—the study of the cosmos—once described the broad philosophical and spiritual endeavor to make sense of existence, to ask who we are, where we are, and why we're here. It is now a branch of mathematical astronomy. So what happened to those bigger questions? Is there nothing else about the universe we need to know?

Instead of detailing the latest astronomical developments, this is a guide to the long history of knowledge that people have gleaned from the stars. It's about what their view of the cosmos told them of the nature of reality and the meaning of life; about the gods and souls, myths and magical beasts, palaces and celestial spheres that we've discarded; about how the scientific view came to dominate; and how in turn that journey still shapes who we are today. It's a tale about people—of priests, explorers, revolutionaries and kings—and it starts not with the Big Bang, nor even with the birth of science, but with the very first humans who looked to the stars, and the answers they found in the sky.

✵ ✵ ✵

Why bother about the celestial beliefs of the past? Archaeologists and historians often don't. We know that science was built on attempts to understand the heavens, but this is rarely a focus for scholars tracing human progress more generally. I think this has created a huge blind spot in our understanding of where we came from. In fact, the patterns people see in the sky have always governed how they live on Earth, shaping ideas about time and place; power and truth; life and death.

We see this in the ancient past: with the eclipse-obsessed Babylonians; the Egyptian pharaohs who built pyramids to guide their souls to the stars; the Roman emperors who fought under the banner of the

Sun. Ideas about the cosmos have shaped the modern world too. These influences are still deeply engrained in our society, even if we've forgotten their origins—in our parliaments, churches, galleries, clocks and maps. Beliefs about the Sun, Moon and stars played a central role in the birth of Christianity, and in Europe's exploration and domination of the planet. They guided the rebellious lawmakers who founded the principles of democracy and human rights, the economists who developed the frameworks on which capitalism depends, and even the painters who produced the first abstract art.

Today, as light pollution envelops our planet, the stars are almost gone. Instead of thousands being visible on a dark night, in today's cities we see only a few dozen (and astronomers fear these will soon be vastly outnumbered by artificial satellites). Most people in the United States and Europe can no longer see the Milky Way at all. It is a catastrophic erosion of natural heritage: the obliteration of our connection with our galaxy and the wider universe. There has been no major outcry. Most people shrug their shoulders, glued to their phones, unconcerned by the loss of a view treated as fundamental by every other human culture in history.

Yet we're still trying to work out our place in the cosmos. Science has been wildly successful: today's five-year-olds know more about the history, composition and nature of the physical universe than early cultures managed to glean in thousands of years. But it has also dissolved much of the meaning that those cultures found in life. Personal experience has been swept from our understanding of reality, replaced by the abstract, mathematical grid of space-time. Earth has been knocked from the center of existence to the suburbs; life reframed as a random accident; and God dismissed altogether, now that everything can be explained by physical laws. Far from having a meaningful role in the cosmic order, we're "chemical scum," as physicist Stephen Hawking put it, on the surface of a medium-sized planet orbiting an unremarkable star.

Critics have fought this mechanistic view of humanity for centuries, often rejecting science wholesale in the process. But now even some high-profile scientists are voicing concerns that until very recently were taboo. They are suggesting that perhaps physical matter isn't all that the universe is, all that we are. Perhaps science is only seeing half of the picture. We can explain stars and galaxies, but what about minds? What about consciousness itself? It's shaping up to be an epic fight that just might transform the entire Western worldview.

With the battle lines drawn, I think we need a shift in perspective, an overview. Here, then, is a book about the cosmos, not a scientific guide but a human one. Rather than give an exhaustive account, I've chosen twelve moments—stepping-stones, if you like—that tell us something about how people through history have seen the sky. In particular, these twelve stories follow the rise of the Western material universe: tracing a path from humanity's earliest expressions in cave paintings and stone circles; through the birth of great traditions such as Christianity, democracy and science; to the hunt for alien life and our recent flights into actual—and virtual—space.

It's a journey that helps to explain who we are today, and perhaps can also guide a future course. It can be hard to see the limits of something when you're embedded in it. I hope that zooming out to survey the deep history of human beliefs about the cosmos might help us to probe the edges of our own worldview and perhaps look beyond: How did we become passive machines in a pointless universe? How have those beliefs shaped how we live? And where might we go from here?

MYTH

THERE'S A CURIOUS PATTERN OF DOTS THAT RECURS IN ART around the planet and throughout history. The number varies, but commonly it's a close-knit group of six circular spots, distinctively arranged in lines of four and two. This motif is seen in far-flung communities, from holes pierced into the gourd rattle of a Navajo tribe to a painting on a Siberian shaman's drum. It even appears in the logo of the Japanese car manufacturer Subaru.

In all of these cases, the dots represent one of the most characteristic features of the night sky: the star cluster Pleiades. This clutch of six or seven stars (the exact number depends on viewing conditions) appears close to the Sun's annual path through the sky, and features in multiple myths and legends: in Cherokee myth, these stars are lost children; the Vikings saw them as the goddess Freyja's hens. They are also a distinctive part of the constellation Taurus. The Pleiades sit just above the shoulder of this celestial bull, with its thrusting horns, prominent eye—the red giant Aldebaran—and another star cluster, the Hyades, splashed in a "V" across its face.

The frequent appearance of this six-spot pattern demonstrates the importance of the Pleiades in societies around the world, as well as the shared human desire to capture aspects of the starry sky in art. But there is more to this story—another example of these dots that seems, frankly, impossible. The cave of Lascaux in southwestern France is famous for its wealth of Paleolithic art: paintings and engravings of animals, thought to be twenty thousand years old, dating from the dawn of humanity. Scholars have argued over their meaning for decades. Meanwhile, barely noticed on the ceiling of its grand entrance hall, are six plain spots that match the Pleiades perfectly. Neatly painted in red ocher, they float above the shoulder of a majestic aurochs bull.

At 17 feet long, "Bull No. 18" is the largest and perhaps most recognizable painting in the entire cave. Its striking similarity to the modern Taurus—it even has V-shaped spots on its cheek—has been known for years. Yet it goes unmentioned in guidebooks and is rarely discussed by mainstream archaeologists. Taurus is one of the earliest constellations to be described: it can be traced back through written sources nearly three thousand years to Babylonian priest astronomers who saw the Pleiades as a bristle on the back of a heavenly bull. But a star map invented by the supposedly primitive hunter-gatherers of Lascaux? The idea was not so much rejected as not talked about at all.

In the last few years, however, experts in fields such as anthropology, mythology and astronomy have begun to argue for a radical reassessment of our Paleolithic ancestors' skills, and the lasting influence of the stories they told. So in this history of humanity's relationship with the stars, let's start with the mystery of Bull No. 18. We'll explore whether the artists of Lascaux could really have painted constellations, and ask why they might have cared so much about the sky. It's a journey that takes us to the heart of what the universe meant to the very first humans who had the ability to imagine, remember, explain and represent. The cosmos they created still shapes our lives today.

✡ ✡ ✡

On September 12, 1940, seventeen-year-old Marcel Ravidat, an apprentice mechanic, walked with three friends into the hills near his village of Montignac in southwest France. According to village legend, there were caves beneath the hills—during the wave of executions that followed the French Revolution, the Abbé Labrousse, owner of the nearby manor, supposedly hid in one—and Ravidat wondered if they might hold treasure. A few days before, he had started to unblock a promising hole in the ground. Now, armed with a knife and a makeshift lantern, he planned to finish the job.

The boys' target was a basin-shaped depression in the ground, surrounded by pine trees and junipers and full of brambles. At the bottom was a small opening that led to a narrow, near-vertical shaft. The boys cleared the thorns—and the remains of a donkey—and dug with their hands to widen the hole to about a foot across. They dropped down stones, and were surprised by how long they rolled and the resonance of the sound. Those brambles had been hiding something big.

Ravidat, the oldest and strongest of the group, dived in headfirst and wriggled several yards through the earth before landing on a conical pile of clay and stones. He lit his lamp, which he'd made from a grease pump and a coil of string, but almost immediately lost his balance and slipped all the way to the bottom. He found himself in a large hall, about 60 feet long, and called for his friends to follow.

They crossed the limestone cavern in near darkness, dodging shallow pools of water on the floor, until they reached a narrow corridor with a high, arched ceiling, like a cathedral vault. Only here did Ravidat raise his lamp, and the boys found their treasure. Covering the white walls was an explosion of life: images from the birth of our species, pulled back into view for the first time in twenty thousand years.

First, they noticed colored lines and strange, geometric signs. Then, moving the lamp around, they saw the animals. There were horses everywhere, golden with black manes, as well as black-and-red bulls, ibexes and a bellowing, antlered stag. Herds climbed the walls and tumbled across the ceiling, some defined and multicolored, others ghostly, as if falling through fog. The boys didn't understand the full significance of what they had found, but they knew it was special, and they celebrated with leaps and cries in the trembling light.

Lascaux Cave (named for that nearby manor) now ranks as one of the most spectacular archaeological discoveries in history. It is just one of hundreds of caves in southern France and northern Spain decorated between 37,000 and 11,000 years ago by anatomically modern humans who first migrated into Europe from Africa around 45,000 years ago, during the last Ice Age. It's a period called the Upper Paleolithic, named for the stone tools in use at the time, and it seems to have hosted an explosion in human creativity. Rock art from around this time is known elsewhere, too—in Indonesia and Australia, for example. But thanks to the complexity, exquisite preservation and sheer volume of its paintings and engravings—nearly two thousand of them—Lascaux is one of the finest examples.

The artists at Lascaux used plant-based brushes or swabs of hair, and a palette of iron and manganese minerals, kaolin clay and charcoal sticks to cover corridors and chambers reaching over 300 feet into the rock. Their creations provide a rare and hauntingly beautiful insight into the prehistoric human mind. Who were these early people? What did they care about, and what triggered them to create art? What was it, in effect, that made them human?

In the decades since the boys' discovery, scholars have come up with a rich parade of answers to these questions. An early idea was that the mysterious figures were simply decoration, "art for art's sake" without any special meaning. Another suggestion was that the animals represented different clans, and that the paintings showed battles and alliances between them. Some experts thought that the paintings were intended as magical spells, to boost the success of hunting expeditions or ward off evil spirits. In the 1960s, scholars took a statistical approach, recording how different types of figures were distributed in the caves and building theories around the patterns they saw—for example, that the horses and bison symbolize male versus female identity.

Then there was Norbert Aujoulat, who perhaps came to know the paintings more intimately than anyone else. A cave enthusiast, he described himself as "an underground man." He would disappear for days at a time on solitary excursions into the French mountains, and helped to discover dozens of subterranean chambers. But he never forgot the first time he saw Lascaux, one winter afternoon in 1970. Since its discovery, the site had opened to the public and closed again: the breath exhaled by thousands of visitors per day, and the germs they tramped in, were damaging the precious paintings. Aujoulat, a twenty-four-year-old local student, joined a private tour guided by Jacques Marsal, one of the four friends led by Ravidat who had discovered the cave three decades before.

To reach the paintings, Marsal led them down a slope through a series of stone-lined entrance halls and doors, built for security, which made Aujoulat feel as if they were approaching the sacred inner space of a temple. The last door was made of heavy bronze and decorated with polished stones. Aujoulat spent only half an hour exploring the treasures beyond that door, but it was enough to set the course of his life. He was bewitched by the overwhelming sense of human presence inside the cave, powerful enough to stretch across so many thousands of years, and he set his sights on understanding how and why the paintings were created.

It was nearly two decades before Aujoulat was able to fulfill his dream. In 1988, as head of the French culture ministry's Department of Parietal Art, he began a monumental decade-long study of Lascaux Cave, from the great bulls circling the ceiling of the entrance cavern to the dense, entangled engravings in a smaller chamber called the Apse. Whereas other scholars had focused on the art, Aujoulat approached Lascaux as a natural scientist, studying every aspect of the cave from the geology of the limestone to the biology of the animals on the walls. He came to the conclusion that everyone else had missed a crucial dimension: time.

When he studied overlapping paintings where horses, aurochs and stags appeared together, he found that in every case the horses were painted first, then the aurochs, and then the stags. What's more, the animals were always shown with features corresponding to specific times of year: the horses with bulky coats and long tails corresponding to the end of winter; the aurochs during the summer; and the stags with prominent antlers, characteristic of autumn. For each species, that was their mating season.

Aujoulat described his findings in a 2005 book called *Lascaux: Movement, Space and Time*. By showing the fertility cycles of important animals, he argued, the cave was a spiritual sanctuary, intended to symbolize creation and the eternal rhythm of life. The cycle of creation represented by the paintings wasn't just an earthly one, however, relating to animals and the weather. It extended to the entire cosmos.

The annual re-creation of life taking place in the Paleolithic world was mirrored, of course, by the cycles of the stars: each season is marked by the passage of the Sun as well as the appearance of characteristic constellations in the night sky. Aujoulat believed this was central to the artists' vision; they were showing, he concluded, how biological and cosmic time were entwined. He compared the cave, with its overhanging walls and paintings that crossed the ceiling, to the "celestial vault," and suggested that the animals weren't being shown on the ground, but in the sky.

That could explain why the animals often appear to be floating—painted at all angles, without any ground line, sometimes even with hanging hooves. If Aujoulat is right, Lascaux Cave is as much about cosmology as it is about biology: rather than copying their immediate surroundings, the artists were synthesizing all of the changes—on the Earth and in the sky—that defined their existence. It was an ode, if you like, to their universe, representing humanity's first ideas about the nature of the cosmos and the origins of life.

Aujoulat was at the heart of the French academic establishment, and his work has been hugely influential. Even so, his ideas about the sky are rarely discussed; without direct evidence, archaeologists find it easier to accept the paintings as a celebration of nature than as a vision of the sky. There are some scholars, though, who think he didn't go far enough, that rather than simply imagining animals in the sky, the artists of Lascaux were painting maps of the stars.

✡ ✡ ✡

In 1921, a French prehistorian named Marcel Baudouin came across a fossilized sponge that was shaped like a penis. The fossil, found in Beynes in north-central France, had a vibrant red patina, which some ancient artist had chipped off in places to create a series of yellow, hoof-shaped dots. "It is the first time I have seen work like this!" Baudouin wrote in excitement. In a paper called "The Great Bear and the Phallus of Heaven," he argued that the pattern matches the northern constellation Ursa Major (the Great Bear), even down to brighter stars being represented by larger dots.

It wasn't possible to date the dots, but he concluded that they were carved in Paleolithic or Neolithic times. Because of the Earth's rotation, the stars of the northern hemisphere appear to circle around a stationary point in the sky directly above the North Pole (known today as the north celestial pole). Baudouin suggested that the fossil was intended to show this pole as a celestial penis, and that the carved dots represented nearby Ursa Major rotating around its shaft.

He was one of the first to see stars in prehistoric art; throughout the 1920s and '30s, several scholars, including Baudouin, also reported constellations in concave depressions called cup marks, dug out of stone monuments and cave walls in locations from southern France to Scandinavia. Their claims were impossible to prove and are now largely forgotten, but decades later, the U.S. archaeologist Alexander Marshack popularized the idea of Paleolithic astronomy in his influential 1972 book *The Roots of Civilization*.

Marshack used a microscope to examine markings on bone fragments made by people in the Upper Paleolithic. One of the first he studied was a 30,000-year-old piece of bone from the Blanchard rock shelter in the Dordogne region of France. It is engraved on one side with sixty-nine disk- or crescent-shaped pits arranged in a snaky line. Marshack showed that the pits were created using twenty-four different types of stroke, suggesting they were carved in groups, on twenty-four different occasions. Rather than simply doodling, someone was keeping track of something; Marshack thought it was the changing phases of the Moon. He surveyed similar patterns on a range of bones, stones and antlers, and argued that the people of the Paleolithic were routinely tracking the sky, using lunar calendars to mark the passing of time. With Marshack's ideas about Ice Age astronomy widely taken seriously, if not proven, it wasn't long before researchers started to look again for prehistoric star constellations, in particular in the chambers of Lascaux.

German astronomer Michael Rappenglück first heard about the idea as a student at the University of Munich in 1984, when he attended a lecture suggesting that Lascaux's paintings might contain star maps. "I was fascinated," he says. Now director of the Adult Education Centre and Observatory in Gilching, Germany, and a former president of the European Society for Astronomy in Culture, Rappenglück has been investigating the theory ever since.

One of the scenes he studied is Bull No. 18. Over long periods of time, constellations shift in the sky because of a wobble in the axis of Earth's rotation; individual stars also follow their own trajectories. So to test how well Bull No. 18 really matches Taurus and the Pleiades, Rappenglück calculated how these stars would have looked around twenty thousand years ago and compared this with measurements taken from photographs of the cave wall. He found that when the bull was created, the Pleiades were slightly higher above the bull's back and that Aldebaran (the bull's eye) was more clearly framed by the Hyades—an even closer match to the painting than they are today.

He's convinced this isn't a coincidence, arguing that our constellation Taurus (which once represented an entire bull, but lost its hindquarters over the centuries to make room for a new constellation, Aries the ram) has its origins in a far older star grouping—let's call it Aurochs—inspired by the giant bulls that the people of the Ice Age hunted for food.

Rappenglück supports his ideas with evidence from anthropology. Societies throughout history have used the Pleiades as a calendar, he points out. Stars circle around the north and south celestial poles each night, but our orbit around the Sun means they follow an annual cycle, too: different stars and constellations "rise" or "set" (first become visible above the horizon at dawn or disappear from view at dusk) at particular times of year. As a distinctive star cluster close to the ecliptic (the Sun's path through the sky), the Pleiades mark the seasons particularly well.

Today, farming communities from Lithuania to Mali to the Andes still mark their agricultural year according to the visibility of the Pleiades. Native American peoples such as the Blackfoot traditionally synchronize their lives with these stars and the life cycle of the bison: when the Pleiades set, it is time to hunt. The Teton Sioux and Cheyenne even name some months after the bison life cycle: November is "the moon of the fertilization of the buffalo cows," while January is "the moon when the coat of the young buffaloes takes on color."

Rappenglück suggests that the artists of Lascaux could have developed a star calendar, with the Pleiades similarly marking key moments in the life cycle of the aurochs bull. He calculates that at the time Bull No. 18 was painted, the Pleiades would have appeared just before sunrise in mid-October, reached their highest point in the sky at the start of spring, and disappeared at the end of August. That means the disappearance and then reappearance of the Pleiades would have defined the mating season of the aurochs, he says, which lasted between August and October. From there it was perhaps a natural step to associate the stars around the Pleiades with the image of a bull. It would have dominated the spring sky to the west of the hilltops that surround Lascaux Cave: a giant, celestial creature with a twinkling red eye and glittering hairs on its back, ready to toss the Milky Way with its horns.

Rappenglück sees possible astronomical associations in other caves too. Another aurochs, nearly four thousand years older than Bull No. 18, in La Tête du Lion cave in the Ardèche, France, has a group of seven dots on its body that he thinks might represent the Pleiades. And in El Castillo cave in Santander, Spain, there's a mysterious group of seven ocher disks dating from 11,000 to 12,000 BC, arranged in a downward-pointing curve and close to a striking frieze of red hand stencils.

After calculating how the sky would have looked at the time, Rappenglück concluded that the dots are a close match for a constellation called the Northern Crown, and suggests that the nearby strip of hands might represent the Milky Way. In 12,000 BC, the Northern Crown never set, but—as the Paleolithic equivalent of Polaris, our polestar—rotated around the northern celestial pole, so it would have been important for marking the direction north. Like the Pleiades, the Northern Crown also features prominently in mythology. A Celtic myth, for example, describes it as the star goddess Arianrhod's home, an icy castle set on a magical, rotating island in the northern sky. Might elements of the tale date from the Paleolithic, when these stars really did trace a circle in the heavens?

Skeptics insist that Rappenglück's ideas can't ever be proved. There are just too many possible combinations—too many sets of dots in European caves and too many stars in the sky. But others argue that the various features of Bull No. 18, in particular, would be an extraordinary coincidence. And Rappenglück is not the only one linking the caves of the Paleolithic with the stories we tell about the stars.

✡ ✡ ✡

It's a long-standing mystery why similar myths often exist in apparently unrelated cultures in different places. Take the story of the Cosmic Hunt, in which an animal is chased into the sky and transformed into a star constellation. Variants of this tale—featuring different stars, hunters and prey—are found all over the world.

In one Greek version of the myth, Zeus tricks the princess Callisto, companion of the goddess Artemis, into giving up her virginity, and she gives birth to a son, Arcas. An incensed Artemis turns Callisto into a bear. Arcas later grows up to be a hunter and almost kills his mother with a spear, but Zeus intervenes, turning Callisto into the constellation Ursa Major and putting Arcas next to her as Ursa Minor (the Little Bear).

Meanwhile, the Iroquois of the northeastern United States tell of three hunters who wound a bear in a forest; its blood stains the autumn leaves. The hunters then follow the bear into the sky and together they become Ursa Major. Among the Siberian Chukchi, the constellation Orion is a hunter who chases a reindeer, Cassiopeia, whereas for the neighboring Finno-Ugric tribes, the pursued animal is an elk.

French archaeologist and statistician Julien d'Huy probes the origins of such stories using the principles behind phylogenetics, a technique developed to glean evolutionary relationships between species by comparing their DNA sequences. Biologists use computer software to analyze similarities in the DNA and construct family trees showing the most likely relationships between species. D'Huy does a similar thing for myths.

Instead of studying DNA, d'Huy analyzed forty-seven versions of the Cosmic Hunt from around the world, splitting them into ninety-three individual components, or "mythemes," such as "the animal is an herbivore" or "a god transforms the animal into a constellation." For each myth, he coded the presence (1) or absence (0) of each mytheme to give a string of 0s and 1s, then used phylogenetic software to compare them and construct the most likely family tree. His results, published in 2016, suggest that the myth originated in northern Eurasia. One branch then spread to western Europe and another reached North America when humans migrated across the Bering

Strait, which was once spanned by a land bridge that connected the eastern tip of Russia with Alaska. That means, he says, that the story must date from before about fifteen thousand years ago, after which the land bridge became submerged.

The original Paleolithic version of the Cosmic Hunt, concludes d'Huy, most likely involves a lone hunter pursuing an elk. The hunt moves into the sky, but before the animal can be killed, it transforms into the Big Dipper or the Plow (the tail and flank of Ursa Major, respectively). Elk were dominant mammals in the forests of northern Eurasia during Paleolithic times, crucial for hunting, and there's evidence that they were important culturally too. A 2017 study of hundreds of animal-tooth pendants found in Estonia, for example, found that elk was the most common mammal represented in the Mesolithic and Neolithic periods (8900–1800 BC), before gradually switching to bears. As the story of the Cosmic Hunt moved around the planet and through history, different peoples would have adapted the tale to fit the animals and constellations most important to them.

Other tales analyzed by d'Huy seem to date back even earlier, spreading out of Africa with the first waves of human migration more than forty thousand years ago. He has compiled a core of "protomyths" that he thinks early humans brought with them as they migrated north and east. Not all of these involve stars. There are dragons: giant, horned serpents that guard water sources and can fly, form rainbows, and produce rain and thunderstorms. But they also include the Pleiades, often as a woman or group of women, set against Orion as man, and the idea of the Milky Way as a river, or a road traveled by the dead.

In other words, the star myths we tell today are not just stories. They're cultural memories passed through generations for thousands of years, that sometimes do reach back to the Paleolithic. D'Huy calls them a "glimpse into the mental universe of our ancestors." That glimpse doesn't directly link the Pleiades to an aurochs bull. But just like the paintings of Lascaux, it overwhelmingly tells of living beings imprinted on the sky.

✡ ✡ ✡

For the native Chumash people of southern California, the universe consisted of three disk-like worlds, floating in a great abyss. At the bottom was the Lower World, inhabited by deformed, malevolent beings. The Middle World, where humans lived, was supported by two giant serpents that triggered earthquakes when they moved. Above that, the Upper World was held up by a great eagle, whose wing movements caused the phases of the Moon.

This cosmos was ruled by the Sun, an old widower who lived in a quartz crystal house in the Upper World and dined on human flesh. Each day he traveled across the sky, carrying a torch and wearing only a feather band around his head. At night, he gambled against Sky Coyote (probably Polaris, the North Star) to determine the fate of the people below. Not surprisingly, the Chumash watched the Sun very carefully. But their knowledge of the Upper World didn't come from just tracking the sky. They knew about it because they had traveled there themselves.

A few centuries ago, the Chumash thrived along the south-central Californian coast, and their journeys give us one more insight into what prehistoric people like the artists of Lascaux might have thought about the heavens. That's because the Chumash lifestyle appears to have been very similar in complexity to that of Upper Paleolithic Europe. They had round grass houses, beautifully carved wooden bowls, fine baskets and plank-built sea canoes, which they used to catch swordfish weighing up to six hundred pounds. The men wore body paint and feather headdresses, the women had skirts of deer or otter skins, and they used shell beads for money.

There were perhaps fifteen thousand of them before the Spanish arrived in the eighteenth century. The soldiers who made first contact in 1769 described large towns, with roofs piled high with grilled fish. In the following decades, however, the population crashed, as the Chumash succumbed to the colonizers and their infections: typhoid, pneumonia and diphtheria.

By the beginning of the twentieth century, the Chumash culture and language had almost disappeared. Then, a linguist called John Peabody Harrington, who worked for the Smithsonian Institution, dedicated his career to tracking down elderly speakers of dying languages across North America, persuading them to share everything they could remember about their heritage.

Eccentric and obsessive, Harrington worked alone. After his death in 1961, Smithsonian curators discovered hundreds of boxes that he had stored in warehouses, garages and even chicken coops throughout the West. Mixed with Native American–made flutes and dolls, dead birds and tarantulas, dirty laundry and half-eaten sandwiches, was what came to be known as the Harrington "gold mine": photographs, sketches, notes and recordings detailing the words and beliefs of cultures that had been thought lost—including the Chumash.

A few years later, Travis Hudson, a curator at the Santa Barbara Museum of Natural History, used thousands of pages of Harrington's notes to reconstruct the most detailed account of astronomical beliefs for any hunter-gatherer community in the world. In his 1978 book *Crystals in the Sky*, Hudson concluded that the Chumash knowledge of the sky was far richer and more sophisticated than Western scholars had ever thought possible.

The Chumash elders interviewed by Harrington spoke of an Upper World filled with powerful supernatural beings. The polestar, Polaris, was Sky Coyote, father of mankind and the being around which the rest of the sky revolved. The stars Castor and Pollux (the Gemini twins) were the Sun's female cousins, while Aldebaran was another Coyote, who followed the Pleiades maidens across the sky. Orion's Belt was Bear, and the Milky Way was a ghosts' road.

The movements of these deities were intertwined with life on Earth. The Chumash knew that when the Sun rose or set at a certain location on the horizon, or when particular stars appeared in a dawn or twilight sky, certain seasonal changes were about to take place: seeds would ripen, deer would migrate, the rain would come.

The winter solstice, the point in the dead of winter when the Sun reaches its farthest point south and the days are shortest, was seen as a critical time for the cosmos. If the Sun couldn't be persuaded to return, darkness would fall and life on Earth would be snuffed out. The Chumash made careful observations to predict the solstice and, on the crucial morning, conducted rituals, often in caves, planting quartz-tipped sun sticks into the ground to "pull" the Sun back onto a northern course.

This knowledge, however, was not for everyone.

These celestial secrets were held by an elite group of astronomer priests called the *'antap*. They formed what was essentially a secret society, led by the Sun priest. They never shared their knowledge with commoners, and wielded great political influence, claiming that they were the only ones who could understand and influence the cosmic system around which Chumash life revolved.

The priests acquired their detailed astronomical knowledge from countless nightly observations, but also with the aid of a plant called *Datura*. It's a hallucinogen (part of the nightshade family) and the priests used it to go on journeys called vision quests. This allowed them to visit the Upper World, where they could contact supernatural guardians such as Coyote, predict and influence the future, and communicate with the spirits of the dead.

It's a practice called shamanism. The term comes from Siberia, where Western travelers in the seventeenth century encountered religious leaders called *saman* among Tungusic peoples, but similar practices and beliefs exist in traditional hunter-gatherer societies all around the world.

Shamans enter trance states to visit an alternate reality or spirit world. During such journeys, they meet and gain power from spirit guides, and this allows them to fulfill a range of roles such as foreseeing the future, harming enemies, controlling the weather and animals, and healing the sick. Trances are induced in different ways—sometimes by hallucinogenic plants such as *Datura* or ayahuasca; by meditation, fasting or sensory deprivation; or by rituals such as drumming or dancing.

Western anthropologists initially rejected shamanism as not even worth studying, dismissing its practitioners as either conmen or mentally ill. But the Romanian historian of religion Mircea Eliade changed that with his seminal study *Shamanism: Archaic Techniques of Ecstasy*, first published in English in 1964. Eliade surveyed the practice of shamanism throughout history, arguing that it is ubiquitous among hunter-gatherer societies from Siberia to North America to Tibet. Because these traditions are all so similar, he argued that they must descend from a common source in the Paleolithic, which spread as people migrated around the planet, just like the myths studied by Julien d'Huy. Shamanism, in other words, was humanity's first religion.

Scholars have since questioned some of Eliade's assumptions. But his work triggered a wave of popular and scientific interest in shamanism. There are now several lines of evidence suggesting that shamanic trances aren't a purely cultural (or imagined) phenomenon, but represent a universal capacity of the human brain. Neuroscientists have measured characteristic patterns of brain activity in shamans undergoing spirit journeys, which share some features with activity in the brain during hypnosis and meditation, suggesting that shamans aren't acting but really do enter a distinct, altered state of consciousness.

Meanwhile, anthropologists have documented the experiences of thousands of westerners in such trance states, usually triggered by drumming, and found that even when people have no idea what to expect, they report very similar experiences to traditional shamans. Western shamans argue that this is because the spirit worlds they visit are real, but scientists tend to see it as evidence that the human nervous system has the ability to generate specific kinds of visions and hallucinations. Both traditional shamans and westerners undergoing spirit journeys often meet and communicate with animals, or transform into an animal themselves. Another key feature is the experience of tunneling down into the ground, or flying up into space, often passing through membranes or barriers to move from one layer to another. These types of visions are commonly reflected in the cosmological beliefs of hunter-gatherer societies: a tiered cosmos, with lower, middle and upper worlds, as seen by the Chumash, is an almost universal theme. Shamans in many different communities believe they can contact the spirits of the Upper World, for example, by flying up to a specific constellation or star. It may have been altered states of consciousness, rather than simple stargazing, that helped to create humanity's first models of the universe.

In their 1998 book *The Shamans of Prehistory*, the South African rock art specialist David Lewis-Williams and the French cave expert Jean Clottes applied ideas about shamanism to Paleolithic sites such as Lascaux. Lewis-Williams had previously studied nineteenth- and twentieth-century rock art of the nomadic San people in South Africa. The San explicitly relate their art to shamanic vision quests, describing the figures as shamans in animal form, for example, or spirit guides.

Lewis-Williams followed up in 2002 with a best-selling book called *The Mind in the Cave*. All human beings have the same nervous system, he argues, and the people of the Upper Paleolithic were anatomically the same as us, so it's probable that they would have experienced the same kinds of hallucinations. In modern Western society, he points out, we tend to dismiss trance states and visions as abnormal or suspect. We value logical, rational thought. But studies of shamanism show that shifting states of consciousness exist, and are highly prized, in pretty much every traditional society on the planet. By seeing cave art only through our own literal lens, perhaps we are missing the point. Entering the deep, narrow caves of France and Spain would have been just like penetrating the nether spirit realm, so perhaps the shamans of prehistory went into the caves on vision quests, just as Chumash shamans did twenty thousand years later, and painted what they saw onto the rock walls.

The theory would help to solve several mysteries about the paintings in Lascaux and other Upper Paleolithic caves. First, it might explain the abstract geometric patterns that are common in these caves, such as dots, grids, zigzags and wavy lines. Such optical effects are commonly seen during the first stages of trance, points out Lewis-Williams (people suffering from migraines often see them too). The Tukano people of South America, who induce trances using *yajé*, a brew made from a psychotropic vine, often paint the geometric symbols they see during visions onto houses or on bark.

It would also help to explain bizarre hybrid figures seen in Paleolithic art, such as a bison-man at Chauvet Cave in southeast France, or the Sorcerer at Trois-Frères cave in the southwest, which has the ears and antlers of a stag, athletic human legs and haunches, a horse's tail and a wizard's beard. In deep trances, people often report seeing images of animals, people and monsters, and can feel as though they are blending with them.

Finally, Lewis-Williams's ideas make sense of images in which the artists incorporated features of the cave walls, as well as cases where people often touched and treated the walls: making hand stencils or finger trails, or even filling cavities with mud and piercing the mud with fingers or sticks. If caves were seen as portals to the underground spirit world, then the cave walls would have been the boundary between the two realities, a membrane through which spirits could appear. "The walls were not a meaningless support," he says. "They were part of the images."

In essence, during such spirit journeys, the physical reality of the cave became entwined with the spirit worlds that existed in the shamans' minds. Each informed the other. People would have entered the cave and painted the visions they saw, physically transforming the walls. At the same time, paintings left by previous visitors would have primed and shaped later visitors' visions. Reality was being revealed to them at the same time they were helping to create it.

Lewis-Williams focuses on caves as a metaphor for the underground realm; he doesn't talk much about the sky. But the evidence from more recent communities, at least, suggests that journeys to the Upper World were crucial too, and were also represented on cave walls. The Chumash priests regularly decorated caves with celestial features including the Sun and Moon; the Tukano painted parallel chains of dots to represent the Milky Way. Rappenglück, for one, argues that interpreting symbols in caves like Lascaux as resulting purely from hallucinations is missing something. They were part of an overall "cosmovision," he says, in which the caves represented not just the Lower World, but the cosmos as a whole.

We can't ask prehistoric shamans directly what that cosmos was like, but after studying the astronomy of the Chumash, Travis Hudson concluded that their universe was "inextricably linked to man and filled with vast sources of powers which influenced all things"; an endlessly recurring cycle of reincarnation "in which matter was neither created nor destroyed, but transformed into life or death."

The beliefs of modern-day Western shamans seem to fit that interpretation. Sandra Ingerman, a practitioner and author based in New Mexico, describes the altered states of shamanism as revealing a different view of reality, in which other living beings are seen "not as objects but as a web of life, where all of life is communicating." It's a web that includes not just animals and plants, she says, but the Sun, Moon and stars.

Meanwhile, Jo Bowlby, who qualified as a shaman among the Q'ero elders of Peru and now runs a healing practice in London, recalls her first experience with ayahuasca. At a nighttime ceremony in the Amazon rain forest, under a blanket of stars, she was offered half a mug of "putrid" drink.

At first, she was horrified to see her hands transforming at lightning speed into every type of animal foot imaginable, finishing with a lobster claw, but then she became overwhelmed by a feeling of pure ecstasy. It was everything and nothing, she says, like being in outer space. And the lesson she learned has stayed with her ever since: "You realize how huge and amazing this universe is. It's an experience of connection, of feeling part of something. We are not separated or isolated. The same energy that feeds the trees feeds you."

✡ ✡ ✡

In September 1940, Marcel Ravidat and his friends at first told no one of their startling discovery at Lascaux. The next day, September 13, 1940, they returned to the cave with better lamps and a rope, setting off at ten-minute intervals to make sure they weren't followed. After further widening the entrance, they explored every corridor, until, far into the cave, just past the densely engraved Apse, they came across a vertical shaft too deep to see down. The boys paused. Who would go first?

Again, it was Ravidat who took the plunge. Heart racing, he climbed down the rope, nervous not because he doubted his own strength but because he feared his younger friends might drop him. When his feet touched the bottom, nearly 30 feet down, he raised his lamp to the walls and saw one of the strangest scenes in all of cave art.

It features a stick man with a bird head and prominent penis—the only human figure in the cave. Often described as "the Dead Man," he lies at a 45-degree angle with his head back and his arms and fingers splayed. Bearing down on him is a bristling bison, head low, horns thrust forward, with a black spot on its shoulder and a series of loops hanging beneath its belly, as if its guts are falling out. Directly beneath the man is a bird perched on a vertical staff.

This bizarre tableau has mystified generations of scholars. But d'Huy and Rappenglück both suggest that the secret to understanding it may lie in the sky. With a slight shift in perspective, it is the man who stands vertical, looking to the heavens as the bird stick and bison follow him upward. D'Huy suggests that the scene might show the Cosmic Hunt, as hunter and beast rise into the sky to become constellations. That would explain why the bison, despite its aggressive position, doesn't appear to be charging forward, he says. The black spot on its withers might be a star, and black marks on the ground beneath could be the bloodstained leaves of the hunted animal, signaling the onset of autumn.

It is no more than a "plausible hypothesis," d'Huy admits. But the shaft scene does look strikingly similar to a Neolithic rock painting from the Maia River in Siberia that is thought to represent an early version of the Cosmic Hunt, in which a hunter takes aim at an elk with the Sun hanging under its belly. Perhaps the loops beneath the Lascaux bison, too, represent not its intestines but the Sun.

Rappenglück, meanwhile, thinks the birdman is a shaman with a staff and the bison is his spirit helper, guiding his journey to the sky. Similar scenes appear in the art of modern-day shamanic cultures, such as the ecstatic shaman in flight to the sky, shown with penis erect and bound to a celestial bull, that appears on a tepee of the Oglala people in North America.

Rappenglück further suggests that the eyes of the Lascaux bison, birdman and bird may correspond to the stars Vega, Deneb and Altair, the "Summer Triangle," among the brightest stars overhead in summer. Twenty thousand years ago, this trio never set but rotated around the northern celestial pole, indicating the time of night like a giant sky clock. Perhaps the people of Lascaux imagined this constellation as a celestial shaman (the Paleolithic equivalent of the Chumash's Sky Coyote) turning each night around the axis of the cosmos. Surrounded by spirit helpers, he ruled and fertilized the sky. Rappenglück interprets the scene as an image of the sky, but also a map for an earthly shaman's own voyage to the celestial pole.

It won't ever be possible to prove what the artist really intended. But the different strands of evidence do seem to converge on one explanation: that this prehistoric scene, far underground in the deepest part of Lascaux Cave, actually represents a journey to the stars. Similarly, the various lines of inquiry described in this chapter—Bull No. 18, the Dead Man, the Cosmic Hunt—seem to me, despite the uncertainties, to add up to an overwhelming broader conclusion: that if we want to understand where we come from as a species, to reach the source of humanity's earliest beliefs and identity, then we have to include a consideration of the wheeling night sky.

Seeing those repeated celestial cycles—night after night, season after season—surely helped to stimulate the very first ideas about who we are and about the nature of reality, ideas that survive in hunter-gatherer communities today. "They had the same questions," Rappenglück says. "What is birth? What is death? Where does the Sun go? What is behind the world?"

The universe that our ancestors came up with in answer to those questions was a quintessentially human one, inspired not just by the sky but by the shifting states of consciousness that our brains can produce. In it, there were no boundaries between living and nonliving, humans and nature, Earth and stars. It was a cosmos that created us as we created it; in which internal experience and external reality were inextricably entwined. We've been trying to separate ourselves from it ever since.

2

LAND

JUST BEFORE DAWN ON DECEMBER 21, 1967, ARCHAEOLOGIST Michael O'Kelly stepped into the darkness of a 5,000-year-old tomb. He clambered through a long, narrow passage toward a burial chamber hidden deep inside the huge mound of stones, and then turned to look back toward the entrance. The visible patch of landscape looked dark and featureless, cut through by a glittering silver river. Flocks of starlings looped across the sky. He checked his watch: two minutes to nine. What happened next would catapult him to fame and change his life forever.

O'Kelly had been excavating this site in Newgrange, Ireland, for the past five years. Workmen realized in the seventeenth century that what appeared to be a small, scrub-covered hill was actually built from ancient stones: a passage tomb, of a type common across parts of the British Isles. But this one was huge—an impressive 300 feet in diameter, with an 80-foot-long passage constructed from great stone slabs leading to a cross-shaped chamber with a high, corbeled roof. Inside and out, the walls were alive with art—elaborate chevrons, diamonds, spirals—picked into the rock using flint-tipped chisels.

Local people said the tomb was a burial place for the legendary kings of Tara, who medieval writers said ruled from a hill nearby.

During his project to excavate and restore the stones, O'Kelly did find human remains, mixed into the earthen floor. But radiocarbon dating showed that the tomb was far more ancient than the stories about Tara. It was built around 3200 BC, centuries before even Egypt's Great Pyramids.

O'Kelly also noted a curious rectangular opening, high above the tomb's entrance, which he called the "roof box." It was partly blocked by a square chunk of crystallized quartz, which seemed to have worked as a shutter; a second chunk had fallen to the floor. Scratches on the stone of the roof box suggested that these shutters had been repeatedly slid open and closed. The opening was too small and high for people to climb through, so O'Kelly was mystified as to its purpose. Perhaps it was a place for offerings, or formed a doorway for the souls of the dead.

Then he considered a third possibility. Another story told by locals was that at midsummer, the light of the rising Sun shone into the tomb, illuminating a distinctive, triple-spiral carving at the back of the burial chamber. O'Kelly couldn't find any witnesses. And he knew that the story was impossible, because the tomb faces southeast, over the valley of the river Boyne, whereas the midsummer Sun rises much farther north.

But the stories were persistent, and O'Kelly realized that the tomb's entrance does point in roughly the right direction to be lit at midwinter, when the Sun rises at its farthest point south. So in the early hours of the morning on the winter solstice in December 1967, he drove over a hundred miles through the darkness from his home in Cork to test the idea. When he arrived, the surrounding fields and even the road below were deserted. He entered the tomb feeling utterly alone.

It was a clear morning, so as he waited in the burial chamber he was hopeful that the sunrise might indeed creep inside. But what actually happened was surprisingly sudden and dramatic. As the Sun's first rays appeared above the ridge on the river's far bank, a thin, bright shaft of light burst through the roof box and struck not the entrance passage but the floor at his feet: a direct hit right in the tomb's heart. The light soon widened to a rich, golden beam about 6 inches across, until the chamber was so bright he could walk around without a lamp and see the roof 20 feet above.

"I was literally astounded," O'Kelly said later. "I expected to hear a voice or perhaps feel a cold hand resting on my shoulder, but there was silence." After seventeen minutes, the Sun passed across the slit and darkness returned. He was deeply moved by the experience, and returned to the tomb every winter for the rest of his life, lying on the soft, sandy floor of the burial chamber as the light shaft danced across his face.

Thanks to O'Kelly's work, the tomb is now a UNESCO World Heritage Site, with tens of thousands of people applying each year for the privilege of standing inside when it lights up. And although his discovery was rare and unexpected at the time, we now know that Newgrange is just one of many stone monuments constructed in western Europe during the Neolithic period (which began with the introduction of farming, and ended with the appearance of bronze tools) that were aligned to events in the sky.

Some are exceptional and dramatic, such as Stonehenge in Wiltshire, England, aligned to the midsummer and midwinter solstices; or a stone circle at Callanish in the Outer Hebrides, Scotland, which captures the nineteen-year cycle of the Moon. But there are many smaller examples, such as the hundreds of simple dolmen tombs* in southern Europe whose entrances face the rising Sun.

What did these stones mean to their Neolithic builders? Why did people go to such efforts to build such monuments, and to relate so many of them to the sky? The answers, as far as we can glean them, reveal aspects of human identity and cosmology at a transformative

* Dolmen tombs are single-chamber tombs in which a large, flat capstone rests on two or more vertical stones.

time in our history, perhaps the ultimate transformation, when our species first adopted agriculture.

The hunter-gatherers of the Paleolithic had existed as an integral part of the natural world, sharing their environment on equal terms with other species. During the Neolithic revolution, people cut those ties and became farmers, controlling and exploiting the land. This shift in lifestyle and mind-set changed humanity forever, setting a trajectory of technological progress that has ultimately made us capable of reshaping not just landscapes, but the entire planet.

The revolution was about more than forging a new relationship with wheat or fields or sheep. It also transformed our wider cosmos: how people viewed the spirit world, and the sky. In fact, there's a case to be made that these new cosmological ideas didn't simply reflect the shift to farming. They caused it.

It's a story that starts not in Ireland, but with humanity's oldest known megalithic monument, built a staggering six millennia earlier than Newgrange, and thousands of miles to the east.

✡ ✡ ✡

In 1994, the German archaeologist Klaus Schmidt was searching for a new project. For the past decade he had been helping to excavate a site in southeast Turkey called Nevalı Çori. It was a hunter-gatherer settlement, inhabited in the ninth and early eighth millennia BC, with houses made from limestone blocks held together with mud. The village included a series of mysterious "cult buildings" (built on the same site over time) that were sunk several feet down into the ground. They were shaped like rounded squares, with stone benches around the edges of the interior, interrupted by T-shaped monolithic pillars. Two further T-shaped pillars stood in the center of each square, decorated with carvings of human arms, like some kind of anthropomorphic beings.

The site was fascinating, a glimpse into the worldview of a society on the verge of transition: within a few centuries, farming would flourish in this region. For the first time in history, anywhere on the planet, people here started to cultivate wheat, and to corral sheep, pigs and goats. But in 1992, the entire settlement was flooded by the construction of the Atatürk Dam. The rest of its secrets were lost forever.*

To find a new site, Schmidt surveyed other prehistoric remains in the area. He came across a 50-foot-high hill, in the foothills of the Taurus Mountains just forty miles from Nevalı Çori, called Göbekli Tepe—"Potbelly Hill"—because of its curves. The hill was strewn with Neolithic flint tools and some broken limestone slabs.

Archaeologists who spotted the slabs in the 1960s had dismissed them as belonging to a medieval cemetery. But Schmidt realized that they matched the T-shaped pillars in the cult buildings at Nevalı Çori. Except that these were gigantic, made from great blocks of stone several yards high. "Within a minute of first seeing it, I knew I had two choices," he later said. "Go away and tell nobody, or spend the rest of my life working here." He chose the second.

Through excavations and geophysical surveys over the next two decades, Schmidt and his team found that the hill is packed with buried pillars and enclosures. There are square chambers similar to those at Nevalı Çori, also dating to the ninth millennium BC. But beneath them is an older layer of much larger circles, up to 65 feet across, dating to the tenth millennium BC. Up to twelve T-shaped pillars around the inner edge of each space were connected by a stone bench. Two more giant pillars—up to 18 feet high and each weighing several tons—stood parallel in the center, with traces of carved arms, belts and loincloths made of animal skins. Other stones here are covered in carvings of animals: spiders, scorpions, vultures, foxes, boars, gazelles.

* Many of the excavation's finds went to the Archaeological Museum in Sanliurfa, though, where they are on display today—including the complete cult building, which has been carefully dismantled and rebuilt inside the museum.

The archaeologist and prehistorian Steven Mithen has said that Göbekli Tepe looks like "an amalgamation of Lascaux cave and Stonehenge" and in time, too, it is a stepping-stone, falling roughly midway between the two.

The discovery of huge stone monuments at such an early date—twelve thousand years ago—was astounding. It takes colossal effort and organization to erect constructions like this, with hundreds of people working together; other sites on such an enormous scale aren't known until thousands of years later. Archaeologists had assumed that hunter-gatherers simply weren't capable of doing it. They figured that the conversion to agriculture, perhaps triggered by climate change or growing populations, eventually made such monuments possible by providing the resources for large, permanent settlements. This led to a more complex society, as well as changes in religious beliefs, which together produced both the ability and the motivation to create giant symphonies in stone.

There were dissenters. The French archaeologist Jacques Cauvin argued in the 1990s that cultural or religious changes must have come first. From a technical point of view, early humans could have started farming long before, "but neither the idea nor the desire ever came to them." Something must have happened, he suggested, to change how they viewed the natural world. But there was little hard evidence for what that shift might have been, or how it happened.

What Schmidt found, however, suggested that Cauvin was right. Here was clear evidence of a complex, organized society, with some form of religion, or at the very least sophisticated mythology, all *before* the invention of farming. What's more, the pillars of Göbekli Tepe were erected at precisely the place where farming was about to originate.

Biologists have pinpointed this small region, between the upper reaches of the Euphrates and Tigris Rivers, as the only place where all seven Neolithic founder crops (chickpea, einkorn wheat, emmer wheat, barley, lentil, pea and bitter vetch) grew together, while genetic studies of hundreds of einkorn and emmer wheat strains have concluded that domesticated versions of both species likely originated from wild strains that grew in the Karacadağ Mountains, just twenty miles or so from Göbekli Tepe.

Large numbers of people, maybe hundreds, would have had to congregate on the hilltop to build Göbekli Tepe. So simply having to feed them all may have created a pressure to develop new and more predictable food sources. Mithen has suggested, for example, that gathering and processing wild grain at or near the site could have led to fallen grain springing up and being gathered again, leading over time to domesticated strains. Rather than being a response to climate change, he concluded, the domestication of wheat "may have been a by-product of the ideology that drove hunter-gatherers to carve and erect massive pillars of stone on a hilltop in southern Turkey."

But the connection might run deeper than that. German archaeologist Jens Notroff and his colleagues, who have continued excavating Göbekli Tepe since Schmidt's death in 2014, see clear evidence of a shifting relationship to the natural world, as suggested by Cauvin. In the cave art of the Paleolithic, people are rarely represented; it's the animals that take center stage. By contrast, the foxes, snakes and scorpions of Göbekli Tepe are reduced to smaller attributes or decorations on those huge anthropomorphic pillars. As the team put it in 2015, "humans are no longer depicted as a coequal part of nature, but are clearly more prominent and 'raised' above the animal world." The art shows, they argue, that people had already begun exerting power over nature: a "mental control" that led to the subsequent physical control of domestication.

Another striking aspect of Göbekli Tepe is an apparent obsession with death. The art here features multiple images of headless people, as well as statues of heads, apparently broken from larger statues. Among animal remains found in the sediment—thought to be the debris from lavish feasts—are hundreds of pieces of human bones. Anthropologists reported in 2017 that most of these are from skulls, and that some are carved with grooves and holes in a way that suggests intact skulls were once hung up for display.

Schmidt interpreted the abstract T-shaped statues as beings from a "transcendent sphere" (naturalistic statues found elsewhere show that the builders could depict realistic humans when they wanted to). And, intriguingly, the circular enclosures appear to have been accessed not via doors or gateways but through small openings in "porthole stones." One of these stones is decorated with a boar lying on its back. The circles, Schmidt suggested, represented the realm of the dead, which could only be entered by crawling through the hole.

In fact, a preoccupation with death, and particularly with skulls, emerges across the region at this time and in the centuries following, with human remains commonly buried inside people's houses. At sites such as Jericho and 'Ain Ghazal in Jordan, dating to the tenth and ninth millennia BC, selected skulls were removed after death and given faces made of plaster, with shells for eyes, before being placed under the floor. At Çayönü in southern Turkey, archaeologists found a building that they called the "House of the Dead," dating to around 8000 BC, with sixty-six skulls beneath the floor and the remains of a further four hundred people. It also held a large, flat stone like an altar, with traces of human and animal blood.

A particularly bizarre example is Çatalhöyük, a 65-foot-high mound on Turkey's Konya Plain, a few hundred miles from Göbekli Tepe. The mound contains mud-brick houses from a settlement that housed thousands of people at its height in around 7000 BC. The closely packed houses were dug down into the ground, and entered by climbing down a ladder through a hatch in the roof.

Inside, the houses were richly decorated with paintings, as well as sculptures of animals that burst out of the walls. There were no doors inside; to move between rooms, inhabitants had to crawl through small portholes. The small chambers were further subdivided into sections, just 3 feet square, which occupied different vertical levels, with their edges marked or guarded by bulls' heads. Human bones were found buried beneath these platforms and in the walls, including a stillborn fetus enclosed in a brick.

The residents seem to have found the walls of their houses highly significant. As well as the objects embedded within them, there were small, undecorated alcoves inside the walls of some rooms, just big enough for a single person to crouch within. And wall sculptures of animals such as leopards and bulls were repeatedly replastered, up to a hundred times.

It seems a crazy way to live: cramped, dark and difficult to move around. Ian Hodder, an archaeologist at Stanford University who has been excavating the site since 1993, has concluded that for the residents of Çatalhöyük, the physical structure of their houses was entwined with their mythical beliefs: "The world was replete with substances that flowed and transformed, and with surfaces that could be passed through."

Archaeologist and rock art expert David Lewis-Williams goes further. He believes that people here were deliberately mimicking the experience of crawling through limestone caves. There are such caves in the Taurus Mountains, just a couple of days' journey to the south, and pieces of stalactite and limestone from these caves have been found in the Çatalhöyük houses. Lewis-Williams has suggested that—just like visitors to the caves of Upper Paleolithic western Europe—the people in these houses saw the walls as permeable interfaces, or portals, to the cosmos's spirit realms. At Çatalhöyük, he says, a house wasn't just a place to live, but "the material expression of a mythical world." It was a model of their universe.

Houses modeled on the cosmos are still known in societies around the world. The Barasana people in the dense forests of Colombia, for example, traditionally live in wooden longhouses called *malokas*. Each one is a miniature universe. The roof is the sky, with a vertical post called "the seat of the Sun" that aligns with the Sun's rays every day at noon. The major horizontal roof beam, oriented east-west, is called "the Sun's path." The floor is the Earth and beneath it is the underworld, where the dead are buried.

In his 2005 book *Inside the Neolithic Mind*, Lewis-Williams argued that a similar principle might also explain other Neolithic monuments and cult buildings found in the Near East, such as the stone circles of Göbekli Tepe. Like Schmidt, he concludes that they modeled the spirit world, the cosmic realm of the dead. These sites incorporate spaces that are sunk into the ground, with human remains often under the floor.

Were the builders only concerned with the underworld, or did they also look up at the sky? Göbekli Tepe is located on a high point in the landscape, so it would have offered a panoramic view of the heavens. Some researchers have suggested that the flat tops of its pillars could have been used to observe the rising or setting of prominent stars, such as Orion's Belt, or were built to commemorate the gradual appearance in the sky (due to precession) of the bright star Sirius. Others have linked some of the animal carvings at the site to specific constellations—for example, proposing that a scorpion depicted underground might represent Scorpio below the horizon.

Notroff isn't convinced: he suspects the enclosures were at least partly subterranean, dug down into the sediment, and had roofs, perhaps made from an organic material such as animal hides. So the site may make more sense as a terrifying journey to a cave-like underworld than as an astronomical observatory. He has experienced these spaces beneath a roof recently built to protect the site, and the shadows make the pillars and carvings look larger and "even more awe-inspiring," he told me. "I can only imagine how all these images of giant scorpions, curling snakes and snarling predators must have affected the young hunter on his first descent into the darkness, maybe with nothing but the flickering light of a torch."

This doesn't mean that the builders of Göbekli Tepe weren't interested in the sky. One of the best-preserved pillar statues wears a carved necklace decorated with a disk and crescent. These symbols are thought to refer to the Moon; it has even been suggested that the necklace identifies this pillar as a "moon-deity." Either way, though, what structures like Göbekli Tepe reveal is a fundamental shift in how people related to the cosmos. It seems likely that the people who constructed these sites had a similar tiered universe—with lower, middle and upper worlds—to that of their predecessors in the Paleolithic and in traditional hunter-gatherer societies today. Lewis-Williams argues that altered states of consciousness were likely still important as a way of journeying between these different cosmic realms.

But such journeys were now occurring in man-made, rather than natural, settings. The residents of Çatalhöyük appear to have copied the cramped passages of natural caves. Elsewhere, the greater control that people now had over these portals allowed the development of simpler, more formulaic designs: circles, pillars, squares. As Lewis-Williams has pointed out, this shift toward purpose-built structures would have allowed for greater social control too, with the emergence of powerful elites and formal rituals—including decorating and displaying selected skulls after death and possibly human sacrifice—that prescribed who could access these other realms and how.

Göbekli Tepe, then, epitomizes two important changes that seem to have happened in parallel just before the adoption of farming, both of which involve societies beginning to separate themselves from, or elevate themselves above, nature. The spirit realms became populated primarily not with animal guides but human ancestors. And instead of using existing caves or natural features as entrances to these other worlds, people started to build their own.

✡ ✡ ✡

It was many millennia before these changes reached Ireland's Boyne Valley. Genetic studies suggest that farming gradually spread throughout Europe from the Near East, carried by migrants who largely replaced the local populations. The new way of life arrived in Greece around 7000 BC and northwestern Europe around 4500 BC. And when these farmers reached the Atlantic coast, they made monuments from giant stones, in an explosion of pillars, circles, tombs and graves from Portugal to Brittany to Sweden.

Different societies expressed this common theme in a variety of ways: in Ireland, a tradition of passage tombs led to Newgrange, one of the most spectacular Neolithic monuments of all. Farmers arrived here around 3750 BC, bringing with them pottery and robust rectangular houses as well as cereals such as wheat and barley. Studies of plant remains suggest that the transition was relatively swift. Within a century or two, cereals were grown across the island, while large areas of forest had been axed or burned.

At the same time, people started building simple stone tombs, with a burial chamber defined by five or six large stones plus a flat capstone on top, all covered with a mound of earth. Over the following centuries, the designs became larger and more complex. Whereas the first tombs were too small to enter, later ones had cairns or mounds up to 65 feet across. Passages inside led to inner chambers decorated with art and corbeled roofs, where rituals could be performed.

Irish archaeologist Robert Hensey, who has studied the development of passage tombs in Ireland, sees these sites, too, as portals. In his 2015 book *First Light: The Origins of Newgrange*, he describes them as "a powerful transcendental network; a chain of monuments which had acted as bridges to other worlds." Through occupying the same space as the bones of their forebears, he suggested, "select individuals could now physically enter the other world, the realm of the ancestors." Just like in the Near East, instead of using natural features of landscape as doorways to other dimensions of reality, people were building their own.

And as at Göbekli Tepe and Çatalhöyük, this journey to the spirit world was deliberately made difficult. Regardless of the size of a tomb, the passage inside was only ever wide enough for one person at once. Reaching the burial chamber often required ducking, crawling or climbing over stones. And as tombs became larger, the burial chambers became more complex, with up to seven recesses, each just big enough for a sitting or squatting adult. Hensey suggests that people might have stayed inside these for long periods, perhaps to facilitate trance states through sensory deprivation (maybe aided by the spooky effect of echoing chants). He notes that in some traditional communities, such as the Kogi of Colombia or the Orokaiva of northern Papua, individuals in training to become spiritual leaders are confined alone, in darkness, for up to years at a time.

At Göbekli Tepe, the link to the sky is unproven. In Neolithic Europe, it's crystal clear; megalithic monuments here often feature celestial alignments. A survey of 177 dolmen tombs in Spain and Portugal, for example, found that every single one faces east, toward a point on the horizon within the arc of the rising Sun. The survey author concluded that each tomb was oriented toward sunrise on a particular day, perhaps when construction began. This fits the idea that such tombs were believed to lead to the underworld: where nature regenerates, and where the Sun appears to go each night before being reborn in the morning.

In Ireland, not all passage tombs have obvious solar orientations. A few have roof boxes like the one at Newgrange, though, strong evidence that the builders wanted sunlight to enter at certain times. There is also an emerging focus not just on the daily rebirth of the Sun but its annual cycle. A 2017 study of 136 Irish passage tombs concluded that more than twenty of them were intentionally oriented toward key dates in the solar cycle, mostly the solstices.

Eventually, between around 3200 and 3000 BC, passage tombs became larger still, often more than 160 feet across, with bigger stones, higher roofs and longer passages. They had other design modifications too, such as art and decorations on the outside (not just inside) of the tombs, public spaces and platforms around the cairns, and flat mounds so that people could stand on top. The empty recesses where people may once have secluded themselves were now filled by ceremonial stone basins.

Together, these changes suggest that the purpose of these sites was shifting away from enabling individual spirit journeys toward public ceremonies, presumably conducted by powerful elites and intended to invoke drama and awe for the watching crowds. The culmination of this tradition was Newgrange, decorated with a gleaming quartz façade: the most impressive passage tomb known in terms of its size, complexity, the quality and quantity of its art and the accuracy of its alignment.

It didn't stand alone, however. This piece of land, famously nestled in a bend of the river Boyne, hosts not just Newgrange but two other passage tombs of similar size, Knowth and Dowth (the latter aligned to the winter solstice sunset), plus around ninety other monuments, including smaller passage tombs, standing stones, timber circles, earth enclosures and a processional way. It was a dramatic ceremonial landscape, which would have required the coordinated effort and resources of hundreds if not thousands of people.

On the morning of the winter solstice, a procession of mourners or worshippers perhaps walked across the river and up onto the ridge before placing human remains in the tomb. At sunrise, the beam of light shone into the burial chamber, symbolizing the journey into the dark underworld. But that may not have been the final destination. Lewis-Williams suggests that people may have imagined the sunlight, with the released spirits of the dead who had been placed in the chamber, then continuing up through the high, corbeled roof and back to the sky, where they would join the Sun "in the eternal round of cosmological life, death and rebirth."

There was a problem, though. No matter how impressive a tomb like Newgrange or Dowth might have looked to the gathered crowds, the main event—the lighting of the burial chamber—could only be witnessed by the handful of individuals inside. Maybe that's one reason the tradition reached a dead end: passage tombs were no longer built after about 2900 BC. The focus shifted instead to a new kind of monument, which took those same illuminations and made them visible to hundreds of people at once.

✡ ✡ ✡

Few ancient monuments have inspired as many different interpretations as the worn, tumbling ruin of Stonehenge, set in the open grasslands of England's Salisbury Plain. Over the centuries, this mysterious giant circle has been described as a druid temple, astronomical observatory, healing center, war memorial and even a landing point for alien spacecraft. But thanks to a series of recent excavations at Stonehenge and beyond, archaeologists are now in a better position than ever before to tell the stones' real story.

The site is unique for the sheer epic size of its stones and the staggering distances they were carried. Giant sandstone slabs called sarsens, weighing 25 to 30 tons each, were probably brought from hills near Avebury, over 20 miles to the north. Smaller bluestones standing among them, weighing up to 5 tons, were brought 140 miles from Wales, one of the most impressive achievements of the entire Neolithic. Adding to the mystery is the monument's famous orientation to the Sun.

Modern excavations and radiocarbon dating show that Stonehenge was constructed in several phases. Just after 3000 BC, a circular earthen ditch and bank (loosely described as a "henge"*) was dug from the chalk using antler picks, with a ring of bluestones just inside and an entrance facing toward the northeast. Several sarsen stones were erected inside the ring and beyond the entrance, leading toward a large, unworked sarsen now called the Heel Stone. Several centuries later, the monument took roughly the form we recognize today, as the

---

* True henges, however, have the ditch outside the bank. Stonehenge is unusual in having its ditch on the inside.

bluestones were rearranged and the giant sarsens added, in a circle of thirty uprights, with horizontal lintels that may have formed a continuous stone ring, 13 feet in the air. At the center, five doorway-like arches called trilithons rose nearly 26 feet, arranged in a horseshoe-shaped arc.

The most complicated astronomical theories suggested for Stonehenge—for example, that it predicted eclipses—have been debunked. But it is indisputable that the trilithon arc, plus an avenue leading away from Stonehenge toward the northeast (it later turns and eventually reaches the nearby Avon River) point toward midsummer sunrise. Thousands now gather every year to watch the midsummer Sun rise over the Heel Stone, but in Neolithic times, the Sun would have risen in line with the avenue itself.

The monument also captures midwinter sunset, in the opposite direction. Whereas the summer alignment is visible from inside the circle, the setting midwinter Sun could have been viewed by a procession approaching along the avenue from the northeast. In fact, the stone surfaces visible from this direction are more carefully dressed, suggesting this midwinter moment was the most important. A rectangle of stones called "station stones" is also possibly aligned to the most extreme rising and setting points of the full Moon, which occur at the solstices every 18.6 years.*

What was it all for? That was the question concerning Mike Parker Pearson, an archaeologist at Sheffield University, U.K., when he took part in a television documentary about Stonehenge in February 1998. Parker Pearson had invited a Malagasy colleague named Ramilisonina to join him. The pair had spent several years working together in Madagascar, where traditional communities still erect standing stones known as *vatolahy* ("man stones") in honor of the dead.

---

* Claims made in the 1960s that the stones incorporate dozens of astronomical alignments, and that the Aubrey holes were used to predict eclipses, are not generally accepted by scholars today.

The day before filming, Parker Pearson took Ramilisonina to nearby Avebury, where there are three Neolithic stone circles. Curious to know what his friend thought of the prehistoric site, Parker Pearson explained to Ramilisonina that archaeologists didn't know why the stones had been erected. "He asked if I had learned nothing from working in Madagascar," Parker Pearson recalled in 2013. "It was obvious to him that such stone circles must be monuments to the ancestors, constructed in stone to represent the eternity of life after death." Perishable materials such as wood, by contrast, belonged to the temporary world of the living.

At first, Parker Pearson dismissed the idea that beliefs from faraway Madagascar could reveal anything new about the purpose of these Neolithic monuments; the idea of Stonehenge as a memorial to the dead had been suggested before. But the next day, during filming at the site itself, he wondered if Ramilisonina's words might help to explain not just these ancient stones, but the entire surrounding landscape.

A few miles up the river from Stonehenge is another Neolithic site, built from earth and wood. Durrington Walls is the largest known henge in the British Isles, an earthen circle that encloses over 42 acres, associated with several large rings of timber posts. Archaeologists had long thought that Durrington Walls was centuries older than Stonehenge, but redating of the Stonehenge stones had just revealed that the two sites could have been in use at the same time. After speaking to Ramilisonina, Parker Pearson wondered if Stonehenge and Durrington Walls might not be two separate monuments after all. Perhaps they were two halves of the same complex: one for the living and one for the dead.

To test the idea, Parker Pearson and his colleagues excavated across both sites from 2003 to 2009. As predicted, the team found evidence of a previously unsuspected settlement at Durrington Walls. It dates to around 2500 BC, when the giant sarsens were erected at Stonehenge. The site overflowed with debris from domestic life, whereas Stonehenge has yielded almost exclusively cremated human remains (archaeologists estimate hundreds of people may have been buried there in the third millennium BC). What's more, the team uncovered an avenue leading from one of the timber circles to the Avon, suggesting the site was linked to Stonehenge by river. They also confirmed several solstice alignments at Durrington Walls—including that this circle and its river avenue both face southeast toward either midsummer sunset or midwinter sunrise—and the remains of lavish, midwinter feasts.

Parker Pearson concluded that this was where the builders of Stonehenge's epic second phase lived. They appear to have traveled from miles around at certain times of year to celebrate their ancestors and perhaps usher the dead from the living world into the eternal afterlife. The midwinter solstice, when the Sun had waned to its lowest point and plant life was dormant, might have been seen, he suggested, as the point at which "the dark world of the dead was closest to the world of the living." Perhaps people gathered then at Durrington Walls to commemorate the recently deceased by feasting and erecting timber posts.

A procession might have started in the midwinter-oriented timber circle at dawn, with people walking down to the river toward the rising Sun. They could have floated downriver by raft or canoe into the realm of the ancestors, perhaps carrying the cremated remains of selected dead, before walking up toward Stonehenge in the afternoon. From this angle, the linteled circle would have presented a solid silhouette against the sky. The Sun would have set directly behind, shining straight through a tight window between the top of that circle and the upper portion of the towering central trilithon. For anyone walking up the slope toward Stonehenge, the last glimmer of sunset would have been held there for a few moments, transformed as at Newgrange into a beam of light framed by stone.

The idea of Stonehenge as a realm of the dead, visited by the midwinter Sun, makes sense in light of theories about passage tombs such as Newgrange. In both cases, the Neolithic builders used the stones to convert their knowledge about repeating patterns of Earth and sky into dramatic moments of sensory perception. Knowing that the solstice falls on a certain day is one thing; collectively witnessing it in the depths of winter would have been quite another: during the time of greatest darkness comes the light. From their knowledge of cosmic cycles, they constructed a dramatic message about eternity—perhaps eternal afterlife—that would last for millennia.

The big innovations of the Neolithic are often said to be stone monuments and farming. Yet both of these can be traced back to a deeper shift, as humans mentally separated themselves from nature, and it became conceivable to manipulate and dominate the natural world. Instead of simply adapting to their environment, people took control, shaping not just individual monuments but eventually entire landscapes to give their beliefs and desires physical form.

It's a revolution begun at and around Göbekli Tepe, but completed six thousand years later by the builders of Stonehenge. Here the animal spirits are gone; human ancestors rule supreme. And the dependence on caves and the underworld has been broken. The farmers of Neolithic Britain constructed a new cosmology, suitable for a larger, more complex society. People now explored their universe not through individual trance states, deep inside caves as at Lascaux or in tombs like Newgrange, but in public arenas dramatically aligned with the sky. Instead of hiding in the dark, they had stepped into the light.

3

FATE

IN DECEMBER 1853, A TWENTY-SEVEN-YEAR-OLD ARCHAEOLOGIST named Hormuzd Rassam was leading excavations near Mosul, now in Iraq, on behalf of the British Museum in London. It was the opportunity of a lifetime, particularly for someone born and raised in the Middle East. But after more than a year of work, he had yet to make a big discovery, and the spot he was desperate to investigate had been promised to a rival team. He had one last-ditch idea, but the timing had to be perfect. So he watched the desert sky, waiting anxiously for the full moon.

Mosul was Rassam's home city. Today, it's known largely as a casualty of the war against terror, left as a pile of rubble and bones after Iraqi forces won it back from ISIS in July 2017. But in Rassam's time, Mosul was part of the Turkish Ottoman Empire, with centuries-old brick walls that enclosed dusty streets, crowded bazaars and mosques with bulging domes and soaring minarets. Rickety, flat-bottomed boats ferried passengers across the Tigris River to the fertile land beyond: cornfields; melon and cucumber beds; and a series of shallow, grassy hills.

Over the past few years, European adventurers digging in these mounds (and at Nimrud, twenty miles south) had revealed a spectacular ancient world. The largest mound, called Kouyunjik, was a mile long. In 1847, the British explorer Austen Henry Layard, with Rassam as his assistant, tunneled into its southwest corner and unearthed the ruins of a great palace, built in the seventh century BC.

The luxurious riverside residence had at least eighty rooms and passages, with stone doorways guarded by huge winged bulls and lions, and walls decorated with around two miles of carved alabaster friezes, showing victorious military campaigns across the Near East. It belonged to the Assyrian king Sennacherib: Layard and Rassam had discovered the great city of Nineveh, capital of the largest empire that the world had ever known.

The Assyrians were famous from the Bible, which tells, for example, of Sennacherib's failed siege of Jerusalem, and describes Nineveh as a wicked city whose inhabitants repented after God sent Jonah there to preach. But before Layard's excavations, no direct trace of this civilization had ever been found. Now, after more than two thousand years, its cities and palaces were emerging from the earth.

When a subsequent excavation was planned in 1852, Layard stayed home to pursue a career in politics, persuading the British Museum to put Rassam in charge instead. Keen to prove himself, Rassam planned to investigate the northern corner of the huge Kouyunjik mound, which he was convinced must hold something else spectacular. But Britain and France were jostling each other for access to antiquities that could be shipped to museums back home, and when Rassam arrived, he found that the British consul in Baghdad, Henry Rawlinson, had handed over digging rights of his favored site to the French.

Rassam dug elsewhere, and by December 1853 his time and funds were running out. He was desperate to explore the site before returning to London, but if he crossed the French and found nothing, the British authorities would likely never trust him again. "So I resolved upon an experimental examination of the spot at night," he wrote later, "and only waited for a good opportunity and bright moonlight for my nocturnal adventure."

He recruited a team of trusted workmen, and on the night of December 20 led them to Kouyunjik. On the second night, they uncovered part of a marble wall, attached to a section of paved floor. The next morning, Rassam excitedly telegraphed Rawlinson and the British Museum with the news that he had discovered another Assyrian palace. But when his team dug further that night, the slabs came to an end after a few feet, surrounded by an ancient rubbish pile.

Rassam was distraught. News of his exploits had already "oozed out" in Mosul, and he feared that the French would soon arrive to stop him, or that the Ottoman authorities would accuse him of looting. On the fourth night, he hired even more men, setting them to work at several sites close to the marble slabs. After a few nail-biting hours, he finally heard a shout—"*Sooar!*"—Arabic for "images." As the men dug a deep trench, a large bank of earth had fallen away, revealing in the moonlight the perfectly preserved image of a muscular, bearded Assyrian king.

The chamber they had discovered turned out to be a long, narrow salon, 50 feet long by 15 feet wide. Its walls were covered with scenes of a lion hunt: the king chasing in his chariot, bow held high; spearing a lion with his attendants; thrusting his dagger through the animal's neck. The reliefs are some of the most exquisite, lifelike art ever discovered from the Assyrian civilization. Rassam was moved by the portrayal of one lioness in particular, "resting on her forepaws, with outstretched head she vainly endeavours to gather together her wounded limbs."

But the biggest discovery was beneath his feet. The floor of this chamber was covered with thousands of broken clay tablets: some completely smashed; others almost whole, up to around 9 inches long. Their surfaces were crowded with tiny, wedge-shaped indentations—a script known as cuneiform, made by pressing the end of a reed into the clay while it was still wet. Rassam really had discovered another palace—built by Sennacherib's grandson Ashurbanipal, the Assyrian empire's most powerful king. And this was his library.

It was a crucial find. We learned in chapter 2 how the origin of farming, around 8000 BC, was a key turning point in human history: people were no longer part of nature, they were beginning to shape and control it. A few millennia later, these same fertile plains between the Euphrates and Tigris Rivers—an area known as Mesopotamia*—witnessed another great human revolution: the invention of writing.

The earliest written tablets known were produced by the Sumerian civilization of southern Mesopotamia, at the end of the fourth millennium BC. Their cuneiform script was later adopted by the Babylonians and Assyrians and spread farther north. By allowing everything from debts and taxes to the will of the king to be permanently recorded, the written word supported the machinery and bureaucracy of ever more complex cities, states and even empires. And, of course, with written records, history can begin. Archaeological remains can hint at what past cultures thought and believed, but words tell us directly.

Ashurbanipal's library is the first systematic insight we have into the mental universe of an ancient civilization. It contained thousands of texts from throughout his empire, which covered all of Mesopotamia and beyond, some of them copies of texts dating back to the third millennium BC. They range from receipts (for oxen, slaves, casks of wine) to prayers to legal documents, literature and medicine: essentially "the forerunners of everything," says Jeanette Fincke, an expert in cuneiform texts who has catalogued the library's Babylonian tablets at the British Museum. "And I honestly mean everything."

What this archive reveals more than anything, though, is a society built around a fascination—if not an obsession—with the heavens. The tablets describe the movements of the Sun, Moon and planets as a divine script, carrying messages from the gods, which shaped

* Mesopotamia, from the Greek, "[land] between the rivers," stretched from southern Turkey down to the Persian Gulf.

behavior and decisions in every area of human life. "When in the month Ajaru, during the evening watch, the Moon eclipses, the king will die," reads one tablet, part of a vast compendium of around seven thousand such omens called *Enūma Anu Enlil*. It's the birth of an idea that has captivated humanity ever since: that our fate is written in the stars.

☼☼☼

The carvings in Ashurbanipal's palace depict him as a bloodthirsty ruler: one relief shows him enjoying a picnic in his garden while the severed head of an enemy king hangs from a nearby tree. In 612 BC, a few years after Ashurbanipal's death, Assyria's enemies got their revenge. A coalition of former subjects, led by the Babylonians, conquered Nineveh and burned the palaces. The heat of the fire made the clay tablets inside bubble and warp, but also baked them hard enough to survive for thousands of years.

In addition to the tablets Rassam found, Layard unearthed crateloads more in the palace that Ashurbanipal inherited from Sennacherib. Between them, the excavators shipped tens of thousands of clay fragments to the British Museum.* Cuneiform tablets had been found before, but the huge scale of the Nineveh finds added urgency to the task of deciphering this strange script.

One of the pioneers was Rawlinson, the British consul. A few years earlier, he had risked his life scaling a cliff face in Persia to copy an ancient text that had been carved there in the mysterious wedge-shaped letters as a message to the gods. Repeated in three different languages including the Babylonians' language, Akkadian, it was a cuneiform version of the Rosetta Stone. By 1860, he and others had

* Rassam and Layard didn't record where they found the different tablets, and the crates became mixed up further after arrival in London, so they are all now treated as one collection.

achieved a working knowledge of the complex symbols, and attempts to read the tablets from Nineveh began.

They reveal Ashurbanipal not just as a military leader but as an obsessive collector of texts who worked tirelessly to gather thousands of them from across his empire. He "wanted to collect the written knowledge and wisdom of the known world," says Fincke. One tablet, for example, contains a message from the king to his agents: "The rare tablets that are known to you and are not in Assyria. Search for them and bring them to me!"

In particular, he targeted Babylonian texts, collecting more than 3,500 of them, dating back up to a thousand years. Although ruled by Assyria since around 900 BC, Babylonia had previously been a powerful empire in its own right. Its capital, Babylon, remained an important cultural and religious center, and the Assyrians assimilated much of the Babylonian worldview.

One of the most famous finds from the library is *The Epic of Gilgamesh*, often described as the world's first story. Thought to have been written in Babylon around 1700 BC but based on Sumerian poems centuries older than that, it describes a young, arrogant ruler—inspired by a real king of Uruk from the third millennium BC—who gains wisdom through a desperate, doomed search for immortality. Hailed today as a literary masterpiece, *Gilgamesh* caused a sensation when it was discovered because it includes a version of the biblical tale of Noah and the Flood, written centuries before the oldest copy of Genesis. (When assistant curator George Smith first deciphered this passage in the reading room of the British Museum in November 1872, he reportedly became so excited that he started taking off his clothes.) The poem is also full of celestial references. In one scene, the king has to outrun the Sun. In another, he and his friend Enkidu defeat the Bull of Heaven—the constellation we now know as Taurus—set on them by the goddess Ishtar (associated with the planet Venus), and throw its severed thigh in her face. Some scholars think it's a mythical explanation for why this constellation, by Mesopotamian times, had lost its hindquarters.

Another Babylonian epic from the library is *Enuma Elish* (which means "when on high"). It's less well known than *Gilgamesh* but arguably just as significant because it is one of the earliest known creation myths, among the oldest surviving attempts to describe how the cosmos came about. It reached its definitive form around 1500 BC, but again was probably based on much older stories. The poem tells how Babylon's patron god, Marduk (Jupiter), defeats the mother goddess Tiamat and the forces of chaos. He tears her in half, "like a dried fish," and from the two pieces he creates the heaven and the earth.

Marduk then brings order to the cosmos, setting the paths of the planets and stars and dividing the year into twelve months of thirty days, entrusting the night to the Moon and the day to the Sun. He unleashes the weather, and causes the Euphrates and Tigris Rivers to run from Tiamat's eyes. Then he builds himself a shrine in Babylon, and with Ea, the god of water and wisdom, creates humankind. Like other early accounts of how people saw their cosmos, it's a rich, epic vision, clearly more concerned with creating meaning than explaining facts. Archaeological evidence suggests that the people of the Paleolithic and Neolithic saw events on the Earth and in the sky as intimately entwined. *Gilgamesh* and *Enuma Elish*, dating from the birth of civilization, reveal a similarly holistic universe in which the terrestrial and celestial reflect and influence each other as two halves of the same coin.

Accordingly, Mesopotamian gods simultaneously inhabited both Earth and sky. Each major divinity existed as a statue in its own home city: Marduk, for example, lived in the Esagila temple in Babylon. Excavations show that the temple was 660 feet long, with huge courtyards leading to an inner shrine, and stood next to a ziggurat, or stepped tower. Priests at the temple attended to Marduk and his divine entourage (also statues), clothing, feeding and entertaining them, and carrying them around the city during religious processions. The twelve-day New Year's Festival was particularly important: during it, said the French Assyriologist Jean Bottéro, "the gods were exalted to not only renew time . . . but the universe itself."

The gods also appeared in the heavens as celestial bodies, with the planets, including Marduk and Ishtar, accompanied by the Moon god, Sin, and the Sun, Shamash. They were thought capable of determining events on Earth, and through their celestial movements, they gave humans clues about what was to come. The priests of the Esagila temple, known as the "scribes of *Enūma Anu Enlil*," were renowned for their ability to decode heavenly messages, with expertise dating back centuries. By interpreting the signs correctly and carrying out the appropriate rituals, it was possible to avoid any dire predicted consequences.

Ashurbanipal's motivation in gathering these texts, then, wasn't purely philosophical. He saw knowledge of the cosmos as vital for his very survival. By far the largest group of Babylonian texts in the king's library concerns omens and divination, particularly relating to celestial events. His master plan, says Fincke, was "to collect as many tablets as possible with instructions for rituals and incantations that were vital to maintain him on his throne and in power."

Watching the sky wasn't the only form of divination: pretty much anything could yield messages from the gods, from sheep entrails, birthmarks, smoke or dice to the call of a particular bird. To avert negative predictions, Babylonians had an arsenal of rituals called *namburbi*, a Sumerian word that means "loosening" or "dispelling": the evil could be untied like a knot. Like the residents of Çatalhöyük, with its cave-like houses, they inhabited a magical world in which there were no boundaries between the physical and spirit realms, and in which everything was at the will of the gods.

Celestial signs were the most powerful, though. A sign observed privately within the home, such as the sudden appearance of insects, might apply to a single person. An omen visible in the street could cover a whole neighborhood. But events in the sky could theoretically be witnessed by everyone, so these heralded the fate of the entire country: its harvest, warfare, politics or king. The tablets from the library detail how priests stationed around Ashurbanipal's empire, but particularly in Babylon, sent him regular reports with information about the celestial events they had observed and advice on what to do.

Their wisdom was collated into the collection of omens known as *Enūma Anu Enlil*. The title comes from the text's first words, meaning "When Anu and Enlil . . . " (Anu was god of the heavens or sky; Enlil, god of the atmosphere, was "lord of the wind.") Compiled around the late second millennium BC, it is essentially a handbook covering the earthly consequences of events from the movements of the planets to the color of the Sun.

"If on the first day of Nisannu the sunrise [looks] sprinkled with blood," reads one tablet, "grain will vanish in the country, there will be hardship and human flesh will be eaten." Another notes that if a solar eclipse takes place while Venus and Jupiter are visible, "the country will be attacked." Among the most important events were lunar eclipses, which often foretold the death of a king. The Moon's disk was divided into quadrants, corresponding to the four regions of the known world: Amurru, Elam, Assyria and Babylonia. The areas darkened by the eclipse revealed which king was to die.

Letters from Ashurbanipal's library describe the chilling way in which Mesopotamian kings avoided this fate. If an eclipse was observed, the monarch temporarily abdicated his position, and a substitute—an enemy, criminal or just the gardener—would be dressed in the king's robes, and placed on the throne with a "girl" or "virgin" beside him as queen. The pair were entertained in luxury for up to a hundred days, enjoying sumptuous banquets, court musicians, and even royal boat trips. Then they were executed, and with the prediction fulfilled, the real king could safely return to his throne.

It's a fascinating glimpse into a civilization ruled by the sky, for whom the celestial dance of the Sun, Moon and planets was literally a matter of life and death. The priests of the Esagila temple weren't simply superstitious fortune-tellers, however. In 1878, a reclusive Jesuit priest started copying more Babylonian tablets from the British Museum's vast stores, and helped to reveal that their knowledge of the sky went far beyond what anyone had dreamed.

✡ ✡ ✡

After the destruction of Nineveh in 612 BC, the Babylonians inherited control of the Assyrian empire, stretching from what is now central Turkey in the north down to the Arabian desert. King Nebuchadnezzar II, who ascended the throne in 604 BC, spent the forty-three years of his reign rebuilding Babylon, until the city surpassed even its former glory. He built a huge palace, and protected the city with moats and walls so thick you could drive around the top in a four-horse chariot. There were eight gates in the walls, the most impressive of which was the Ishtar Gate, through which a 65-foot-wide processional street led into the city, ending at the Esagila temple. The gate and parade route were lined with glazed blue bricks, decorated with fierce yellow-and-white animals: dragons, lions and aurochs bulls.

Next to the temple, Nebuchadnezzar rebuilt the city's ziggurat (previously destroyed by Sennacherib) so it was taller than ever before. Thought to have reached 300 feet high, it had stairs around the outside and was topped with a shrine to Marduk, again decorated with vibrant blue bricks. The tower was called Etemenanki, which translates as "House of the Foundation of Heaven and the Underworld," and it was full of mythological and cosmological significance. There's a clear resonance with the Ark in *The Epic of Gilgamesh*, for example: both were divided into seven stories and covered an area of one *ikû*, around 300 feet square.

This structure was reflected in the sky too. *Ikû* is what the Babylonians called the great square of our modern constellation Pegasus, points out Andrew George, who studies Babylonian culture at the School of Oriental and African Studies (SOAS) in London. According to George, Etemenanki was "a structure founded in both levels of the universe at once—one whose hugeness . . . transcends the gap between them." It was the home of Marduk and the ultimate source of Babylon's—and the king's—security and power. (In the Bible, for which Nebuchadnezzar is the wicked king who sacked Jerusalem and exiled the Jews, it became the Tower of Babel.)

Much of this was confirmed by German archaeologists who conducted the first scientific excavations of Babylon from 1899. By the time they arrived, however, there were hardly any clay tablets left. These had already been removed by Rassam, who dug through large areas of the city in the 1870s, and by local people digging illegally, who sold them to antiquities dealers.* Thousands of tablets were ultimately bought by the British Museum, where they caught the attention of a trainee priest named Johann Strassmaier.

Born in rural Bavaria in 1846, Strassmaier joined the Jesuits—an all-male Catholic order—at age nineteen. A few years later, Bavaria became part of the newly united Imperial Germany, and the Jesuits were targeted by Otto von Bismarck, the country's first chancellor. Bismarck saw their deference to the pope as a challenge to his secular government, and in 1872 he banned them from teaching or working in Germany. Strassmaier emigrated to a Jesuit college in England, where he specialized in the study of languages. He was ordained in 1876, and two years later moved to a Jesuit-owned house in Mayfair, London, walking distance from the British Museum.

But he couldn't escape the tensions between secular and religious worldviews; scholars in London were clashing over a series of revolutionary scientific discoveries that appeared to undermine ideas in the

* As at Nineveh, the findspots of these tablets weren't recorded, and many of them were damaged during recovery and transport.

Bible. In 1859, the biologist Charles Darwin had set out his theory of evolution by natural selection, challenging the biblical account of how species were created. Then in 1872 came the Flood tablet, causing some to claim that this crucial episode in the Old Testament was simply a reworked Mesopotamian myth. With a strong tradition of academic scholarship and an interest in defending the accuracy of the Bible, the Jesuits wanted to be part of the debate over the finds pouring out of Mesopotamia. Strassmaier was assigned to study cuneiform tablets at the British Museum, and set about teaching himself Akkadian.

Strassmaier was a small man, affable and kind, with a round face and "a nose that can not be easily forgotten." He originally planned to write a book on the history of the Semitic languages, but was dismayed by the vast number of tablets that lay unread and eroding in the museum's stores. "How can a history of these languages be written," he remarked to a colleague, "whilst 60,000 cuneiform tablets remain uncopied and untranslated?"

So he embarked on a schedule that he kept for almost twenty years, arriving at the museum's student room at ten o'clock each morning and working through until four o'clock without any breaks. In that time, he copied the symbols from thousands of tablets, producing neat ink drawings on sheets of paper folded in half. Whereas curator George Smith had read texts from Nineveh, Strassmaier focused on the tablets coming out of Babylon. They mostly dated from the time after Nebuchadnezzar, between the fifth and the first centuries BC, during which time the city fell to the Persians and then to the Greeks.

At first, Strassmaier diligently copied economic records such as bills and contracts, the texts most scholars thought too boring to bother with. But he soon noticed large numbers of tablets with few words, just numbers. What text there was—planet names, for example—hinted that the subject matter was astronomical. The numbers made no sense to Strassmaier, so in 1880, he asked fellow priest Joseph

Epping, who had been his math teacher in Germany (and was now based in the Netherlands), for help.

Epping was reluctant at first. He couldn't read cuneiform, and though astronomy was "not totally alien" to him, he wrote later, the task seemed too daunting: "I did not believe to be such a computational artist, that I could solve an equation, that had so large a number of unknowns, and so little a number of knowns." But Strassmaier handed over his drawings, and eventually Epping started wrestling with the mystery numbers, looking for patterns that might reveal their meaning.

He started on a fragment with seven columns of numbers that cycled up and down. It took him months to make sense of it. As Epping worked, other cuneiform scholars were just getting an inkling of the Babylonians' facility with mathematics, using a number system based on 60 (which we still use every time we write a time in hours, minutes and seconds or an angle in degrees) to tackle algebra, fractions and even quadratic equations. But still, what he found was a bolt from the blue.

In 1881, Epping announced that the numbers represented steps in a calculation of the dates and times of a series of new Moons, covering the years 104 to 101 BC. Another text included a similar table for the positions of Venus and Jupiter. The calculations were impressively accurate, even taking into account the subtle variations in the apparent speed of the Moon and planets through the sky (caused by their elliptical orbits). Although Greek and Roman writers often refer to the Babylonians' astral wisdom, no one had expected that alongside their magical omens and prayers, the scribes of Marduk had developed a new type of knowledge about the cosmos. Epping called his discovery a "precious historical treasure." The priests really could foretell the future, using accurate formulas to predict celestial events decades in advance.

As more tablets have been catalogued and read, historians can now trace a gradual progression in the priests' abilities. *Enūma Anu Enlil*, the handbook found in Ashurbanipal's library, contains a series of omens that list risings and settings of Venus,* dated to the second millennium BC. Some of the numbers seem to be based on observations, but others have been corrected to fit a pattern. The scheme isn't very accurate, but shows that the Babylonians were already trying to describe the heavens using mathematical rules. The later tablets from Babylon (and also some written by temple priests in the city of Uruk) show that from around the eighth century BC, the priests started keeping more systematic records, watching the sky each night and writing down everything they saw. These "astronomical diaries" also included notable terrestrial events, from the level of the Euphrates River to the prices of wool, barley and sesame to reports of monstrous births.

Within a few generations, the scribes started to notice "great cycles": periods after which particular types of events roughly repeat. Ishtar repeats her wandering path after eight years, for example, and Marduk after seventy-one years, while eclipses follow an eighteen-year cycle. By checking what had happened during previous great cycles, they could monitor signs in the sky without even needing to watch.

Then, around 400 BC, came another a jump in sophistication. The priests invented the zodiac by dividing the ecliptic (the path through the sky followed by the Sun, Moon and planets) into twelve equal segments of 30 degrees, naming each one after a nearby constellation, such as GU.AN.NA ("Bull of Heaven") or MAS.TAB.BA GAL.GAL ("Great Twins"). This gave them an accurate coordinate system for recording and computing events in the sky. Shortly afterward, they came up with arithmetic methods to describe the repeating cycles recorded in their diaries.

These were based on finding "period relations," which express different astronomical cycles in terms of one another. For example, each

* Its first appearance before sunset and last appearance before sunrise.

planet moves through the zodiac at a characteristic speed (its "tropical cycle"), but superimposed on this is a zigzag pattern in which it sometimes stops or temporarily reverses direction (its "synodic cycle").* Venus can be described pretty well by a very simple relation—in eight years, it goes through eight tropical cycles and (almost exactly) five synodic cycles—while others are far more complex. The final step was to incorporate the subtle variations in speed that occur throughout these cycles, by adding or subtracting different values over time according to set rules.†

It's very clever math, says James Evans, a historian of astronomy at the University of Puget Sound in Tacoma, Washington. The scribes no longer needed to rely on long lists of past observations; they now needed just a small set of numerical parameters to define the behavior of each celestial event.‡ Epping had uncovered the moment when humanity transitioned from simply experiencing phenomena in the sky to explaining them.

And that wasn't the only surprise hidden in the crumbling clay.

---

* This is because Earth, and the other planets, are all orbiting the Sun. Mercury and Venus are closer to the Sun, so they always appear close to it in the sky. When they move behind the Sun (as seen from Earth), they appear to be going backward. The other planets (Mars, Saturn, Jupiter) are farther from the Sun than we are, so sometimes we overtake them on the inside.

† Rather than tracking celestial bodies through the zodiac, the priests were most interested in computing the times and positions of key events—such as a new Moon or lunar eclipse, or the moment at which a planet stops or changes direction—as these were what triggered omen predictions.

‡ Researchers are still finding surprises in the tablets. In 2016, historian Mathieu Ossendrijver, from Humboldt University in Berlin, found that the priests were using geometric techniques in their astronomy too. He reported a Babylonian tablet that recorded a calculation of the distance Jupiter had traveled in the sky at different dates, using a method equivalent to plotting its velocity against time and then calculating the area under the graph. This method was previously thought to have been invented by European astronomers in the fourteenth century AD.

✡ ✡ ✡

In 336 BC, more than a thousand miles northwest of Babylonia, a young prince named Alexander ascended to the Macedonian throne. Over the next five years, he carved out a huge empire, winning Greek states, then Asia Minor, then Egypt. And in October 331 BC, after a decisive battle against Persian forces on the plains near Nineveh, he marched his armies to Babylon.

According to the later Roman historian Quintus Curtius Rufus, lots of the inhabitants climbed the city walls to watch Alexander the Great arrive. But most went out to meet him as he approached the blue-glazed gate. Officials carpeted the ceremonial road with flowers and lined it with silver altars, heaped with perfume. They also sent out gifts—herds of cattle and horses; lions and leopards in cages—and showed off their cultural treasures with a procession of musicians, wise men and the scribes of *Enūma Anu Enlil*.

Surrounded by his armed guard, Alexander entered the gate by chariot and went straight to the royal palace. Taken by the city's beauty and antiquity, he made it his new capital. His victory ushered Babylon into the Greek world—and brought the scribes into contact with the astronomers and philosophers of the West. Their two views of the cosmos could not have been more different.

Whereas the temple priests saw celestial events as written on a flat tablet, Greek scholars were interested in three dimensions: they wanted to know how the solar system was arranged. And while the Babylonian belief in omens meant precision mattered above all else, the Greeks had little tradition of accurately observing the sky. They based their models on lofty philosophical ideals.

In the fourth century BC, the dominant figure in Greek thinking about the universe was Alexander's tutor Aristotle. His fundamental assumption was that since the heavens are divine, they must be structured in the appropriately perfect and efficient way: as a series of spheres. He proposed a spherical Earth at the center of the cosmos, surrounded by concentric circles or spheres that carried the orbits of the Sun, Moon, five known planets and fixed stars.

The only imaginable heavenly motion was constant speed in a perfect circle, but that couldn't explain why the planets sometimes stop and change direction. In the third century BC, Greek astronomers came up with an elegant solution: the planets moved in small circles called epicycles at the same time they were tracing a larger loop around the Earth. Off-center orbits were suggested to explain the varying speed of the Moon and Sun. These geometric theories included no accurate numbers; the principle was what mattered. Until the second century BC, that is, when an astronomer named Hipparchus changed everything.

Born around 190 BC, Hipparchus worked on the island of Rhodes and seems to have conducted a one-man revolution of Greek astronomy, essentially transforming this philosophical art into a practical science. He made extensive astronomical observations, and is credited with compiling the first star catalog. He also criticized his peers for their sloppiness, arguing that their models of the cosmos were useless if they didn't accurately match what actually happened in the sky.

His attitude, according to James Evans, "represented a radically new way of regarding the world—at least among the Greeks." Hardly any of Hipparchus's work survives directly, but the astronomer Ptolemy later reported that Hipparchus used astronomical observations to derive accurate numbers—period relations—to describe the cyclic behavior of the Sun, Moon and planets. Then Hipparchus used the new math of trigonometry (and possibly even invented it; he was the first we know of to use such techniques) to plug these numbers into the existing geometric models.

"Hipparchus turned a broadly explanatory geometric model into a real theory," says Evans. He wasn't able to fully explain the motions of the planets using Aristotle's perfect circles. But for the first time, the Greeks could calculate the position of the Sun or Moon in the zodiac for any given date.

Ptolemy was the other great astronomer of the ancient Greek world. Working in Alexandria in the second century AD, he built on Hipparchus's work in a monumental text called *Almagest.* In it, he set out a logical, mathematical explanation for all of the movements seen in the sky, derived step by step from observations. It included the planetary theory that eluded Hipparchus: Ptolemy suggested that epicycles move through the sky at a constant speed not as seen from Earth or from the center of their orbit but from a third point, which he called the equant. Though complicated, this scheme was impressively accurate, and the *Almagest* proved to be one of the most influential science books ever written, defining a view of the cosmos that lasted for fifteen hundred years.

For much of recent history, then, this chain of events was thought to explain the origin not just of Western astronomy, but of scientific thinking in general, part of the so-called Greek miracle, as Evans puts it, "as if the Greeks had in one swoop invented science, along with history, poetry and democracy." But in 1900, Joseph Epping's colleague and successor, Franz Xaver Kugler, read something unexpected in the Babylonian tablets, the full significance of which would not be realized for many decades to come.

Kugler, from a landowning family in Königsbach, Bavaria, was square-jawed, determined and difficult. Another ex-student of Epping's, he was appointed as math professor at a Jesuit college in the Netherlands, and taught himself Akkadian in order to take over analysis of Strassmaier's drawings in 1897, a few years after Epping died.

Kugler had a strained relationship with Strassmaier, complaining that his colleague's frequent linguistic suggestions were not helpful for his astronomical analysis. He was also a scathing critic of Panbabylonism, a school of thought that emerged in the late nineteenth century and argued that the Hebrew Bible was directly derived from Babylonian culture and mythology, and that the Babylonians had developed highly sophisticated astronomy as early as the third millennium BC. It was Kugler who worked out much of the detail of the astronomical theories Epping had unearthed. And he noticed something odd about the period relations that the Babylonians used to calculate the behavior of the Moon.

This was the priests' most complex theory. To fully describe the Moon and predict all-important lunar eclipses, they had to combine several different lunar cycles: the Moon's variation in speed (anomalistic month); its progression through its phases (synodic month); and the time it takes to travel between the "nodes" where it crosses the Sun's path (draconitic month). To do this, the Babylonians ultimately used a cycle of nearly 350 years, from which they derived the average length of the synodic month as precisely 29.5306 days.* Kugler noticed that the numbers in this theory were *identical* to those used by Hipparchus. In other words, Hipparchus didn't derive the numbers in his theories from observations at all. He took them from the astronomers in Babylon.

In fact, over the last few decades it has emerged that pretty much all of the numbers on which Hipparchus's theories were based, including his period relations for the planets, came from Babylonian tablets. Historians already knew that some aspects of Babylonian math and astronomy had filtered to the Greeks, including the zodiac signs and the base-60 number system (which Hipparchus was one of the very first Greeks to use). But the Babylonians were still seen as primitive stargazers, inferior to the scientifically minded Greeks. The French Assyriologist George Bertin, for example, responding to Epping and Strassmaier's findings in 1889, insisted that even if the Greeks had adopted some of the priests' terminology, it was the Babylonians who had learned astronomy from the Greeks: "The Babylonians . . . soon discovered the accuracy of their new masters in science."

The discovery of Hipparchus's numbers embedded in older Babylo-

* Instead of a decimal number system, the Babylonians used a sexagesimal system (one in which the base number is 60, as we still use for degrees and time today). Both the Babylonians and Hipparchus used the same figure of 29;31,50,8,20 days, which converts to 29.5306 days (29 days, 12 hours and 44 minutes). The modern value is also 29.5306 days.

nian models turns that view upside down, proving that the fundamental ingredients for his theories came from the temple tablets. (More evidence is still being uncovered. In 2017, for example, Australian researchers claimed that a Babylonian tablet from the second millennium BC contains a trigonometric table: perhaps the priests helped to inspire Hipparchus's invention of trigonometry too.)

Hipparchus's reliance on Babylonian astronomy is so extensive, in fact, some scholars think he must have visited the Esagila temple himself and worked with the priests there, copying observations and equations from their tablets and converting them into Greek. More than that, though, contact with the priests may have transformed his very approach. After the hand-waving philosophical discussions back home, Hipparchus must have been "shocked," says Evans, to discover that the Babylonians were accurately predicting future positions of the Sun, Moon and planets in the sky. No wonder he made it his mission to make the Greek models just as precise.

In Hipparchus, then, two opposing worldviews collided. The Babylonian arithmetic progressions yielded precise predictions but included no three-dimensional structure, while the Greeks had geometric models but no accurate numbers. Neither approach on its own could produce a complete description of the sky. When they came together, the science of astronomy was born.

Of course, that wasn't all the Babylonians helped to forge. Entwined with astronomy from the beginning was the parallel discipline of astrology.

✡ ✡ ✡

In September 1967, French archaeologists excavating near a Roman sanctuary at Grand in northeast France found broken fragments of ivory at the bottom of an ancient well. Along with pottery, jewelry, fruit stones and shoes, the team eventually recovered nearly two hundred pieces from two pairs of ivory tablets, smashed and discarded around AD 170. Their surfaces still hold traces of gold leaf and colored paint, and they're beautifully carved with a circle of figures still intimately familiar today, from a crab and a scorpion and two scaly fish. They were used for casting horoscopes.

Before Alexander conquered Babylon, the Greeks had plenty of ways to foretell the future, from dream specialists to temple oracles, but there was no particular tradition of reading someone's fate in the sky. Without the ability to calculate the positions of the Sun, Moon and stars, the idea of casting a horoscope simply didn't exist. But sometime in the second century BC, after contact with Babylon, a craze for astrology swept through the Greek and Roman world. It reached throughout the Roman Empire but was particularly popular in Greco-Roman Egypt. Elaborate zodiacs start appearing on Egyptian temple ceilings. And papyrus fragments found in ancient rubbish dumps have yielded hundreds of briefly scrawled horoscopes noting details of the sky at the moment of a person's birth. James Evans suggests these were astrologers' notes, summarizing information about a client that would be displayed during a consultation on a board like the tablets from Grand. These have portrait busts of the Sun and Moon carved in the center, surrounded by a zodiac circle. Around that are thirty-six decans, groups of stars that the ancient Egyptians used to divide up the sky.

A narrative poem called the *Alexander Romance* (a fictional version of Alexander the Great's life which exists in several versions and was written around the second century AD) includes a passage describing how a similar tablet was used. In the story, the last native Egyptian pharaoh, Nectanebo II, travels to the Macedonian court after being defeated by the Persians, posing as an astrologer as part of an elaborate plan to trick Queen Olympias—who would later give birth to Alexander—into sleeping with him. He tells the queen that the horoscope reveals a ram-horned god will visit her during the night; Nectanebo subsequently disguises himself as this deity.

During the consultation, Nectanebo uses a "princely and costly board" made of ivory, ebony, gold and silver, decorated just like the Grand tablets. He opens a small ivory case and carefully pours out gemstones to represent the celestial bodies—crystal for the Sun, sapphire for Venus, a blood-red stone for Mars—placing them on the board to show their positions in the heavens at the moment of the queen's birth. In Greco-Roman times, wealthy clients probably had such consultations in temples and sanctuaries, says Evans. For everyone else, street astrologers may have cast horoscopes in squares and markets, drawing their charts in sand trays or on the ground.

The inspiration for astrology based on birth charts and zodiac signs—the kind we recognize today, popular in New Age websites and self-help books—is often credited to the ancient Egyptians; classical writers say it was invented by a seventh-century pharaoh called Nechepso. That's a false story. Greek astrology did incorporate traditional Egyptian elements, not least groups of stars called decans, originally used to tell time at night. The Greeks added features such as the "horoscopic point"—the part of the ecliptic rising at the time of one's birth, seen as so important that the entire chart was ultimately named after it. But as with mathematical astronomy, the fundamental ingredients of Western astrology came from Marduk's priests.

Since around 400 BC, the Babylonian scribes had been branching out. Instead of just giving predictions for king and country, they made forecasts for individuals based on the position of celestial bodies in the sky at the time of birth. Epping and Kugler deciphered the first Babylonian "horoscopes"; a few dozen are now known. One of the earliest, dating from 410 BC, records the birth of a child on the fourteenth night of the month Nisannu, when Jupiter was in Pisces, Venus was in Taurus, and the Moon was beneath the "Horn" of the Scorpion (the stars of our constellation Libra). "Things will be good for you," the tablet says.

While the Babylonians' astronomical techniques enabled Greek astrology, the desire to study horoscopes was in turn a key motivation for Greek astronomers. Hipparchus, for example, wrote a now-lost treatise on astrology, with the historian Pliny the Elder remarking that Hipparchus "can never be sufficiently praised, no one having done more to prove that man is related to the stars and that our souls are a part of heaven."

Ptolemy was also an advocate. Alongside the *Almagest*, he wrote another epic and hugely influential work, *Tetrabiblos*, in which he summarized the methods of astrology and tried to arrange them in a logical system. Personal qualities "which concern the reason and the mind are apprehended by means of the condition of Mercury," he wrote, "and the qualities of the sensory and irrational part are discovered from . . . the Moon." Ptolemy differed from the Babylonians in that rather than seeing celestial signs as divine warnings, he believed powers emanating from stars and planets, such as "humoral shifts," could trigger effects on Earth, influencing everything from the weather to personality and health. But he too was driven to achieve mathematical accuracy, at least partly because he wanted to read human secrets in the stars.

It took more than a thousand years before scholars in western Europe superseded Ptolemy's system and constructed our modern, heliocentric view of the heavens. In 1543, Copernicus suggested that the Sun, not the Earth, was at the center of the cosmos, a theory supported by Galileo when he turned his telescope skyward and found, for example, that Venus had phases like the Moon. Then in 1609, Johannes Kepler banished epicycles and equants when he realized that celestial orbits are not circular but elliptical.

For these founding fathers of astronomy, the idea that the stars influence our fate was still embedded in their motivation and worldview. Galileo regularly made astrological predictions for rich clients, and drew up horoscopes for his illegitimate daughters. Kepler hoped to strengthen and reform the discipline, describing himself as "throwing out the chaff and keeping the grain." He discounted the idea that cultural inventions like names or zodiac signs could affect earthly events. But he firmly believed that different qualities of light from the various planets could influence climate and health, and he suggested that just like human beings, the Earth had a soul, sensitive to the harmonies of the stars.

Ultimately, though, astrology was incompatible with the scientific revolution. In 1641, the French philosopher René Descartes famously separated mind from body, consciousness from the material world, part of an inexorable shift in the West towards physical causation as the only acceptable type of explanation. Astronomy and astrology had to go their separate ways: the former making sense of the universe based on objective measurements; the latter emphasizing intangible connections and subjective meaning. There could only be one winner. Without an obvious physical mechanism by which distant celestial bodies might affect our lives, the intellectual standing of astrology slowly collapsed.

While scientific astronomy soared, astrology was left to "stumble along," as Nicholas Campion, director of the Sophia Centre for the Study of Cosmology in Culture at the University of Wales, puts it: "a system that is dislocated from its cosmology." For many scientists, it's a threat that needs stamping out. In the UK, celebrity physicist Brian Cox has described astrology as "undermining the very fabric of our civilization," while biology and skeptic Richard Dawkins complained that it is "shriveling and cheapening the universe."

And yet despite (or perhaps because of) the lack of scientific support, the interest in zodiac signs and horoscopes persists. In fact, astrology is now reportedly rising in credibility and popularity, particularly among millennials seeking guidance, escape and even ambiguity in a stressful, ultra-rational world. Astronomy and astrology might seem complete opposites, even enemies. But in a way they are twins, reflecting two essential sides of our nature, and born of the same fundamental human desire to see patterns, order and meaning in the sky.

✡ ✡ ✡

In February 323 BC, Alexander once again approached Babylon with his army. The city's envoys included an astronomer priest named

Bêl-apla-iddina, who warned Alexander that according to the celestial omens, his life was in danger if he entered the city. The details given by different ancient writers vary: according to one version, the priest advised that Alexander should approach Babylon facing east to avoid catching sight of the setting Sun. But the marshy terrain was too difficult for his troops, so the king turned back, and arrived facing westward after all.

Once in Babylon, Alexander made plans for his next military campaigns—he wanted to attack Arabia in the south; Carthage and Italy in the west. Several Greek and Roman writers recount a story that in May that year, while the king was away from his throne—Diodorus says he went for a massage, Plutarch says he was exercising—an escaped prisoner entered the palace, crowned himself and sat on the empty throne. Alexander's Babylonian advisers told him to put the man to death. The classical authors seem mystified by this strange event. But recent scholars have suggested that the priests, concerned for Alexander's life, may have tried to enact the "substitute king" ritual to save him.

A few weeks later, Alexander attended a drinking party, and afterward fell seriously ill. An astronomical diary for the evening of June 11, 323 BC, is the only contemporary record of what happened next. "The king died," wrote the scribe. "Clouds." Aged just thirty-three, Alexander was gone, and the career of one of the world's greatest generals was over.

The city's fate was sealed. Alexander had made Babylon his capital and was rebuilding Marduk's temple and the Etemenanki tower. But after his death, his kingdom was split between his generals. Seleucos, who took Mesopotamia, built himself a new capital and forced Babylon's citizens to move there. Only the priests stayed, diligently recording their nightly observations in the abandoned city.

The region was next conquered in 125 BC by the Parthians, from modern-day Iran, and shortly afterward was subsumed into the Roman Empire. Within a few centuries, Babylon, like the great cities of

Nineveh and Uruk, lay buried under the sand, forgotten until the nineteenth-century exploits of Layard and Rassam. It was the end of humanity's first civilization, a flourishing of armies and empires, temples and towers, myths and magic, which created many of the foundations from which our own society is built. Its dying breath—the very last cuneiform tablets ever found—date from the first century AD. They are astronomical almanacs, forecasting future events in the sky.

As far as we can tell, people have recognized star constellations and followed the annual cycles of the Sun, Moon and stars as far back as the Paleolithic. By the Neolithic, they were starting to shape their cosmos, building monuments to create and capture key moments and effects. But the ability to keep written records—and the administrative system that writing supported—offered the opportunity to vastly extend that control. In a centuries-long effort, the Babylonians transformed a wandering, whimsical sky, plaything of the gods, into a predictable, mathematical universe.

Many civilizations through history have developed mathematical models to describe the sky. Chinese emperors employed teams of astronomers to draw up sky maps and predict events such as eclipses. The Mayans, whose leaders associated themselves with heavenly bodies, counted celestial cycles that lasted millions of years. But the scribes of *Enūma Anu Enlil* were the first we know of to move from an analog cosmos to a digital one; the first to swap the messy complexity of reality for the simplicity and power of numbers.

4

FAITH

ON OCTOBER 28, AD 312, TWO ARMIES CLASHED JUST OUTSIDE Rome, with seismic consequences for the future Western world. Two brothers-in-law, Maxentius and Constantine, were fighting for control of the western Roman Empire, which stretched from Britain to North Africa. Maxentius held Rome, while Constantine advanced across the Alps from Gaul. On the eve of the battle, Constantine set up camp a few miles north of the city walls.

What happened to Constantine that day, described by ancient authors such as Eusebius, bishop of Caesarea, has acquired legendary status as one of history's great turning points. At around noon, the power-hungry emperor saw a divine vision: a flaming cross of light above the Sun, emblazoned with the words "Conquer by this." It was enough to convert Constantine from paganism to Christianity. He ordered the sign—made from the first two letters of Christ's name, *chi* and *rho*, superimposed—to be painted on his soldiers' shields.

Maxentius, meanwhile, claimed protection from Mars, the Roman god of war. Rattled by his rival's military victories, Maxentius had tried everything he could think of to halt Constantine's progress, from conducting temple rituals and sacrifices to reading animal entrails and omens in the sky. He prepared for a siege behind the capital's impregnable walls, stockpiling grain and destroying the stone arches of the Milvian Bridge, Constantine's route across the Tiber River into Rome. But he wasn't sure the citizens would remain loyal; at chariot races in the Circus Maximus on October 27, the crowds shouted Constantine's name. The next day, after consulting a collection of oracles called the Sibylline Books, Maxentius chose instead to meet his rival in open battle.

With the bridge disabled, Maxentius's men crossed the Tiber on a temporary platform made from wooden boats, meeting Constantine's army a few miles north of its banks. The attacking forces, though much smaller in number, pushed Maxentius's troops back. With no escape route, thousands of his soldiers were forced into the river and drowned, until, according to one account, the water could barely force its way through the piles of bodies. Maxentius perished too as he tried to flee, weighed down by his armor. Constantine fished his body out of the mud and paraded his severed head through the streets.

The victory gave Constantine undisputed control of Rome's western territories. Fighting under the Chi-Rho symbol, he later gained its eastern lands too—from Macedonia as far as Syria and Egypt. After generations of instability and civil war, he finally united the Roman Empire. Although the west fell within a century or two as Rome's political influence over its vast territory gradually disintegrated, the eastern empire, ruled from a new capital, Constantinople, lasted for another thousand years.

The significance of Constantine's victory goes far beyond geopolitics. Throughout his reign, the emperor also broke with centuries of religious tradition, single-handedly transforming his chosen faith from a minor, persecuted sect into a hugely powerful church. His conversion paved the way for Christianity, rather than the old planet-based gods, to become the dominant religion not just for Rome but for the entire Western world. The Battle of Milvian Bridge, then, marks a key moment in an even greater clash in human history: between the sky worship of early civilizations and the monotheistic religions that dominate today.

To commemorate his victory over Maxentius, Constantine built a huge stone arch near the Colosseum in Rome. It still stands, spanning the grand ceremonial route taken by emperors as they entered the city in triumph, and it features a giant inscription, originally cast in bronze, that attributes Constantine's success to "divine inspiration." Many historians have taken that phrase as referring to his epic moment of conversion, and the flaming cross in the sky. But over the last few years, scholars like Elizabeth Marlowe, an art historian at Colgate University in Hamilton, New York, have pointed out that the marble sculptures and reliefs that cover the arch include no Christian symbols.

Instead, they show the Roman sun god, Sol. On the eastern side of the arch, Sol rises from the ocean in his four-horse chariot, balanced on the west side by the descent of Luna, goddess of the Moon. Sol is identifiable elsewhere on the arch, too, from a band of light rays around his head—known as a radiate crown—and a raised right hand; in several places Constantine mirrors this pose. What's more, Marlowe has shown that the arch was carefully offset from the road so that for approaching crowds, the view beyond it centered on a colossal bronze statue dedicated to the Sun. Far from affirming Constantine's Christianity, she says, "the favored deity is unambiguously Sol."

In other words, the emperor's famous conversion isn't everything that it first appears. But then, neither is the victory of monotheism over the gods of the sky.

✡ ✡ ✡

Most early societies worshipped the sky in some form, or associated their gods with celestial bodies. There are earthly gods too, of course, representing everything from animals and ancestors to rivers and crops. But the vast majority of religions—from all periods of history, anywhere in the world—have a prominent role for celestial beings. The very word "deity" derives from a root—*dšiuš* in Hittite; *dyaus* in Sanskrit (which later became "Jupiter" in Latin and "Zeus" in Greek)—that meant "shining in the sky."

According to the twentieth-century Romanian historian Mircea Eliade, who surveyed hundreds of religions around the world, the sheer size and power of the sky drives spiritual experiences. Simply by being there, the heavens reveal how tiny we are in the cosmos at the same time as putting us in touch with the vast, unimaginable whole. "The sky, of its very nature, as a starry vault and atmospheric region has a wealth of mythological and religious significance," Eliade wrote. "Atmospheric and meteorological 'life' appears to be an unending myth."

Some sky gods are associated with specific celestial bodies, such as Babylonia's Marduk and Ishtar, or the Egyptian sun god Ra. Others are supreme creator beings that are embodied by or live in the heavens. The goddess Mawu, worshipped in Benin, West Africa, wears the blue sky as a veil and clouds for clothes. Debata, known in Sumatra, releases lightning when he opens his mouth to smile. Qat, supreme being of the Banks Islands in Melanesia, created the dawn when he cut into night's darkness with a red obsidian knife. But with the rise of the three great monotheistic religions—Judaism, Christianity and Islam—this wealth of celestial personalities was swept away. Their divine dramas were replaced, for much of humanity, by the concept of one unchanging God.

It's a revolution that started in Canaan, a region centered on Palestine, between the Jordan Valley and Mediterranean Sea. Ruled by the Egyptians for much of the Late Bronze Age, a people called the Israelites emerged here after around 1250 BC. Texts and archaeological evidence suggest that early on, the Israelites worshipped celestial bodies among a pantheon of gods led by Yahweh (linked by some scholars to the Sun) and his wife, Asherah (associated with trees, and later Venus).

The region split into two kingdoms: Israel in the north and Judah in the south. The Assyrians destroyed Israel in 722 BC and exiled thousands of its people; in 586 BC, the Babylonians did the same to Judah. After Israel's fall, a religious group emerged in Judah, centered in Jerusalem, that recognized Yahweh alone as the sole creator of the universe: a deity who couldn't be imaged, and who forbade worship of all other gods. Surveys of religious texts from the period suggest this was initially a minority view, but that such "Yahweh- alone" beliefs then strengthened among the exiles in Babylon.

In 538 BC, Babylon was in turn conquered by the Persians, who helped the exiles to return home, and rebuild their Temple of Jerusalem. With Persian support, this monotheistic group came to control Judaean religious institutions, and they assembled and edited the documents that became the Hebrew Bible. (It has been suggested that the Persians were sympathetic because they recognized in the exiles' religion aspects of their own Zoroastrianism, with its supreme creator god, Ahura Mazda, opposed by the hostile Angra Mainyu. Early Judaism was certainly influenced by Zoroastrianism, particularly its notion of the cosmos as an epic struggle between good and evil.)

Although historians' understanding of these events is murky, scholars generally agree that the rise of Yahweh among the Jews at this time was linked to their experiences of exile and loss. They also saw the powerful empires of Assyria and Babylonia destroyed within one lifetime, points out David Aberbach, a professor of Jewish Studies at McGill University in Montreal. Perhaps that encouraged them to see material gods and territorial identity as weak and transient, he says, and to choose instead "an abstract and indestructible God, rather than gods of wood and stone."

All nations claimed that their own chief god was supreme, but this new deity was different. Unlike the Greek chief deity, Zeus, say, who despite being the most powerful god in the universe still faced limits to his actions and could be thwarted by other gods, Yahweh was transcendent: not within the cosmos but above it, and no longer bound by its rules. According to Eliade, this "notion of God's 'power' as the only absolute reality" was "the jumping-off point for all later mystical thought and speculation on the freedom of man."

It was an idea so powerful, it would change the spiritual face of humanity: Judaism, Christianity and Islam are now followed by more than half of the world's population. Initially, Yahweh had little influence beyond Israel. Then, in the first century AD, a Jewish sect emerged, led by a teacher from Nazareth called Jesus Christ, who claimed to be the son of God. Converts such as Paul the Apostle brought the story of his death and resurrection to Rome. The faith was slow to take hold, though. When Constantine, the son of a high-ranking Roman army officer, was born in Naissus, modern-day Serbia, in AD 272, the tide of change was still lapping at the shore.

✡ ✡ ✡

Constantine grew up in what the Cambridge historian A. H. M. Jones described as "evil times." Roman rule had expanded fast since it first rolled into Greek territories in the second century BC. Five hundred years later, the huge, unmanageable empire was on the brink of collapse, threatened by invasions, civil wars, famines and plagues, with rival armies proclaiming new emperors almost too fast to keep count.

Religion was also in flux. The traditional Roman gods had long since merged with the Greek pantheon: Rome's chief god, Jupiter, was identified with the Greek Zeus, while the goddess Aphrodite took on attributes of Venus, and the Greek solar deity, Helios, became associated with the Roman sun god Sol. These planetary gods occupied a central position in daily life, not just for religious rituals and sacrifices but fortune-telling too, as astrology, imported from Babylon, became popular across the Greco-Roman world.

By the first century AD, the planets even ruled over the calendar, with the introduction of a seven-day week—that also probably originated in Babylon—starting with Saturn (Saturday), followed by the Sun (Sunday), Moon (Monday), Mars, Mercury, Jupiter and Venus.* As Rome's empire grew, however, so did its pantheon, as the Romans happily adopted the gods of the provinces they conquered. The new faith of Christianity had to carve out a place in a society already

* Archaeologists have seen the planetary calendar everywhere from ruined baths of the first-century emperor Titan to the wall of a house buried by the volcano Vesuvius in AD 79.

"crammed full of deities," as one historian put it, from the mother goddesses Artemis of Ephesus and Isis of Egypt (associated with the star Sirius) to the Persian sun god Mithras.

It took the iron hand of Diocletian, an ex-commander of the imperial bodyguard, to restore order. After becoming emperor in AD 284, he emphasized Rome's traditional gods, and punished the Christians who refused to worship them. He also divided the sprawling empire into two parts, each led by senior and junior partners in a four-emperor system known as the Tetrarchy. Diocletian and Maximian (Maxentius's father) ruled the east and west, respectively. They both retired in AD 305; Constantine's father, Constantius, took over from Maximian, and when he died in England the next year, his army proclaimed Constantine his successor. But the Tetrarchy system had created multiple claimants for the various thrones, including Maxentius, and over the next few years, Constantine had to negotiate his way through a series of battles and betrayals as the rivals vied for power.

In AD 310, after he defeated Maximian (who came out of retirement to support his son) in Marseille, France, Constantine ordered his army to run off the main road to visit a temple sanctuary.* There, according to one contemporary account, the emperor had a divine vision in which he was promised military victories and a long rule. The bearer of the message wasn't a flaming cross, though, but the classical god Apollo, who was often associated with the Sun.

Roman leaders traditionally insisted that their power came from Jupiter, a deity with origins in the older Indo-European sky god Dyaus. But some claimed links with a supreme solar deity, seeing themselves as a kind of conduit that reflected the Sun's rays on Earth. According to Jonathan Bardill, author of *Constantine, Divine Emperor of the Christian Golden Age*, it's a tradition that goes back to the Greek-speaking Ptolemaic kings who ruled Egypt after Alexander the Great, in turn

---

* This was probably the sanctuary at Grand, origin of the astrological tablets mentioned in chapter 3.

influenced by the ancient Egyptian pharaohs and their worship of the sun god Ra. In the first century BC, Julius Caesar wore a radiate crown; his heir Augustus (Rome's first emperor) erected an obelisk brought from Egypt as a giant sundial. In the first century AD, Nero erected a colossal bronze statue of himself as the Sun.

These efforts did not generally end well. When Marcus Aurelius Antoninus, a teenager from Emesa, Syria, succeeded the throne through family connections in AD 218, he was a priest at his local temple, where worship of the Syrian sun god Elagabalus centered on a large, cone-shaped meteorite. He took the stone to Rome and worshipped it daily, "dressed in silk robes, a lofty tiara, and cheeks painted red and white." He was assassinated four years later, and his mutilated corpse dragged through the streets. Aurelian, who became emperor in AD 270, had slightly more success when he tried to replace Jupiter with the cult of Sol Invictus, the "Unconquered Sun."* Its central festival was Sol's birthday, December 25, a few days after the winter solstice, when the Sun resumes its journey north and the days start to lengthen toward spring. Aurelian was assassinated too, five years into his reign, but the cult of Sol Invictus survived him.

It's not clear why Constantine chose to follow his ill-fated predecessors with his solar vision; perhaps he wanted to distance himself from the traditional religion of the Tetrarchy. But from AD 310, all of Constantine's mints produced coins that featured Sol, described as "companion of the emperor," in his characteristic pose: standing naked with his right hand raised and wearing a radiate crown. When Constantine set out to liberate Rome in AD 312, says Bardill, "it was with the sun-god as his guardian."

Did he then have a second vision that converted him to Christianity? Some researchers have suggested that the "flaming cross" Constantine reportedly saw was a solar effect called a sun dog, which occurs when sunlight is refracted by ice crystals in the atmosphere

* Historians disagree about whether Sol Invictus was a rebranding of Elagabalus, or the traditional Greco-Roman Sol, or a completely new solar deity.

and can make the Sun appear cross-shaped. But there doesn't necessarily have to be a meteorological explanation. Roman leaders often used reports of dreams and visions to encourage their troops or claim divine support. (When Augustus entered Rome following Caesar's murder, it was said that a circular rainbow formed around the Sun.) Constantine's cross in the sky is probably just another version of the Apollo story, retold later with a Christian spin.

Nonetheless, after Constantine defeated Maxentius, his religious policies did shift. Christians had been targeted on and off since Nero's time, culminating in Diocletian's "Great Persecution," during which anyone who refused to sacrifice to the pagan gods was imprisoned or executed. Constantine reversed that in AD 313, allowing the empire's inhabitants to worship whatever god they chose. In 321, he decreed that the Christian day of worship (Sunday) would be an official day of rest for Roman citizens, and after he gained sole control of the empire in 324, he started using Christian symbols on his coins. He removed or repurposed pagan statues and built a series of important Christian churches, including the Church of the Holy Sepulchre in Jerusalem, over the supposed locations of Jesus's crucifixion and empty tomb, still seen as Christianity's most holy site.

Constantine also supported Christians financially and involved himself intimately in the running of their church. In May 325, he convened hundreds of bishops from across the empire for the First Council of Nicaea, the first attempt to agree on a common doctrine for all of Christianity. The early Christian bishop and historian Eusebius describes him sitting in the middle of the great hall in purple-and-gold robes "like some heavenly messenger of God, clothed in raiment which glittered as it were with rays of light." He bullied the squabbling bishops into near unanimity, helping to build a powerful, unified, "Catholic" church.*

This conventional history, however, is mostly gleaned from accounts

* The term "Catholic" comes from the Latin *catholicus* (or *katholikos* in Greek), meaning "universal."

by Christian authors such as Eusebius. Other sources reveal that there's more to the story, just as with Constantine's vision and victory arch. For years after his defeat of Maxentius, for example, Constantine kept making coins that featured Sol (though he eventually stopped around 324). And although Christians observed a rest day on Sunday, in his decree Constantine didn't refer to it as "the Lord's Day" as Christians did. He introduced the law "in veneration of the Sun."

In AD 330, at his new, supposedly Christian capital of Constantinople, Constantine erected a giant statue of himself on a 120-foot-high column of purple porphyry. It depicted him naked with a radiate crown, facing east toward sunrise. The statue was felled by high winds in 1106, but literary sources preserve the inscription: "For Constantine who is shining like the Sun."

In other words, Constantine didn't give up his solar beliefs. But how could he follow both religions, when Christianity expressly forbids the worship of other gods? Some clues come from his letters, in which he describes, for example, the saving power of God's "most brilliant beams," and says that God "held up a pure light" through his son. Historians such as Bardill argue that Constantine never really converted from paganism to Christianity at all. Instead, he simply joined the two, seeing the Christian God as a sort of supreme solar deity, whose rays were spread on Earth by Christ. By deliberately blurring the boundaries between the two faiths, he could embrace his new beliefs without having to give up the old.

He wouldn't have been the first, of course. For centuries, Christians had been drawing on the power and radiance of our neighboring star.

✡ ✡ ✡

In the Old Testament, the prophet Malachi refers to the coming Messiah as "the Sun of Righteousness"; Christ later described himself as "the light of the world." Before Jesus was crucified, his Roman captors mocked him by giving him a radiate crown made of thorns. But during the first couple of centuries after Jesus's death, as Christians in the Roman Empire tried to attract followers and distinguish themselves from their parent Jewish faith, they increasingly borrowed the rituals and trappings of Sun worship.

Instead of praying toward Jerusalem, as the Jews did, they faced east toward sunrise. And instead of keeping the Jewish Sabbath, they moved their main day of worship to Sunday, in line with pagan sun cults. In the second century, the Christian author Tertullian denied that this choice of day had anything to do with its solar connotations. But by Constantine's time, Eusebius was happy to recognize a direct link, explaining that "the Saviour's day . . . derives its name from the light, and from the Sun."

The major Christian festivals were also scheduled according to the Sun's movements. Easter, the celebration of Jesus's resurrection, was originally based on the Jewish Passover (itself a direct descendant of the Babylonian new year festival, Akitu). Like Akitu, Passover was celebrated on the first full moon after the spring equinox; under Constantine, bishops at the First Council of Nicaea voted to move Easter to the following Sunday. And from at least the fourth century AD, Jesus's birth was celebrated on December 25, the birthday of Sol Invictus, which pagans marked by lighting candles and torches, and decorating small trees.

The result, points out Oxford historian Marina Warner, was that Christ's life became intimately identified with the annual cycle of the Sun; his birth is still celebrated just after the winter solstice (when the Sun begins to rise toward the spring) and his return from the dead just after the spring equinox (when the Sun finally triumphs over darkness, and the days last longer than the nights). Other Christian imagery reinforced the metaphor. As early as the first century AD, the twelve apostles were widely regarded as representing the twelve signs of the zodiac, through which the Sun passes in the sky.

Meanwhile, the Moon became linked with first the Christian Church, and eventually the Virgin Mary. In her book *Alone of All Her Sex*, Warner notes that in the regions where Christianity first took hold, the Sun represented ferocious energy and power. It was the gentler moonlight, associated with precious moisture-giving dew, which nourished life. She argues that this inspired the idea of the grace of God mediated through Mary just as the light of the Sun reflects off the disk of the Moon. "Had Christianity not taken root in the sun-baked east," says Warner, "the astral images it employs might have been very different."

Using the Sun as a symbol for God made switching to Christianity relatively easy, because converts didn't have to give up their familiar rituals and festivals. But it also meant that as Christians furiously denied links to paganism, with many choosing to die rather than sacrifice to pagan gods, aspects of solar worship were becoming embedded in their faith. As the archaeologist Jacquetta Hawkes put it: "With the malicious irony so often apparent in history, even while they fought heroically on one front, their position was infiltrated from another."

Constantine took that merging even further, though, identifying Christ not only with the Sun, but with himself. Just as Christ spread divine rays of light on Earth, so did the emperor. One fifth-century description of his sun statue in Constantinople, for example, details how even the city's Christians laid sacrifices at its base: "They venerate it with incense and candles, and they worship it like a god." Whether intended or not, Constantine's original choice to follow Sol was a political "masterstroke," says Bardill. It provided the bridge he needed to link sky worship with Christianity, allowing both pagans and Christians in his empire to unite behind one ruler and one supreme solar god.

This blending had profound consequences not just for Constantine's image but for how Christians saw their savior. The earliest known depictions of Jesus come from a private house converted into a Christian church around AD 235, in the city of Dura-Europos in today's Syria.

Described as the only surviving church walls from before Constantine's rule, they show a figure in a simple tunic, healing a paralyzed man; walking on water; tending his sheep. After Constantine's time, though, typical Catholic depictions were quite different. In the fifth-century church of Hosios David in Thessaloniki, Greece, for example, Christ sits on a celestial throne made of rainbows, dressed in purple robes, with a golden halo and his right hand raised.

Although now associated with Christianity, the halo (specifically, a disk behind the head called a nimbus) was originally used by pagan cults such as Mithraism to depict the divine nature and radiance of the sun god. Then Constantine depicted himself with one on his victory arch in Rome—a first for a Roman emperor. Only after that did Christians start using halos too, and in doing so triggered a transformation in which Christ took on more and more imperial features. Thanks to Constantine, the humble teacher became a cosmic emperor, ruling over the universe with the radiance of the Sun.

It's an image that remained powerful throughout the Middle Ages. Scholars argue about whether Christians who borrowed the halo intended to show their Messiah as a powerful emperor, or whether they simply hoped to communicate the radiance and lucidity of the Sun. Either way, once it was adopted, the image took hold, says Thomas F. Mathews, an art historian at New York University: "He *became* what people pictured him to be."

✡ ✡ ✡

Worship of the Sun and stars didn't just shape ideas about Jesus Christ. It is also at the root of modern Western beliefs about heaven and the fate of human souls. In a children's book called *What's Heaven*, author and former California First Lady Maria Shriver describes the afterlife as "a beautiful place where you can sit on soft clouds . . . when your life is finished here on earth, God sends angels down to take you to heaven to be with him." The idea that after we die, our disembodied souls float up to live with angels in the sky is popular among many Christians today. It would have shocked the ancient Israelites.

According to J. Edward Wright, director of the Arizona Center for Judaic Studies, who has studied early descriptions of heaven, people in ancient Israel believed—as Paleolithic and Neolithic societies seem to have done—that the cosmos had three tiers: a flat Earth, with the underworld beneath it and the sky above. The Hebrew Bible (repurposed in the Christian Old Testament) describes the sky as a tent or canopy stretched over the Earth, but also as a solid "firmament," with a stone floor and storehouses for meteorological phenomena such as wind, snow and hail. These descriptions use the imagery of earthly cities and palaces: heaven has entry gates, for example, and a central throne room, from which Yahweh rules the universe, surrounded by a council of holy beings.

But just as commoners weren't welcome in the royal palace, this heaven was not for ordinary humans. The Hebrew Bible doesn't say much about what comes after death, notes Diarmaid MacCulloch, a historian at Oxford University and author of *A History of Christianity: The First Three Thousand Years*. "What it does say . . . suggests that human life comes to an end and, for all but a few exceptional people, that is it." Neighboring societies had similar beliefs. In *The Epic of Gilgamesh*, the innkeeper Siduri counsels the hero to give up his search for immortality, telling him that the gods keep eternal life for themselves.

Similarly, the epics of Homer—our earliest literary record of Greek thought, dating from around the eighth century BC—also include no recognizable heaven for the vast majority of people. The true "self" was the physical body, and although a soul or *psyche* was thought to survive after death in a dark, dusty underworld, this was a mere shadow of the living person. In the *Odyssey*, it's a fate that horrifies the hero Achilles: "Never try to reconcile me to death," he tells Odysseus, adding that he would rather be a poor man's servant on Earth than be "lord over all the dead that have perished."

After the sixth century BC, however, that changed. When Greek philosophers broke from mythological accounts and looked instead for physical explanations of the cosmos, their models also fed into religious beliefs, not just in Greece but in the Near East as well. We heard in chapter 3 how Aristotle set out a system of concentric celestial spheres carrying the Sun, Moon and planets around the Earth; this idea inspired the "seven heavens" described in many Jewish and Christian (and later Islamic) texts. But in an earlier and even more fundamental shift, Aristotle's teacher, Plato, revolutionized ideas about the soul.

One of Plato's most famous teachings, written in the fourth century BC, is a story in his dialogue *The Republic* in which prisoners are chained in a cave, facing the wall, so that all they can see are shadows on the rock. The captives believe that these shadows represent real things, but in fact they are just reflections of reality, which exists in the light outside the cave. Likewise, Plato argued, material things that we perceive in our lives, and believe to be real, are mere reflections of the unchanging ideas or "forms" that lie behind them. It's a philosophy in which matter is secondary to consciousness, and physical objects are derived from ideas.

Not surprisingly, then, Plato thought that souls are more important than bodies. In another dialogue, *Timaeus*, one of his characters describes a benevolent god who shaped the chaos of the cosmos into an ordered system of celestial spheres, crafting first its soul, and only then its physical form. Plato suggested that humans, too, have immortal souls, which originate in the divine realm of the stars.

He described a scheme in which each soul descends through the planetary spheres toward Earth, where it becomes joined with a physical body at birth. When we die, our soul is released from that body and either it is reincarnated or, if we're virtuous enough, ascends through the spheres to return to its place in the stars. "We ought to fly away from earth to heaven as quickly as we can," he wrote, "and to fly away is to become like God." (It's an idea later echoed by the painter Vincent van Gogh in a letter to his brother in 1888: "As we take the train to get to Tarascon or Rouen," he mused, "so we take death to come to a star.")

We each have within us a divine spark, a perfect soul inside our frail, mortal body that can travel to the heavens and live eternally among the stars. It's a remarkable, inspiring idea. So where did Plato get it from? That's a story that was hidden for thousands of years.

✡ ✡ ✡

On the edge of the Egyptian desert south of Cairo, near the village of Saqqara, there's a collection of half-ruined pyramids. Part of a vast royal burial ground for the ancient capital of Memphis, they are smaller and for the most part younger than the famous pyramids of nearby Giza. They were built from blocks of limestone around a core of rubble, but the limestone itself has long since been stolen. Once around 200 feet tall, they now resemble low, crumbling hills.

On January 4, 1881, two brothers from Berlin—Heinrich (Henri) and Émile Brugsch—visited these rubble piles, hoping to explore the underground chambers hidden beneath. The ancient entrance passage of the first pyramid they tackled was sealed by a heavy granite trapdoor, so they squeezed on their bellies through a narrow tunnel cut by looters centuries earlier. Henri was terrified that the huge, ruined stones suspended precariously above would fall and crush them, but eventually the brothers dropped unscathed into a subterranean corridor. "What a surprise awaited me, what reward to my efforts!" Henri wrote. "Wherever I looked, right and left, the smooth limestone walls were covered in innumerable texts."

The hieroglyphs were expertly carved and arranged in columns. Stooping low, stepping over stones and boulders, the brothers clambered along the corridor and into a wider chamber. It had a peaked limestone ceiling, which was painted black and covered with white, five-pointed stars. Here, too, the walls were covered with hieroglyphs.

In the dim candlelight, they read the same name over and over: Merenre, beloved of the Sun.

During the nineteenth century, just as in Mesopotamia, the colonial powers were enthusiastically combing Egypt for ancient treasures. Near Luxor in the south, European explorers were opening royal tombs dug into cliffs in the Valley of the Kings. Although the tombs had almost invariably been emptied by looters, the art and inscriptions painted on their walls yielded invaluable information about the history of this mysterious civilization. Disappointingly, though, the chambers of the much older pyramids at Giza were completely blank. Auguste Mariette, the aging Frenchman in charge of Egypt's antiquities department, became convinced that all of the pyramids were "mute."

It wasn't even worth the trouble of opening the smaller ones, he argued. The French government disagreed, and in the summer of 1880, a team of local workmen burrowed into one of the Saqqara pyramids and reported finding hieroglyphics inside. Mariette refused to believe it, insisting they must have entered a nobleman's tomb by mistake. But in December that year, when he was seriously ill in Cairo (suffering complications of lifelong diabetes), news came of a second pyramid apparently full of texts. With his health fading fast, he sent his longtime colleagues, the Brugsch brothers, to check.

On the morning of January 4, Henri and Émile took a train south from Cairo, then a two-hour donkey ride, to reach the newly entered pyramid. To the west of the star-painted room was another chamber, also with a peaked, starry ceiling. The walls here, too, were covered in columns of rich inscriptions. And in the corner they saw a sarcophagus made from red-speckled granite, its lid shoved back. It was carved with more hieroglyphics, which Henri roughly translated: "The Great God and Lord of the Light Zone, Living Like the Sun."

On the floor beside the sarcophagus lay the embalmed body of a young man. Originally wrapped in fine linen, the mummy's bindings had been torn off by looters, and the shreds were strewn around it like cobwebs. The body had been stripped of amulets and jewelry, but from the high quality of the embalming, with the mummy's delicate facial features still recognizable, Henri concluded it belonged to the pyramid's owner, King Merenre.

The brothers returned to Cairo, with Henri determined to take the 3,000-year-old pharaoh to Mariette that same evening. "Perhaps, I said to myself, it will afford the dying friend a last pleasure," Henri wrote, "to be able to see with his own eyes the mummy of one of the oldest kings of Egypt and indeed of the world." They put the mummy in a wooden coffin and strapped it to a donkey for the two-hour ride to the station, then heaved it into the baggage car of the Cairo train, telling the surprised guard they were accompanying an embalmed mayor of Saqqara. Damage to the rails meant the train stopped short of the city, and as the sun set, they had to walk the last few miles.

"To lighten the load, we left the coffin behind and held his dead Majesty at the head end and at the feet," recalled Heinrich. "Then the Pharaoh broke through the middle and each of us took his half under his arm." They finally delivered the mummy to Mariette (who was reportedly horrified by the sight of the battered king) just a few days before the old man died.

It wasn't the mummy, though, but the newly discovered "Pyramid Texts," as Heinrich called them, that turned out to be the most important find. Similar inscriptions have now been discovered in ten Saqqara pyramids owned by kings and queens of the fifth and sixth dynasties, dating from the twenty-third and twenty-fourth centuries BC. They don't contain historical details, but they are the oldest and most extensive source of information we have about ancient Egyptian religious beliefs. They show that while other Near East civilizations held that heaven was just for the gods, Egypt was different.

The Egyptians' view of the cosmos was complex and full of mixed metaphors. They imagined the sky as an ocean, crossed by the Sun god Ra in his celestial barge, but also as the belly of an enormous falcon, or of the goddess Nut, either in the form of a cow or a woman, who arched over the Earth god Geb on her hands and feet, eating the celestial bodies each night and giving birth to them in the morning. Life was a daily cycle, which revolved around Ra's repeated death and resurrection. At sunset, Ra died and passed into the netherworld below the Earth. During the night, he merged with the body of Osiris, depicted as a mummy lying deep in the netherworld. Through this union, he received the power of new life, and at sunrise was reborn.

The pyramids were built to enable the pharaohs to mirror this journey. They were made from stone to last for eternity, says British Museum Egyptologist John Taylor, and were intended as "not simply the resting place for the corpse, but the interface between the world of the living and the resurrected dead." It's an intriguing echo of theories about Neolithic monuments such as Stonehenge, which was in use at about the same time.

As with the Sun, the pharaoh's soul was believed to merge each night with his physical body—the mummy in the tomb—in order to be reborn the next day. The Pyramid Texts were a collection of spells and incantations to help the process. They describe various stages, culminating in the king rising up with the Sun and taking his place among the stars. "I row in the sky in your boat, Sun," declares one inscription, while another reads: "A footpath to the sky is laid down for me, that I might go up on it to the sky."

The Saqqara pyramid chambers were laid out to reflect this daily journey, with the resurrected king initially traveling east out of the burial chamber toward the rising sun. The exit corridor then turns north, perhaps to point toward the circumpolar stars, which rotate around the northern celestial pole. The Egyptians associated these constellations with immortality, calling them the "Imperishable Stars" because they never set. In several places, the Pyramid Texts name them as the deceased king's ultimate destination, although other stars feature prominently too, including Orion's Belt (associated with Osiris), Sirius (the goddess Isis), and the Lone and Morning Stars (both thought to be the planet Venus, which appears as a lone, bright star at dawn and dusk).

Although the bigger, older Giza pyramids have no inscriptions, the kings who built them may well have shared similar beliefs. The Pyramid Texts are thought to have already been ancient by the time they were written down. And Giza's three main pyramids all face due north, toward the celestial pole. The oldest and largest of all, Khufu's pyramid, built in the twenty-sixth century BC, is oriented to within a twentieth of a degree. That's "maniacal precision," says Italian astrophysicist and archaeoastronomer Giulio Magli, who has studied the pyramid, and strongly suggests an "obsessive interest in the circumpolar stars." This pyramid also has internal shafts running north and south from the main burial chamber. Astronomers have calculated that at the time the pyramid was built, the two shafts pointed accurately toward the circumpolar stars, and to the highest rising point of Osiris (Orion's Belt). Perhaps they were "symbolic pathways," suggests Magli, to propel the king's soul into the sky.

The Pyramid Texts were only available to the king and his family, who had their own pyramids. In later centuries, though, similar spells were written inside coffins and on papyri for nonroyals too. The Book of the Dead, which emerged around 1600 BC, appears to have been commonly used, and included spells to help the owner pass a test in which his or her heart was weighed to judge whether they deserved to enter heaven. The Egyptians didn't invent the idea of a divine realm in the sky—that seems to be an almost universal belief—but they are the first we know of to see this as an ultimate destination for virtuous human souls.

Egyptian beliefs are often seen as a historical dead end: a lost religion, fascinating but barely relevant to today's concepts of the afterlife. But historian Nicholas Campion, director of the Sophia Centre for Cosmology and Culture at the University of Wales, argues that in fact, this is where it all started; that it was probably the Egyptians who inspired the concept of the immortal soul in Greece. The Greek author Herodotus said as much, Campion points out, writing that the Egyptians were "the first people to put forward the doctrine of the immortality of the soul." That's not proof in itself—Herodotus got a lot of things wrong—but the link is plausible. In the sixth century BC, Greece and Egypt came under the common rule of Persia, providing the opportunity for Greek philosophers to mix with Egyptian priests.

Ancient biographies of the mathematician and philosopher Pythagoras—who is thought to have been among the first in Greece to propose the idea of an immortal soul and who heavily influenced Plato—claimed that he studied at temples in Egypt before opening a school in southern Italy. The Egyptians are often sidelined in the history of astronomy; from a scientific point of view they were nowhere near as advanced as their Mesopotamian neighbors. Nonetheless, says Campion, they were fundamental to the development of Western cosmological ideas. While the Babylonians provided the math, he says, the Egyptian contribution was metaphysical: "the inclusion of the soul."

Thanks to Pythagoras and especially Plato, the idea that our souls belong in the stars became popular throughout the Greek and then the Roman world, including the belief that contemplating the cosmos therefore draws us closer to God. "Observe the movement of the stars as if you were running their courses with them," advised the second-century Roman emperor Marcus Aurelius. "Such imaginings wash away the filth of life on the ground." Plato's ideas also contributed to a variety of "mystery" cults, still popular in Constantine's time, such as Gnosticism, Hermeticism, and Mithraism. All promised to impart secret knowledge about how to prepare the soul for its journey to heavenly realms. Whereas the Jewish God was passionate and interventionist, these religions worshipped "the one": an unchanging, immaterial deity that radiated knowledge and light.

Platonic concepts seeped into Judaism. The idea that the faithful could expect to join God in heaven is hinted at in the Hebrew Bible twice, but only in later books. Ecclesiastes, composed in the Persian period, is unconvinced, asking: "Who knows if a man's spirit rises upward?" But Daniel, written after the conquests of Alexander the Great, comes down in favor: "Those who are wise will shine like the brilliant expanse of the sky, and those who lead many to righteousness will be like the stars." Centuries later, Islam inherited similar ideas about an eternal afterlife in the sky.

Plato's biggest influence, though, was on the newborn faith of Christianity. "By the time Christians were beginning to construct their own literature, their writers clearly found such talk of the individual soul and resurrection completely natural," says MacCulloch, "and it became the basis of that Christian concern with the afterlife which sometimes has bordered on the obsessional."

✡ ✡ ✡

Shortly after Easter in AD 337, while preparing to invade Persia, Constantine fell ill. He visited a spa at his mother's city of Helenopolis, then, too sick to make it back to Constantinople, he traveled instead to nearby Nicomedia, where he summoned a group of bishops. He switched his royal purple gown for a pure white robe and was finally baptized, just a few days before he died.

His body was taken home in a golden coffin, and placed in the sumptuously decorated Church of the Holy Apostles, where he had prepared himself a sarcophagus. It was surrounded by twelve empty tombs, which were intended to hold the remains of Jesus's disciples.* Some historians see this as perhaps the clearest indication that Constantine really did equate himself with Christ. Others have suggested that as the Apostles symbolized the twelve signs of the zodiac, the emperor could equally have been presenting himself as the Sun. Constantine's official memorial coin, issued shortly after his death, featured him rising to the sky, like Sol, in a four-horse chariot and radiate crown.

* The apostles' tombs and relics were removed by his son Constantius II. Constantine is now seen not as a deity, but as a saint.

Most likely, it was both at once. Constantine fixed Christianity into history as a major global force; scholars agree that he ensured the future of the Catholic Church and is largely responsible for the reach of Christianity around the globe today. But to do it, he embedded within the faith the image of his divine, radiant Sun. It's a story that reflects the broader religious ambiguities of his time, during which the changing fortunes of rival empires brought together clashing worldviews about the nature of the universe and our place in it. In many cases, monotheism didn't eliminate old beliefs about gods in the sky as much as it assimilated and adapted them. Christian concepts such as heaven, the soul and an eternal afterlife were woven over centuries, out of ancient and multicolored threads from Egypt, Persia, Israel, Greece, Rome and beyond.

Not in every case, though. In the eighteenth century, the political writer Thomas Paine, known, among other things, for his searing attacks on religion, described Christianity as "a parody on the worship of the Sun, in which they put a man whom they call Christ, in the place of the Sun, and pay him the same adoration which was originally paid to the Sun." The early Christians certainly did take symbols and rituals from the rival Sun cult. But there is a fundamental difference between the two that Paine's criticism doesn't acknowledge, an aspect of Plato's thought that mainstream Christianity did not ultimately accept.

Plato's creator existed within the universe, fashioning the celestial spheres from the material he had available. "Out of disorder he brought order," Plato wrote in *Timaeus*, shaping a world that was "as far as possible a perfect whole and of perfect parts." By contrast, the Jews believed in a transcendent God above and beyond the cosmos, who made the world from nothing. "The heaven is my throne, and the earth is my footstool," says the Hebrew Bible's book of Isaiah.

In general, Christians stuck with the latter view. Eusebius, for example, was happy to liken Constantine to the Sun: "As the sun . . . liberally imparts his rays of light to all," he wrote, "so did Constantine, proceeding at early dawn from the imperial palace, and rising as it were with the heavenly luminary, impart the rays of his own beneficence to all who came into his presence." But at the same time, he made clear that even the Sun is not divine in itself, but just part of God's creation. Constantine might have been loyal to Sol, but for Eusebius, the ultimate focus of worship was outside the universe. The emperor's support of Christianity, he said, meant that men were no longer "to look with awe upon the Sun or moon or stars and attribute miracles to these, but rather to acknowledge the one above these, the invincible and imperceptible Universal Creator, having learned to worship him alone."

That's the position still held by the Catholic Church. Guy Consolmagno, chief astronomer at the Vatican Observatory, put it clearly in 2013: "The God I believe in is not of the universe," he wrote, "but existed before the universe began; not a part of nature, but supernatural." As discussed earlier in this chapter, it's a belief that allows for an all-powerful God, unlimited by the universe's existing rules or resources. But that's not all. It also has profound consequences for the cosmos itself.

Plato's cosmos was a living, intelligent creature, divine in its own right, with its own soul that spread through all of reality. Consequently, everything within the universe shared in that soul: from animals and people, with our mortal bodies, to the stars, which Plato described as "divine and eternal animals." Largely thanks to him, such beliefs were common in the classical world. Five centuries later, the Roman historian Pliny the Elder described the Sun as "the soul or . . . the mind of the whole world." Earlier societies with celestial gods, such as the Egyptians and Mesopotamians, whatever their views on the postmortem fate of human beings, would also have seen the universe as an interconnected, living system. When they made the Sun, Moon and planets into gods, they were really worshipping the cosmos itself.

The significance of the switch to monotheism, then, goes beyond the reduction from many gods to one. It's really a transformation in the kind of universe in which we live. The fifth-century theologian Augustine of Hippo cemented the mainstream Christian view. He had huge respect for Plato: "None come nearer to us [Christians] than the Platonists," he said. But he dismissed the idea of a cosmos infused with God's soul. "Who cannot see what impious and irreligious consequences follow," he said, "such as that whatever one may trample, he must trample a part of God, and in slaying any living creature, a part of God must be slaughtered?" In effect, says Campion, Augustine "rejected the notion of the universe as a living creature . . . Instead, some parts of the world were no longer alive and were distinguished from those that were."

I'm not sure that Constantine himself would have seen it that way. But his conversion marks the moment, for Western civilization at least, when humanity rejected the cosmos as a divine, living being, as all that there was. Instead, it became merely the product of a separate creator. Where once humanity's fate was determined by the movements of the celestial bodies and the stars gave home to the gods, now we were no longer at the center of a universe that encompassed everything. It was possible to imagine stepping outside it and looking down.

Our religious beliefs remain steeped with the influences of the Sun, Moon and stars. But one more tie with the cosmos was cut.

5

TIME

IN THE ARCHIVES OF OXFORD UNIVERSITY'S BODLEIAN LIBRARY, there's a small but unusually chubby book. Known simply as *MS Ashmole 1796*, it consists of around two hundred pages of calfskin, bound between leather-covered wooden boards. Written in the fourteenth century, its square, black lettering is adorned with painted capitals and interspersed with neat diagrams showing complicated mechanical arrangements of axles and wheels.

The book was bequeathed to the university in the seventeenth century by the collector and astrology enthusiast Elias Ashmole, and lay unstudied for hundreds of years. In 1965, an Oxford astronomy historian named John North finally translated its Latin script, along with a note in the margin, *Hic est liber sancti Albani*, which translates as "This is a book of St Albans." He realized that hidden among these pages was something special: a description of a spectacular invention to which a monk called Richard of Wallingford, the abbot of St. Albans monastery from 1327 to 1335, dedicated much of his life.

The invention was a colossal clock, installed high beneath the great south window of the abbey's church. But Richard's construction was no ordinary timekeeper. It was a sophisticated, self-moving model of the cosmos, unlike anything else known when it was built, and so far ahead of its time that when the librarian and antiquities specialist

John Leland saw it two centuries later in 1534, he described it as a marvel still without equal in all of Europe. The clock demonstrated the workings of the universe, Leland said, from the oceans to the heavens: "One may observe the course of the Sun, the Moon, or the fixed stars, or again one may regard the rise and fall of the sea."

No trace of the clock now remains, and for modern historians reading those words it was hard to know what Leland really saw—until North realized in the Bodleian Library that he was holding Wallingford's instructions for building this very machine. The secrets he discovered reveal not just an impressive invention, but a crucial moment in human history.

This chapter, then, tells the story of why and how the monks of medieval Europe chased down time, and how in doing so, they transformed humanity. As we'll see, their efforts to track daily cycles ever more accurately were ultimately successful beyond their wildest dreams. Yet they also destroyed the very thing they pursued. Until this point in history, time was a sign of divine cosmic order, as shown to us through the cycling motions of the Sun, Moon and stars. The invention of mechanical clocks unleashed a very different kind of time, powerful enough to weaken our bond with both God and the universe, and set the foundations of a new way of life.

✡ ✡ ✡

Richard was born in 1291 or 1292, the son of a blacksmith, in the town of Wallingford, Berkshire, about fifty miles west of London. His father died when Richard was ten, and within a year or two, he was adopted by the head of the local Benedictine priory, "on account of his loneliness and aptitude and his great promise." The prior later paid for him to study at nearby Oxford University, where he apparently neglected theology, preferring astronomy and math.

The Benedictine order (with members nicknamed "Black Monks," for the color of their habits) was popular throughout Europe. In 1314, aged about twenty-two, Richard traveled to the monastery in St. Albans and took his vows to become a monk himself, probably because he needed the abbot's financial support for his studies. Founded in the eighth century,* St. Albans was one of the most powerful monasteries in England, boasting swaths of land and a huge church (with a nearly 300-foot-long nave, reportedly the longest in the Christian world).

After being ordained as a priest in 1317, Richard returned to Oxford, where he studied for nine more years and wrote several influential treatises. One described his invention of an astronomical instrument called the albion, a series of metal disks that could be used to calculate the positions of the planets in the sky. Another covered astrology; influenced by Ptolemy's *Tetrabiblos*, it explained how to forecast the weather—events such as floods, droughts, storms and tides—from the arrangement of heavenly bodies (he was right, of course, about the tides).

In September 1327, Richard visited the abbey at St. Albans and his life changed again: the abbot, Hugh Eversdone, died while Richard was there, and Richard was voted his successor. According to the abbey chronicles, he was reluctant to accept; perhaps he realized just what a challenging situation he was getting into. The abbey was heavily in debt. Part of the church had collapsed and still wasn't completely repaired. And England was split by civil war.

In January that year, King Edward II had been deposed and imprisoned by his French wife, Queen Isabella, and her lover, Roger Mortimer; the pair then ruled on behalf of Isabella's young son, Edward III. In the months that followed, there was political instability along with widespread social uprisings as different factions fought for power. The violence was particularly fierce in monastic towns like St. Albans, where the crisis fueled existing public resentment against the abbot.

The monasteries were among England's great landowners, and they effectively controlled the lives of the townsmen and peasants who

* There had previously been a church at this site, built to commemorate Saint Alban, who was martyred during Emperor Diocletian's persecution of the Christians in the fourth century AD.

occupied their property. The commoners had to pay steep tolls and taxes, and were forbidden to hunt or poach on the abbot's estates. In January 1327, the citizens of St. Albans attacked and laid siege to the abbey, demanding a wide-ranging charter of rights that included representation in Parliament. Abbot Hugh had previously enjoyed the support of Edward II; now, under pressure from Queen Isabella, he was forced to grant almost all of their requests—a humiliation that may have contributed to his death.

By September, the buoyant townsmen were rebelling further—destroying nearby woodlands, for example, and invading the abbey's warrens and fishponds. Richard would have a fight on his hands. First, though, he had to travel to the pope's court in Avignon, France, to have his election as abbot confirmed. On his trip, Richard would have seen the grand cathedrals and bridges of continental Europe, as well as read new manuscripts and met some of the leading scholars of the day (the philosopher and theologian William of Ockham, to name one, was at the Avignon court at the same time). It must have been an inspiring journey. But on the night Richard arrived back at the abbot's manor, in April 1328, he felt a burning sensation in his left eye. He had developed leprosy.*

✡ ✡ ✡

Almost exactly a thousand years had passed since Constantine unified the Roman Empire behind the new Christian capital of Constantinople. In the centuries after the emperor's death, civilization flourished in his eastern territories (later known as the Byzantine Empire), as well as in the Islamic world, where scholars translated and built on the knowledge of classical antiquity. But the western part of the Roman world never recovered from the shift in influence to the

---

* Richard believed he had leprosy, although some historians have suggested that he might have been suffering from tuberculosis, or syphilis.

east. Partly because of a huge influx of Germanic migrants from the north, the emperors in Constantinople were unable to hold control over these distant regions. Rome finally fell in AD 476, marking the end of antiquity and the start of the Middle Ages.

Europe was thrown into violent disarray as the social, economic and political infrastructure that had existed for centuries broke down, and long-distance trade collapsed. The main points of light—fortified, stable centers of continued power and (some) learning—were the Christian monasteries. Religious hermits in the Egyptian desert had started to form communities as early as the third century, and the practice soon spread into the West. Life in these monasteries came to be regulated by strict schemes such as the sixth-century "rule" of St. Benedict (who founded the Benedictine order). And what governed the monks' daily existence, perhaps more than anything else, was an obsession with time.

Day and night were divided into strict time slots, filled with rounds of study and manual work punctuated by regular, communal prayers. This "temporal discipline," as Harvard historian David Landes describes it, distinguished western Christianity from the other monotheistic religions. In Judaism and Islam, as well as the eastern Christian churches, daily prayers were conducted according to natural cues such as sunrise, noon and sunset. But in western Christianity, especially monastic Christianity, there was a growing focus on regularity and punctuality. Extra rounds of worship were introduced, several of which were defined not by watching the Sun, but by counting the hours of the day. Benedict described seven daily prayer times: as well as morning, evening and night, the monks had to pray together at the first, third, sixth and ninth hours (called prime, terce, sext and none).*

Benedictine rule was adopted by several other monastic orders, and "canonical hours," as they were called, soon became the official policy

---

* The count started at sunrise, or around six a.m. Centuries later, *none* shifted from midafternoon to midday, giving us the modern word "noon."

of the Catholic Church. The strictness and efficiency of such schedules enabled the monks to be highly productive, as well as helping their way of life to survive despite the bloody, turbulent times. But there was a deeper motivation behind the close attention to the clock.

In a society where everyone had their role to play—knights defended the realm; peasants grew food—the monks were responsible not just for their own souls but those of all Christians. Their vocation, according to Landes, was "to pray and pray often, and in so doing to save that multitude of the faithful whose worldly duties or inconstancy prevented them from devoting themselves entirely to the service of God." That duty revolved around the daily cycle of collective prayer, which was marked by the sound of bells. It was crucial to be on time, to avoid cutting short the worship and to ensure that everyone's prayers could be synchronized: chanting out loud, together, was thought to make the efforts more powerful. Good timekeeping, then, was more important even than life and death. The spiritual salvation of humanity depended on it.

But how to keep time accurately enough? The ultimate arbiter of time had always been the motion of the cosmos itself. The first humans marked days and seasons by the movements of the Sun; months by the cycles of the Moon. The ancient Babylonians and Egyptians, followed by the Greeks and Romans, used sundials to partition the day and the rising of stars to tell time in the dark. Although astronomers divided days into twenty-four equal hours as we do today, wider society kept so-called seasonal hours, in which daylight and darkness were each divided into twelve, so their length varied through the year.* Terrestrial methods such as sand timers and water clocks (in which a float was attached to a pointer on an hour scale) were used to measure shorter time periods—everything from senators' speeches to sessions with prostitutes.

In the Middle Ages, though most areas of scholarship regressed, the

* Except the Babylonians, who divided the day into twelve double hours (they also made use of both equal and seasonal hours).

art of timekeeping flourished, as Christian monks and scholars looked for ever better ways to ensure that prayers happened on time, particularly when it was dark. In the 570s, for example, the bishop Gregory of Tours gave detailed instructions for watching the rising of certain stars and measuring the passage of time by chanting psalms. This still meant, however, that someone had to stay awake all night.

As the political situation in Europe started to stabilize in the tenth and eleventh centuries, contact with the Muslim world, particularly in Spain, boosted Western scholarship. At around this time, monks started using water-driven alarms. Although described as clocks, these didn't tell the time as such. Instead, they were attached to a simple mechanism that after a certain number of hours would repeatedly strike a small bell, to get the ringer of the monastery's main bells out of bed (as immortalized in the popular children's rhyme "Frère Jacques").

The earliest detailed account of such a clock is in a tenth- or eleventh-century manuscript from a Benedictine monastery in northern Spain, but they soon became common around Europe. When a fire broke out at the abbey in Bury St. Edmunds in 1198, the monks ran to the clock for water. These alarms weren't particularly accurate, though, and they had to be reset from scratch each evening. Then, toward the end of the thirteenth century, came a revolution.

All previous clock designs—such as sand timers, burning candles and water clocks—had attempted to create an even, continuous flow. After all, that's apparently the nature of time itself. But it's actually very difficult to keep something moving at a perfectly constant speed. In the thirteenth century, clockmakers started experimenting with descending weights attached to a cord or chain wrapped around a barrel, the rotation of which drove the mechanism of the clock. But the weights fell too quickly and their speed wasn't even: they accelerated as they fell to the floor.

The crucial breakthrough was to add an oscillating device—called an escapement—that alternately blocked and released the moving train of wheels, so that the weight fell in controlled, regular chunks. It was, says Landes, "among the most ingenious inventions in history." Instead of trying to measure time as a continuous flow, an escapement divides it into regular beats—the "ticks" of the clock—which can then be counted. Until that moment, the Islamic world and China had produced the world's most sophisticated clocks: from rugged gearwheels driven by fast-moving streams, described by the Andalusian engineer al-Murādī; to a clocktower, powered by water or mercury, designed by the Chinese imperial tutor Su Sung. From then on, Europeans would be the global masters of time.

The exact origin of the escapement is lost to history. In 1271, this device apparently did not yet exist; the English astronomer Robertus Anglicus wrote that clockmakers experimenting with lead weights to make a wheel turn once each day "cannot quite perfect their work." Just a few years later, church financial records reveal that a new type of clock had appeared, which was expensive, often built or repaired by specialist craftsmen, and erected high on a church wall.

The earliest mentions are all from English churches, starting with an Augustinian priory in Dunstable, Bedfordshire, where in 1283 the monks made themselves a clock to go over the *pulpitum* (the massive screen that divides the choir of a church from the nave). Following soon after were Exeter, Ely, Canterbury and St. Paul's in London—where a clockmaker called Bartholomew was paid 281 rations for his work—all before the end of the century. (The first records from the rest of Europe appear slightly later—for example, an iron clock built in Milan in 1309.) The first literary allusion to a mechanical clock also appears at around the same time, in the thirteenth-century French poem "*Le roman de la rose*":

And then through halls and chambers

Made his clocks chime

By wheels of such cunning

Ever turning through time

These clocks didn't necessarily have dials or hands; their purpose was to ring the bells for prayers. The monks, however, weren't the only ones chasing machines that could turn themselves.

☼ ☼ ☼

In the first century BC, the Roman author Cicero wrote about two "spheres" built by the great mathematician Archimedes. Both had been taken by the Roman general Marcellus when he sacked Archimedes's home city, Syracuse (in what's now Sicily), in 212 BC. He donated one of the objects—a solid globe marked with stars and constellations—to a temple in Rome. But the other one was so special that Marcellus kept it for himself. To build it, wrote Cicero, Archimedes must have been "endowed with greater genius than one would imagine it possible for a human being to possess."

This gem was a mechanical model of the cosmos, which showed the movements of the Sun, Moon and five known planets as seen from Earth. Cicero said that Archimedes "had thought out a way to represent accurately by a single device for turning the globe those various and divergent movements with their different rates of speed." When the sphere was moved, "the moon was always as many revolutions behind the sun on the bronze contrivance as would agree with the number of days it was behind it in the sky."

For much of history, Cicero's account wasn't taken particularly seriously. No such devices survived from antiquity and his description (written, after all, in a fictional dialogue) went far beyond what ancient Greek craftsmen were thought capable of. But in the last few decades, researchers have been studying a mysterious bronze device, recovered by sponge divers in 1901 from a first-century-BC shipwreck near the Greek island of Antikythera. Its corroded inscriptions, gearwheels and dials are so intricate that some writers initially claimed it was an elaborate hoax (or even made by aliens). It is genuine, however, and its battered workings have taken more than a century to decipher.

Researchers now realize that it was a machine to model the motions of the heavens, just as Cicero had described.

Originally held in a wooden case roughly 12 inches high, the Antikythera mechanism had a large bronze dial on the front that showed the varying movements of the Sun, Moon and planets in the sky. A revolving black-and-white ball revealed the Moon's phase, while inscriptions listed star risings at different times of year. On the back were two spiral dials: a 235-month calendar (with a 4-year inset dial showing the timing of athletic events, including the Olympic games) and a 223-month cycle for predicting eclipses.

In other words, it was a portable cosmos: a universe encapsulated in mechanical form. The varying speeds of the different pointers were calculated using complicated trains of bronze gearwheels (more than thirty survive, but it probably originally had many more) with differing numbers of teeth—the cogs that today we call "clockwork."

Nothing close to the sophistication of the Antikythera mechanism appears again in the historical record for well over a thousand years—not, in fact, until Richard of Wallingford's clock. But the idea of using gearwheels to model astronomical motions was not completely forgotten. In the Science Museum in London, there's a sixth-century Byzantine sundial with a mechanical calendar attached that uses eight gearwheels to display the positions of the Sun and Moon in the sky. An identical calendar appears on a thirteenth-century astrolabe from Isfahan, Iran. There are hints that such knowledge was eventually transferred back to Europe: in 1232, for example, ambassadors of the sultan of Damascus presented a jewel-studded planetarium to the Holy Roman emperor Frederick II.* How it worked isn't known, but a

* Not to be confused with the Roman Empire, the Holy Roman Empire was a varying territory in western and central Europe ruled by Frankish and then German kings from 800 to 1806.

contemporary manuscript that might refer to it talks about "a device of certain wonderful wheels."

All of these devices had a severe limitation, however: they had to be turned by hand. Water has been used throughout history to drive astronomical displays. For example, the Roman architect Vitruvius, writing in the first century BC, described an impressive type of water clock that turned a large bronze disk engraved with stars.* But hydraulic forces aren't powerful enough to drive complex geared mechanisms. In the thirteenth century, as scholarship in Europe once again thrived, theologians and philosophers became intensely interested in creating a model of the cosmos that could turn itself.

In 1248, for example, the Franciscan friar Roger Bacon wrote a treatise arguing that science and technology could achieve far greater wonders than magic. He described inventions such as submarines, flying machines, curved lenses and gunpowder, before moving on to something that he said was more valuable than any of these: a self-moving astronomical sphere. Such a device didn't yet exist, but Bacon believed it was possible. In later works, he hinted that it could be powered by magnetism,† and referred to "one of the greatest of secrets in the experimental sciences, or indeed anywhere . . . a body or instrument which is to move with the motion of the heavens, and transcend all instruments of astronomy."

In other words, the first inventors trying to build perpetual motion machines weren't trying to get free energy or to "cheat nature." They

* Such a clock is thought to have filled the Tower of the Winds, an octagonal building built in the first century BC, which still stands in the Roman marketplace in Athens. Fragments of bronze dials like those described by Vitruvius, dating from the second century AD, have also been found in Salzburg, Austria, and Grand, France.

† He may have been referring to the work of the French scholar Pierre de Maricourt, also known as Peter Pilgrim, who had been experimenting with natural magnets and concluded (wrongly, though it was an ingenious idea) that the north and south poles of a magnet align with the celestial poles. He suggested that a spherical magnet angled to the appropriate point in the sky could therefore be spun around as the Earth turned.

were trying to make a miniature cosmos that would turn itself. Such a model, said Bacon, would be more valuable than the treasure of a king, for it would hold the secret of the workings of the universe. So when the inventor of the escapement first sliced up time just a few years later, the implications reached far beyond improving the accuracy of prayers. Two great mechanical traditions—timekeepers and astronomical models—were about to collide.

✡ ✡ ✡

When Richard of Wallingford returned from Avignon to St. Albans, his main task was to take back control from the townspeople and remove the rights they had won from Abbot Hugh. One particular point of conflict was the milling of flour. The peasants were legally required to pay for use of the abbey's horse mill. But they had started building their own smaller hand mills, meaning a loss of profit for the abbey.

Richard's strategy involved aggressive legal attacks on the town's most influential citizens. He accused them of moral crimes such as adultery, then wined and dined the judges and built influence with the new king, Edward III. By May 1332, the townsmen were forced to surrender their charter of rights—and their eighty or so hand mills. Richard confiscated the millstones and had them cemented into the abbey's parlor floor.

Meanwhile, his disease was advancing. He was physically weak, blind in one eye, his face was horribly disfigured, and he could barely talk. But his victories seem to have earned him the respect of his monks. He survived two plots by rivals to overthrow him, and embarked on various building works at the abbey. All of these events, though, were just distractions from Richard's main plan. He was determined to build on the recent revolution in timekeeping to create something that the world had never seen: a magnificent celestial clock.

MS Ashmole 1796 isn't signed, but notes in the margin suggest it was copied from Richard's papers soon after his death, by a scribe at St. Albans.* The book contains a range of math and astronomy texts, including several known to have been written by Richard of Wallingford. Mixed up with these, though, were some jumbled, unedited drafts not known from any other source: instructions for building his clock. These are the earliest detailed description of any mechanical clock known, the closest insight we have into the origins of an invention that changed the world. From them, North was finally able to reconstruct the details of this "cosmic machine."

The notes describe how a falling weight, controlled by an oscillating escapement, drove a hefty iron gearwheel, which rotated once each day with the Sun and was connected to a bell-striking mechanism that rang every hour. This escapement consisted of a semicircular piece of metal that was pushed backward and forward by alternating pins on two parallel wheels. It differs from the so-called "verge and foliot" escapement that became common later and which, until North's discoveries, was assumed to be the original design.†

Richard doesn't seem to have invented this part of the mechanism himself; he writes as if it were already known. But intriguingly, the arrangement he describes is identical to the back-and-forth hammer mechanism used to strike his clock's bell. North concluded that this type of escapement must have started out as an oscillating bell-ringing device, presumably created for a water clock, before someone realized that it could also be used to regulate the motion of a falling weight. From the peals of a bell to the ticks of a clock: that was the

* By the late fourteenth century, it belonged to a monk called John Loukyn, who was a subsacristan at St. Albans. The sacristan or subsacristan (origin of the modern word "sexton") of an abbey was responsible for the church's sacred equipment, including the bells, and later for the clocks.

† Verge and foliot escapements have pallets at either end of a vertical spindle that are pushed backward and forward by teeth at the top and bottom of a single wheel. The design of escapement described by Richard is otherwise known only in drawings made by Leonardo da Vinci in Italy at the end of the fifteenth century.

leap of lateral thinking that enabled Europe to jump ahead of all the centuries of scholarship of the Byzantine and Islamic worlds.

MS Ashmole 1796 also describes how many astronomical features were incorporated in the clock. The 24-hour Sun wheel was connected to a slightly faster second wheel, which drove a 6-foot iron display dial in time with the circling stars. The iron disk was marked with a map of stars, including the constellations of the zodiac. In front of it, a fixed grid marked the positions of key locations in the sky, including the horizon, hour lines and local meridian.* It would have been a spider's web of graceful lines and curves: the result of projecting the three-dimensional dome of the sky onto a flat surface.

Richard wasn't the first to incorporate such a dial into a self-moving astronomical display. A rotating star map also features in the astronomical water clock described by Vitruvius. And financial records from Norwich Cathedral in the 1320s describe construction of a 90-pound iron disk that turned with the sky. But Richard achieved this with breathtaking accuracy, using complex mathematical tables to calculate the appropriate numbers of teeth, and the sizes and relative angular speeds of the gear trains he used. North worked out that if the clock's main Sun wheel turned once every 24 hours, then the iron star disk would have made one revolution in 23 hours, 56 minutes and 4.12 seconds—an error of just 0.03 seconds per day.

Richard didn't stop there, though. He also added an ingenious Sun pointer. Overall, this made one circuit per day. But as discussed in chapter 3, the Sun's speed isn't constant; it appears to speed up and slow down as it travels through the sky. Richard accounted for this using what appears to be a completely new invention: an oval-shaped gearwheel. It was made up of four separate circular arcs, the proportions of which Richard had calculated to precisely represent the varying solar speed. It drove the Sun pointer so that at any given moment,

* The local meridian is the great circle that passes through the point in the sky directly above the observer as well as the celestial poles.

the location at which the pointer crossed the ecliptic line on the star map corresponded to the Sun's exact position in the sky. North, an expert in the history of cosmology as well as in medieval science, had never seen anything like it. The oval wheel was "a combination of mathematical and mechanical genius," he said, "perhaps without equal in the entire history of medieval or Renaissance rational mechanics."

A second pointer showed the Moon's position, while a rotating black-and-white sphere showed the Moon's phase (useful for planning nighttime journeys, as well as scheduling the bloodletting procedures that were part of the monks' health regimen). The clock even calculated the occurrence of lunar eclipses. The maker of the ancient Antikythera mechanism—by far the most sophisticated astronomical device known until this date, with many similar features—didn't attempt this; he simply used information from Babylonian predictions to mark expected events on a monthly calendar. Richard's clock, by contrast, derived eclipses automatically from the motions of the Sun and Moon (based on Ptolemy's theories), and used a rotating black disk to cover the Moon ball by the appropriate amount.

Finally, in line with the abbot's astrological interests, his device included celestial influences on Earth, with one dial showing the rise and fall of the tide at St. Albans's nearest port, London Bridge. This monthly cycle was relatively easy to calculate, with the pointer driven by a gearwheel that turned once every thirty days.

Richard's surviving notes don't discuss the planets, but the antiquarian Leland later said that the clock did show them. A Wheel of Fortune mentioned by Leland is also missing from the fourteenth-century text. This could refer to the wheel of the Roman goddess Fortuna. Popular in medieval times, it represented changing fortunes, with a beggar at the bottom, for example, who rises to become a king before falling off his perch again. But this would be an odd reference in a Christian monastery. North suggested instead that this feature may have displayed an astrological doctrine known as "the place of fortune," which predicted an individual's life span from celestial positions at their time of birth.

Little is known about the process of constructing the clock, though abbey records say it took many years and went hugely over budget. Detailed records from a clock built a few decades later in France reveal the possible scale of operations. It was commissioned in 1356 by Peter IV, one of the kings of Aragon, at his castle in the Pyrenees. The master clockmaker Antonio Bovelli brought with him ten skilled assistants from the pope's court in Avignon, who filled the castle courtyard with furnaces and smithies, using wooden templates to hammer the huge iron wheels into shape. The clock's frame weighed nearly a ton and its bell weighed three tons, and when the parts were lifted into the clock tower by crane, such a large crowd gathered to watch that the carpenters had to erect scaffolding to hold them all.

Bovelli's clock was much simpler than the one at St. Albans, taking only nine months to build. Even so, it never really worked as intended. Within thirty years, men with hourglasses rang the great bell. Richard, on the other hand, seems to have pushed beyond anything else known at the time to create a spectacular demonstration of the divine order and logic of the cosmos, which was still running two centuries later. The rediscovery of his notes proved that he was not only a "great abbot," said North, but "the most original English scientist of the later middle ages."

His extraordinary clock has smashed modern assumptions about what astronomers in fourteenth-century Europe were capable of. It also marks one of the most important turning points in the history of technology.

✡ ✡ ✡

Over the next few decades, a tradition of great clocks swept across Europe. It was a period of economic as well as scholastic revival; trades such as textiles and wheat were expanding, and society was rebelling against the old feudal order. Towns from Strasbourg to Milan were becoming lively marketplaces, with an organized workforce and the rise of a new merchant class.

The accelerating pace of life and work, particularly in towns, was already driving a greater awareness of time. Life was punctuated by town bells, marking everything from the start and end of work shifts and market trading hours, to street cleaning and closing time at the pub. At first, these public bells followed the seasonal canonical hours that rang from the churches. But with the invention of mechanical clocks, that started to change. Although the earliest clocks were in churches and cathedrals, they soon appeared in public squares too.

Following Richard's lead, the devices that sprang up across the continent had complex astronomical displays. They also often included moving figures. The 1350s cathedral tower clock in Strasbourg, Germany, featured a gilded cockerel that flapped its wings and crowed at midday (a tradition that survives in today's cuckoo clocks). On the astronomical clock in Prague, built in 1410, a skeleton shakes an hourglass; the designer of the fifteenth-century clock in Mantua, Italy, boasted that it showed the proper time for activities from surgery and dressmaking to tilling the soil. These devices weren't just about striking the hours. They were models of creation, concerned with the meaning and purpose—even the morality—of time, in heaven and on Earth.

The cosmological nature of these early devices shows how time was still seen as embedded in the wider universe; the ticks of the clock were inseparable from the cycles of the sky. But this characteristic was also key in enabling the seismic social changes that were to follow. These clocks weren't mere playthings or tools; they had divine authority. For the people whose days they regulated, these displays that echoed the majestic cycles of the heavens must have been awe-inspiring. The ability of Richard's clock to predict eclipses, for example, would have seemed magical to the inhabitants of St. Albans, proof of a direct link to celestial realms. The abbey's horse mill, subject of so much strife, relied on simple cogs and gears.* In his clock, Richard took that mundane gearwork and connected it to the heavens, and to God.

With the rise of public clocks, the control over time, and therefore over workers' daily lives, gradually moved from the monasteries to the towns. In St. Albans, for example, the townsmen built their own clock in the fifteenth century, as a protest against the abbey's control over the working day. There was also a change in how days were divided. Monastic water clocks had kept seasonal or canonical hours (they were reset each day, with different hour scales for different times of year). But there's no simple way to do this with an automated geared mechanism. So with the introduction of mechanical clocks, there was a shift from canonical to equal hours.† And rather than simply ringing on the hour, the new clocks marked each hour with an appropriate number of chimes—a practice called "hour-striking" (of which Richard's clock is the first known example) that was widespread by the end of the fourteenth century.

The introduction of equal hours reinforced the process of secularization by cutting ties with the monastic schedule, as well as starting to detach timekeeping from the seasonal patterns of the Sun. Hour-striking meant that everyone within earshot became aware of time not just as a series of intermittent bells, but as a regular, cumulative process throughout the day. And as time became more accurate, it was also less negotiable. Life was increasingly ruled not by events or natural cues, but the inexorable march of the clock.

Historians generally pinpoint this period as laying the ground for capitalism and the Industrial Revolution: according to the influential twentieth-century historian Lewis Mumford, "The clock, not the

* The mill would have used large, crudely cut wooden gears, rather than the precisely made iron wheels of the clock, but the underlying principle—toothed gearwheels that transferred forces by meshing together—was the same.

† In which each day is divided into 24 hours of equal length, regardless of the time of year.

steam engine, is the key-machine of the modern industrial age." The precisely made gears and technological expertise necessary for clock-making led to the development of machinery that allowed automated mass production.* But more important was the increased regulation and coordination of work and workers, and the change in attitudes toward time.

John Scattergood, professor emeritus of medieval English at Trinity College Dublin, analyzed literary references to clocks in the Middle Ages and found that with the emergence of mechanical clocks, writers started using clock-related metaphors to emphasize a new set of virtues, such as constancy, punctuality and exactness. This wasn't universally welcomed. In the fourteenth century, a clock interrupted poet Dafydd ap Gwilym's dream of a beautiful girl: "Alas the clock beside the dyke . . . awakened me," he complained. "Let its mouth and tongue be vain with its two ropes and its wheel, the stupid balls which are its weights, its four-square-case and hammer, with its ducks who think it day, and its restless mill wheels. Churlish clock with foolish chatter . . ."

In general, though, wider society soon shared in the work ethic and productivity of the monasteries. The activities of everyone from mill-workers to merchants were synchronized and regulated by the mechanical beat of the clock—although money, rather than spiritual salvation, was the new driving force. In 1433, a Spanish trader named Leon Battista Alberti wrote that the first thing he did every morning was to list the tasks that he needed to do that day. "So many things: I count them, think about them and to each I assign its time," he wrote. "I'd rather lose sleep than time."

But this shift in attitudes to time runs deeper than an increased appreciation for punctuality. With the malleable flow of lived experience now chopped into regular, measurable pieces, people also started

---

* For example, features such as epicyclic and differential gears, in which smaller wheels ride around on larger wheels, were first used to model the planets' motions in astronomical clocks. They later became a key component of automated looms (and more recently of car axles and 3D printers).

thinking in a more mathematical way. David Landes argues that regular hour-striking encouraged counting and simple arithmetic (calculating how many hours until the end of a shift, for example) in a society that was previously largely innumerate. He also points out that before about the thirteenth century, units of measurement tended to be variable and dependent on physical objects, such as the English foot, or the French *pouce*, or inch, which means "thumb." The switch from seasonal to equal hours, he suggests, encouraged the concept of abstract measures—a standardized unit that exists *in itself*—something that was vital for growing bureaucracy and trade. Meanwhile, clocks themselves gradually became smaller and simpler, often no longer models of the cosmos, but more purely about telling "the time" (though their direction of travel—clockwise—still reflects the Sun's path through the sky*).

For Mumford, this newfound ability to think in abstract measures and numbers was the real revolution wrought by clocks. "Men became powerful to the extent that they neglected the real world of wheat and wool, food and clothes," he wrote, "and centered their attention on the purely quantitative representation of it in tokens and symbols." Quantity was no longer simply one indication of value but its very definition. From the striking of the hours emerged the economic seeds of our modern way of life.

✡ ✡ ✡

The revolution in timekeeping also profoundly influenced the thinking of scientists and philosophers, in ways that overlapped with and reinforced the social changes already underway. First, from building astronomical clocks that modeled the cosmos, it was a short step to suggesting that the universe was itself like a clock. One of the first writers to make the comparison was the Italian poet Dante, in his

* As viewed from the northern hemisphere.

Paradiso, written between 1316 and 1321 (while Richard was studying at Oxford), which tells of Dante's journey through the spheres of heaven. When Dante reaches the sphere of the Sun, he sees the workings of God's cosmos, and compares it to a monastic clock striking the hour for morning prayer: "a glorious wheel . . . sounding the chime with notes so sweet."

The idea of the universe as a kind of machine, governed by predictable rules, dates back to antiquity, but in medieval times, the endlessly beating escapement of astronomical clocks made the concept irresistible. In 1364, the Italian engineer Giovanni de' Dondi built a complex astronomical clock, partly to prove that Ptolemy's system of concentric spheres wasn't just a mathematical model but an accurate representation of how the heavens really work. In 1377, the French philosopher Nicole Oresme suggested that the cosmos was like a timekeeper that worked whatever the season, was never fast or slow and never stopped: "The situation is much like that of a man making a clock and letting it run and continue its own motion by itself."

For these medieval scholars, though, the universe wasn't a dry, lifeless contraption, as we might associate with clocks and machines today. In *Paradiso*, although Dante compares the universe to a monastic clock, he describes it also as God's bride: when the cosmos chimes for matins she is singing a love song to her husband and creator. But by the seventeenth century, philosophers such as René Descartes* pushed the metaphor to its logical conclusion, arguing that not just stars and planets but animals, too, are simply automata: mechanisms driven by predetermined rules, like the Strasbourg cockerel. Only humans were different, because of an added "soul."

The driving force of the universe was no longer God's love but cause and effect; to understand something, it had to be explained by a

---

* Descartes seems to have been fascinated by clocks. He designed his own clock mechanism, as well as automata ranging from a magnet-powered tightrope walker to a clockwork dog that would spring at a partridge.

physical mechanism. According to the philosopher and science historian Stephen Toulmin: "Any 17th-century scientist who was satisfied with less was reproached by his colleagues, as invoking 'miracles' and 'occult qualities.'" This "mechanization" of the universe was what finally made astrology unacceptable, and it laid the foundation for all future scientific thought. Meanwhile, following the general shifts in attitudes to time among wider society, clocks also transformed scientists' ideas about the nature of time itself.

The notion of regular seconds ticking by might seem obvious, but it isn't inevitable. Anthropologists have reported several indigenous societies, such as the Nuer people on the banks of the Nile in Sudan, the Ainu culture of the Russian island of Sakhalin, and the Amondawa tribe of Amazonia, Brazil, who lack any concept of abstract "time."* For most of human history, the movements of the Sun, Moon and stars didn't just define time but created it; without their motion, there simply was no time. As clocks became more accurate, though, all of that was to change.

Improvements came initially from making better parts. In the fifteenth century, clockmakers introduced the minute hand and devised a way to power clocks not with falling weights but using a coiled spring. This allowed much smaller clocks, even watches. Astronomers, who wanted more accurate timekeeping in order to record the observations more precisely, pioneered these developments, and in the seventeenth century they came up with an even more impressive advance.

The key insight came from the Italian astronomer Galileo Galilei. One problem with escapements on early clocks was that they could run at any speed. Clock owners carefully arranged weights on the rotating arms to achieve the desired pace but any slight shift threw the clock off course. Galileo worked out that a pendulum—a weight

---

* The Nuer, for example, have rough lunar months, named after (and dependent on) activities carried out at different times of year, such as *Jiom*, the period when cattle camps are formed. The Amondawa recognize morning and afternoon, and the dry and rainy seasons, but not months or years. Neither culture has a word for "time," or any independent unit of it.

suspended on a cord or wire—always swings at a set speed, regardless of its mass or the amplitude of the swing.* So in theory, a pendulum designed to beat once per second would always do so each time it was released. In 1656, the Dutch astronomer Christiaan Huygens† (working with the clockmaker Salomon Coster) built the first pendulum-driven clock. The new design led to a spectacular improvement in performance, from at least fifteen minutes variation per day to just a few seconds.

In the past, clocks and watches had to be adjusted frequently. The reference point, the "true time," was the Sun. Many early watches had miniature sundials inside the cover, for just this purpose. The minor variations in the Sun's speed through the sky (which create a discrepancy of up to sixteen minutes between "Sun time" and "clock time") hadn't been large enough to matter. But now clocks were more accurate than the Sun. The devices initially built to imitate the cosmos had transcended it, with profound consequences for human thought.

Poised to take advantage of this development was the physicist Isaac Newton, working on his theory of gravity. His pioneering work *Principia*, published in 1687, revolutionized the scientific view of the cosmos, and remains one of the most important scientific publications ever. In it, Newton set out three basic laws of motion and explained how these, combined with his law of gravitational attraction,‡ could account for all of the movements of the solar system from wandering

---

* This is only strictly true for a particular kind of arc, known as a cycloidal arc, which differs slightly from the path taken by a weight on a wire. But as long as the pendulum swings only on a narrow section of arc, the error is very small (the longer the pendulum, and therefore the smaller the section of arc, the smaller the error).

† Huygens was an expert in lens grinding and telescope construction. He's famous for, among other things, detecting the first moon of Saturn in 1655 and the true shape of Saturn's rings in 1666.

‡ Newton's law of gravity states that every particle in the universe attracts every other particle with a force that is directly proportional to the product of their masses and inversely proportional to the square of the distance between them.

planets and distant moons to once-in-a-lifetime comets and even the ocean tides. From a dizzying cosmic array of seemingly independent objects and motions, he described an interconnected universe held together in gravity's web, all explainable by one simple equation.

Crucial to Newton's thinking about motion was the idea of absolute space and time: together forming a mathematical grid, essentially, upon which the actual objects and movements of the cosmos are superimposed. He separated time from the Sun's motions, for example, introducing the notion of "true and mathematical time," which "flows equably without relation to anything external."* One of Newton's major inspirations for this was the accuracy of the newly invented pendulum clocks. He had frequent contacts with the Astronomer Royal, John Flamsteed (who introduced pendulum clocks in London's Greenwich Observatory when it opened in 1675), and is thought to have owned a pendulum clock made in the 1680s. In the *Principia*, Newton discussed how pendulum clocks proved that scientists needed to distinguish between normal, or relative, time and the absolute time necessary for precise observations or calculations regarding the cosmos.

There was some initial opposition, for instance from the German philosopher Gottfried Leibniz, who felt that Newton was introducing mysterious, imperceptible concepts and forces that he couldn't explain. Leibniz argued that we cannot know something exists unless we perceive it, and we don't perceive time, only events. Time, he insisted, is simply the order in which things happen. This is actually much closer to how most physicists now see time than Newton's view (as we'll see later in this book). But for several centuries at least, Newton's concept of abstract time, flowing regardless of events, was the accepted and unquestioned scientific view.

As pendulum clocks spread, the distinction between Sun time and true or "mean" time filtered into wider society. From the 1670s, con-

* We'll return to the concept of abstract space in chapter 6.

version tables giving the difference for each day or week in the year (known as the "equation of time") were often pasted inside clock cases, so that clocks could be continually adjusted to follow the Sun. But as people ordered their lives more and more by their clocks, this process was reversed. Soon it was sundials that came with conversion scales: it was the Sun, not the clocks, that needed correcting.

One by one, big cities left solar time behind—mean time was made standard in Geneva in 1780; London in 1792; Paris in 1816—and the idea of regular, absolute time, existing independently of the Sun, became common sense. At first, people still felt that noon should be when the Sun was highest in the sky, so each town or city followed its own mean time. But in the nineteenth century, railway schedules made that impractical, leading to the imposition of hourly time zones. The twentieth century brought the invention of Daylight Saving Time, during which the entire clock is shifted to make more convenient use of daylight, and the popularity of air travel means we're now familiar with hopping multiple time zones in a day. The idea of time as an independent, abstract flow is so self-evident, we find it hard to imagine it any other way.

Clocks have since become ever more accurate, as increasingly ingenious mechanical escapements gave way to oscillating quartz crystals, and then to lightning-fast atomic vibrations. And although the circular dials that once modeled the cosmos haven't quite given way to digital displays, our definition of time no longer relies on astronomy. In 1967, scientists formally severed the link between time and the heavens when they redefined the fundamental unit of the second. Traditionally determined by the Earth's rotation as 1/86,400 of a day, the second is now described in terms of a specific number (over 9 billion) of oscillations of a caesium atom.*

* The figure of 1/86,400 comes from 60 (seconds) x 60 (minutes) x 24 (hours) = 86,400. A second is now defined as 9,192,631,770 oscillations of a caesium-133 atom. Physicists do still have to insert occasional leap seconds into the calendar, however, to ensure that clocks stay aligned with day and night.

Split-second timings far too fast for the human brain to perceive now underlie aspects of society from communications systems, power grids and financial networks to movie footage and sports results. Meanwhile, the perception of passing time—of spending it, saving it, wasting it—infuses virtually every moment of our waking hours. The importance of punctuality and the consequences of being late are among the first things that children learn, and we're rarely out of sight or earshot of some form of clock. "Our responsiveness to these cues is imprinted," says Landes, "and we ignore them at our peril."

Ironically, it seems that the more precisely we measure time, the less of it we feel we have. Nearly half of all Americans, for example, say they feel as though they lack enough time in daily life. This sense of "time famine" is a commonly cited reason for not engaging in leisure activities or helping others, and has been linked to unhealthy eating, sleep problems and stress. Our willingness to submit to the clock is what makes the complexities and opportunities of modern urban life possible. But there's a price to pay for dividing our lives into smaller and smaller parcels.

✡ ✡ ✡

In the early 1330s, King Edward III visited the abbey at St. Albans. When the king saw the clock, he gently rebuked Abbot Richard for neglecting repairs to the church while pouring the abbey's resources into a mere timepiece. Richard replied that future abbots could rebuild the monastery, but no one else after his death would be able to finish the clock. He knew his own time was running out. His disease was worsening, and by 1334, he was seriously ill. In 1335, according to the abbey chronicles, a great thunderstorm set fire to his chamber, and from then on, "he had scarcely a day without pain to the end of his life." He died on May 23, 1336, aged about forty-four.

The clock he built was apparently still working when the librarian John Leland saw it in 1534. It didn't last much longer. King Henry VIII had just split with the Roman Catholic church to form his own Church of England and was in the process of claiming the wealth of England's monasteries for himself. Leland was commissioned to tour the buildings before they were sold off to save any manuscripts or valuables they held. "The break with Rome marks not only a break in history but one in historical knowledge," notes North, "and we owe much to Leland for making this break less serious."*

But he couldn't save the clock. The abbey's church was eventually sold to the St. Albans townsmen for 400 British pounds, and the clock itself disappeared from history. A 1631 survey of English church monuments, for example, makes no mention of it; presumably it was broken up for its valuable iron parts. The townsmen might have used some of the components for their own clock, renovated in the time of Henry VIII. The clocktower still stands today, and North has pointed out that its great medieval bell is about the right size to have once belonged to the abbey, and that its inscription, *Missi de celis habeo nomen Gabrielis* ("I am heaven sent, in Gabriel's name"), suggests it came from a church. The sound of Richard's clock may ring out still.

Certainly, the implications of the technology Richard pioneered still resonate through the modern world. The social and philosophical changes inspired by mechanical clocks helped to create the scientific worldview that defines modern Western society, as well as driving the breathtaking economic and technological advances that propelled Europe out of the doldrums of the Middle Ages and allowed it to dominate the planet. Meanwhile, these self-turning machines caused one more fundamental split with the universe that we inhabit. Time is now embodied not in the cycles of the cosmos, but in our ever more accurate clocks.

* One of the manuscripts Leland preserved from the St. Albans monastery was the *Gesta Abbatum*, a chronicle of the abbots' lives, without which we would know next to nothing about Richard of Wallingford.

OCEAN

THE CREW OF THE HMS *ENDEAVOUR* HAD BEEN SAILING FOR two months without sight of land when, at the end of March 1769, they saw clumps of seaweed and land-nesting birds: shearwater gulls, terns and great black frigates. Then came a few palm-covered islets that they passed without stopping, and as the sun rose on April 11, their destination appeared at last: the green, crinkled peaks of Tahiti.

The sturdy, flat-bottomed ship was built as a collier in Whitby, northeast England—a "floating coal bucket" as one author described her—before being renamed and refitted for this pioneering Pacific voyage. Previously carrying a crew of twelve, her hold was converted to provide cramped quarters for more than ninety men and three years' worth of supplies. She left Plymouth in August 1768, packed to the brim and commanded by a man who had also started his career in the Whitby coal trade: Lieutenant James Cook.

The crew sailed south through the Atlantic via the Portuguese colonies of Madeira and Rio de Janeiro, before stopping to meet the sealskin-clad people of Tierra del Fuego at South America's tip. Battered by hail and snow, they rounded treacherous Cape Horn at the end of January 1769, and began the long, daunting haul across the Pacific.

As they plowed through thousands of miles of open water, Cook's men looked forward to their arrival with growing excitement. The Spanish had long sailed trading galleons across the Pacific, but large parts of this vast ocean were still unexplored, and the rival British and French empires had just begun to send ships there in search of new land and trade opportunities. One of them, the *Dolphin*, arrived home to Britain a few months before the *Endeavour* left. The *Dolphin*'s crew, the first Europeans ever to visit Tahiti, brought back stories of a tropical paradise where they had feasted on fruit and pork, and where a nail or bead could buy sex with beautiful, bare-breasted girls.

As the island grew on the horizon, it appeared to be everything they had hoped. The dramatic forested peaks were fringed by fruit trees, coconut palms and black sand beaches, which in turn gave way to turquoise lagoons and bays enclosed by a ring of coral reefs. On April 13, the *Endeavour*'s crew anchored in Matavai Bay, on Tahiti's northern coast, and were immediately surrounded by a crowd of canoes, filled with locals waving plantain shoots and offering coconuts, roasted breadfruit, Tahitian apples and fish.

During his three epic voyages, Cook would discover more about the Pacific than any European before or since. But for me, his first visit to Tahiti remains the most fascinating of his career. He was accompanied on the *Endeavour* by a team of artists and scientists, led by the young aristocrat and naturalist Joseph Banks, ready to record the new landscapes and species that awaited them. These scholars had no idea that they would also discover something else: a vibrant culture impossibly scattered across the ocean, and an understanding of the cosmos so alien from their Enlightenment worldview that its secrets are still being deciphered. It's a story that has rewritten European assumptions about the limits of human navigation and exploration in a vast ocean, and about our relationship with the stars.

But all of that would have to wait: Cook had orders. He and Banks greeted the local chief, a muscular old man named Tuteha (the men exchanged gifts, though Banks was sorry to have to decline offers of sexual hospitality on account of the thatched houses being "intirely [*sic*] without walls") then got to work, on a narrow spit of sand on the island's northern coast. Cook chose a site that could be protected by the ship's cannons and surrounded it with a strong wooden fence, guarded night and day by musket-bearing marines.

He erected a huddle of tents inside and, aided by the astronomer Charles Green, filled them with instruments from the finest makers in London, including a brass quadrant and sextant, at least four telescopes and two clocks. This camp, marked by the Union flag, was where Cook would carry out a mission vital for the security, pride and wealth of his nation; a pioneering task seen as crucial in Britain's efforts to explore and map—and therefore dominate—the planet. He called it Fort Venus.

✡ ✡ ✡

We have always relied on the Sun and stars to pin down not just time but place; for most of human history, knowing where we are on the planet has been inseparable from knowing where we are in the cosmos. As the Earth spins on its axis, our bearings are defined by the sky: east and west by the rising and setting Sun; north and south by the celestial poles around which all of the stars turn. Meanwhile, the positions of the heavenly bodies we see at any moment reveal where we are on the surface of the globe. Landmarks guide local journeys, and a magnetic compass can provide a bearing; but until the last few decades, the only way to know our position on the Earth as a whole has been to look up.

Many animal species take advantage of celestial cues, from dung beetles that sight off the Milky Way for a direct route back to their nests, to migrating birds—and even seals—that steer by the stars. The ability to decipher the map of the sky has also shaped human migration, battles and trade since ancient times, particularly for ocean voyages out of sight of land. As far back as the second millennium BC, the Minoans of Crete are thought to have used the constellations Ursa

Major and Ursa Minor (the Great Bear and Little Bear), which circle the northern celestial pole, to guide warships and trading fleets to distant ports; in Homer's *Odyssey*, composed around the eighth century BC, the wandering hero keeps the Great Bear on his left in order to sail east toward home.

The stars that mark north and south vary as the Earth wobbles slowly on its axis, a cycle that lasts twenty-six thousand years. There is currently no star at the precise southern pole (though the Southern Cross circles it); but for the last few centuries, the northern pole has been pretty well defined by Polaris, at the tip of the Little Bear's tail. This star—directly above the Earth's North Pole—barely moves as the rest of the night sky turns, and it has been crucial for seafaring, known in English for example as the North Star, Lodestar, Steering Star, or simply "the Star."

As well as pointing the direction north, Polaris shows how far north you are—known as latitude—by its height in the sky; most early navigational instruments are based on this principle. If its altitude is 50 degrees, for example, then your latitude is 50° north. On the equator (0° north), Polaris appears on the horizon; at the North Pole (90° north), it is directly overhead. The first navigators may have estimated the star's height using a stick held at arm's length; from the end of the ninth century, Arab traders judged its angle using a *kamal*, a wooden board with a knotted string attached to its center. In the fifteenth century, Portuguese navigators armed with the mariner's astrolabe (which measures altitude using a heavy brass ring marked with degrees) pushed southward across the Atlantic to Madeira, the Azores, the Cape Verde islands and Sierra Leone. In doing so, they kicked off the so-called age of discovery, the period of European global exploration and expansion that over the next few centuries would shape the modern world.

In the late fifteenth century, tables became available to calculate latitude during the day from the midday height of the Sun, and the mariner's astrolabe was gradually replaced by the more accurate cross-staff and backstaff (in which an adjustable bar is moved over a straight metal rod marked with degrees). Meanwhile, Portuguese explorers were joined by the Spanish—who sponsored the transatlantic expeditions of Columbus from 1492 and the first expedition to circumnavigate the globe in 1521—and later the English, French and Dutch. From the sixteenth century, goods such as sugar, spices, silk—and slaves—were carried on a web of trade routes across the Atlantic and Pacific.

All of this, of course, was achieved without being able to determine longitude, the distance traveled east or west. Whereas the night sky easily reveals north-south differences, the Earth's rotation means that when traveling east or west, there's no fixed point in the sky to use as a guide. Sailors had to estimate how far they had traveled from the speed that water flowed past the ship; or use tricks such as sailing east or west along lines of fixed latitude until they hit their destination. But this meant taking inefficient routes, and countless lives lost due to navigational errors.

The sixteenth-century Dutch astronomer Gemma Frisius was first to realize that although we think of longitude in geographical terms, as a position in *space*, with respect to the sky it's actually a function of *time*. Observers at differing longitudes are at different points in the Earth's daily rotation, so at any set moment, they will each see the Sun or stars at a different point in the sky; the greater the difference in longitude, the greater the difference in the time of day. Frisius suggested that as long as a traveler knows the time at his starting point and at his current location, he can find his change in longitude by calculating the difference between the two.*

This was easier said than done, however. Mechanical clocks weren't reliable enough to keep time over long periods even on land, let alone through the swaying motions and temperature and humidity fluctuations of an ocean voyage. Another suggestion was to use the

* The Earth spins at 15 degrees per hour—making a 360-degree rotation in a 24-hour day—so if you are sailing west from London, for example, and the Sun reaches its highest point (12 p.m.) as a watch set to London time reads four p.m., you have traveled 4 x 15 = 60° west.

sky itself as a clock. The Moon travels relatively fast across the background stars, so if it were possible to predict its future position relative to certain stars or the Sun at set times each day, for a reference location of known longitude, sailors anywhere in the world could use observations of the Moon to work out the time back home.

The problem here, too, was accuracy. The Moon's path through the sky is horribly complicated, affected by the gravity of both the Earth and Sun, so it wasn't possible to predict its movements anywhere near precisely enough for longitude calculations. Any nation that could gain an edge in navigation stood to profit hugely in terms of territories and trade, and finding longitude at sea soon became the most pressing intellectual concern of the day.

It would take two hundred years to find an answer. During that time, various European rulers offered giant sums of prize money for anyone who solved the problem, and both France and Britain established state-funded royal observatories. Although now thought of as scientific institutions, King Charles II's instructions in 1675 to Britain's first Astronomer Royal, John Flamsteed, were clear: he must dedicate himself to "rectifying the tables of the motions of the heavens and the places of the fixed stars, so as to find out the so much-desired longitude of places for the perfecting the art of navigation."

The pendulum clocks that revolutionized time measurement on land didn't work on moving ships, but they did allow much more precise astronomical observations, and when combined with Newton's theory of gravity, published in 1687, European astronomers could soon accurately predict the paths of all the major heavenly bodies—except, that is, for the meandering Moon. Determining longitude on land was by now fairly straightforward, but neither clocks nor astronomical tables nor observing instruments were accurate enough to do it at sea.

As the start of Cook's first Pacific voyage neared, however, both approaches—clocks and astronomy—were finally on the verge of becoming practical. In 1759, after more than thirty years' experimentation, the Yorkshire carpenter John Harrison came up with a device he called H4: a small clock, or chronometer, that could keep time on a ship (a story popularized in Dava Sobel's 1995 best-seller *Longitude*). With the newly invented sextant, which allowed more accurate measurements of altitude at sea than ever before, navigators could compare local time at noon with the time back home.

But the H4 was still being tested and argued over when Cook set sail in 1768. To find Tahiti, he relied instead on the Moon. Astronomers and mathematicians had refined Newton's lunar tables, and in 1767 Britain's fifth Astronomer Royal, Nevil Maskelyne, published the very first *Nautical Almanac*, containing precomputed lunar distances from the Sun at three-hour intervals, for the Royal Observatory's location in Greenwich, London.* The almanac allowed a navigator with a sextant and watch (used to keep local time from noon the previous day) to derive his longitude in as little as half an hour, and Cook was one of the first to use it.

The tables were by no means perfect, however, with the potential for errors of up to 1.5 degrees, which on the equator corresponds to more than a hundred nautical miles. That was because of a gaping hole in astronomers' understanding of the solar system, which it was Cook's mission to fill.

✡ ✡ ✡

Once settled on Tahiti, Banks and his party set about documenting life on the island, from painting the wildlife (sitting in mosquito nets to stop swarms of flies from eating the paint) to listening to local tunes on the nose flute. Banks also hosted a constant stream of islanders in his tent, including, on April 28, a woman he described as "about 40, tall and very lusty," though if she had once been handsome, he said, "few or no traces of it were left." A crew member who had visited

* The *Nautical Almanac* has been produced continuously ever since, and Greenwich was ultimately selected as the reference against which all global longitudes (and times) are measured.

Tahiti with the *Dolphin* immediately recognized her as Purea, a local chief who the *Dolphin*'s crew had thought was the queen of the island, but who had now been toppled by Tuteha.

Meanwhile, Cook was waiting for his big day. Until then, one of his major concerns was to control the fate of the ship's iron nails. The *Dolphin*'s mission had been seriously threatened by the trade of nails for sex, with the undisciplined crew literally pulling apart the ship in search of iron, and the locals reluctant to part with valuable provisions when items they desired could be bought in other ways. Cook drew up strict trade rules, punishable by flogging, stating that iron and other necessary items were not to be bartered for anything but essential supplies (a spike nail for a small pig; a white glass bead for ten coconuts). His other challenge was to resist the Tahitians' impressive and prodigious talent for theft. In one incident, two of Banks's colleagues were relieved of an opera glass and snuffbox; in another, a local man was killed when he tried to snatch a musket. Dependent on the locals for supplies, Cook was desperate to avoid more bloodshed.

By May 1, Fort Venus was finished, and the crew finally brought the astronomical instruments ashore to be stored in the heavily guarded compound. But the next morning when they came to unpack the huge brass quadrant, its case was empty—a loss that could have jeopardized the mission. Cook came up with a nonviolent solution, telling his men to seize Tuteha's and Purea's canoes, and keep the vessels in custody until the quadrant was produced. Meanwhile, Banks hurried off with a small search party to confront the prime suspect, closely followed by Cook and a group of armed guards.

After threatening an agitated crowd with his two pocket pistols, Banks was able to retrieve the quadrant, albeit in bits. He and Cook returned to Fort Venus with the brass pieces wrapped in grass, satisfied with the outcome of their excursion. When they arrived, however, Cook was dismayed to find Tuteha, forcibly restrained and convinced he was about to be executed. It took several days for Cook and the chief to make up (and food supplies to be restored) with a display of wrestling and a roasted pig.

Purea's canoe had also been seized, and the next morning she sent an adviser to the *Endeavour* to retrieve her property. He stayed all day, fascinated by the European ship and particularly the artists and scientists on board, and he stayed that night, too, sleeping under the canoe's awning. Banks doesn't seem to have been impressed at first, simply noting cattily that the man was "not without a bedfellow tho the gentleman cannot be less than 45." But from then on, the visitor spent more and more time with Banks's group, becoming an adviser and guide, if not a friend. A high priest and perhaps the most learned man on Tahiti, he was an expert in religious lore, medicine, astronomy and navigation. His name was Tupaia.

✡ ✡ ✡

Much to Cook's relief after a night of changeable weather, the morning of June 3 dawned without a cloud in the sky. He, Green and their botanist companion, Daniel Solander, had just a couple of hours to make final checks on their equipment, including a state-of-the-art pendulum clock in a tall wooden case and a series of chunky brass telescopes, their tripod stands perched on barrels dug deep into the sand. Just after 7:21 a.m. came the moment they were waiting for, when a faint black dot—the planet Venus—began to creep across the fiery disk of the Sun. This tiny pixel carried a huge message for those who knew enough to decode it: the size of the entire solar system.

The past two centuries had seen a cataclysmic change in scientists' understanding of our place in the universe. Ptolemy's system of Earth-centered spheres endured for a millennium and a half, but in 1543, the Polish astronomer Nicolaus Copernicus finally turned it inside out, suggesting that all of the planets, including Earth, were in fact orbiting the Sun. By the early seventeenth century, telescope observations by astronomers such as Galileo and Johannes Kepler provided overwhelming evidence that Copernicus was right. Instead of gazing out from a privileged position at the fixed heart of the cosmos, we were clinging to just one of many planets hurtling around our nearest star.

This shift in perspective had huge implications for humanity, which we'll return to in later chapters. But to pin down our position on the surface of the globe, describing the layout of the planets wasn't enough. Solving the complex interplay of gravitational effects on the Moon—vital for improving the lunar tables used to calculate longitude—required knowing not just the motions of the solar system but its absolute size, and in particular the distance from the Earth to the Sun.

Attempts to measure the scale of the solar system date back to Greek astronomers such as Hipparchus, who pioneered a method known as parallax. Based on the principle that when objects are seen from different angles, closer ones appear to shift in position more than farther ones, he used naked-eye observations made hundreds of miles apart to estimate the distance of the Moon from Earth. In the seventeenth century, astronomers—armed with telescopes and Newton's math—returned to Hipparchus's method. It requires measuring tiny apparent shifts in the sky, virtually impossible with the distant, fiery disk of the Sun. But Kepler had shown that there is a simple mathematical relationship between the size of a planet's orbit and the time it takes to complete one revolution. This enabled astronomers to work out from the periods how all the distances in the solar system are related: find just one and they would know them all.

In 1716, England's second Astronomer Royal, Edmond Halley, suggested how to measure a planet's parallax more accurately than ever before, by watching Venus as it passed in front of the Sun—an event known as a transit. Observers in far-flung places would see Venus sweep across the Sun's disk along two different chords, so they could calculate the planet's parallax by timing exactly how long it took for the transit to occur in each place.* It was an ambitious plan, requiring observers all over the globe, each of whom would need to time

* This calculation would also account for the apparent shift in position of the Sun. This is why Venus transits work better for calculating parallax than Mercury transits (which happen more often), because Mercury is closer to the Sun, so its observed shift in position relative to the background Sun would be

the hours-long transit to the nearest second. And the next one wasn't due until 1761.

When the time came, France and Britain were on opposite sides of the Seven Years' War, a battle over global territory that involved every major nation in Europe. Both sent out teams of astronomers anyway, but the fighting, plus cloudy weather and bad planning, thwarted many of the observations and the results were inconclusive. Once the war ended in 1763, planning began in earnest for the next transit six years later, which would be the last chance for more than a century. With the rival nations competing for prestige and power, chasing Venus became the space race of the eighteenth century. "It behoves us . . . to profit as much as possible by the favourable situation of Venus in 1769," said the Oxford astronomer Thomas Hornsby when setting out Britain's plan, "when we may be assured the several Powers of Europe will again contend which of them shall be most instrumental in contributing to the solution of this grand problem."

The British-sponsored teams headed for Norway and Canada, with Tahiti a last-minute addition after the *Dolphin* brought news of its discovery. On board the *Endeavour* were two astronomers: not just Green, who had observed the 1761 transit, but also Cook himself, who had learned maritime astronomy while surveying the North American coast.* By June 3, the pair had already spent weeks using their sextant and newly repaired quadrant to determine the precise longitude and latitude of Fort Venus (information just as vital as the transit measurement). Then they switched to their telescopes, carefully watching for the precise moments that the spot of Venus entered and left the Sun's disk, while listening to the ticks of the clock.

As a backup in case of clouds, Cook also sent colleagues from the

much smaller. (The other planets have larger orbits than Earth, so they never pass between us and the Sun.)

* Cook's observations of a 1766 solar eclipse, for example, were published by the Royal Society and helped to establish the longitude of Newfoundland.

Endeavour to watch the transit from two nearby locations, including Banks, who led a group to the little island of Mo'orea. From his journal we know that Banks invited the local chief to watch events through his telescope; "3 hansome [*sic*] girls" from the island were later sent to Banks's tent, and were easily persuaded to stay the night. Cook reported no such distractions at Fort Venus, but still, he was a little disappointed by his results. An optical phenomenon known as the "black drop effect"* blurred the exact moments that the transit began and ended, and the group's timings varied by as much as thirteen seconds.

Other observers had the same problem, and mathematicians later crunching the data from around the globe† came up with different results, so the Earth–Sun distance wasn't actually improved upon until the next transit of Venus in 1874. Some historians argue that Cook should nevertheless have been proud of his results. The astronomer Hornsby, using observations from Tahiti and Canada, a French team in Baja California (most of whom died of yellow fever shortly after the transit‡), as well as teams from Russia and Scandinavia, calculated a mean Earth–Sun distance of 149,623,007 kilometers (nearly 93 million miles). That's within a whisker of the accepted modern value§ achieved by bouncing radar signals off Venus and Mars.

Meanwhile, European explorers continued to use the stars to chart the planet. Even as chronometers came into use, navigators still had

---

* The explanation for this effect has been controversial; it is now thought to be caused by light diffracting, or bending, around the edge of the planet.

† Some 151 observers at seventy-seven stations around the world reported their results to various societies in around six hundred papers.

‡ The French astronomer Guillaume Le Gentil didn't have much better luck: he just missed the 1761 transit because of the war, then waited eight years in the Indian Ocean area for the 1769 transit only to have his view blocked by clouds, a disappointment that almost sent him mad.

§ The distance between the Earth and Sun is known as one astronomical unit (AU). Its value shifts slightly with time and reference point so in 2012, astronomers agreed to fix the definition at exactly 149,597,870 km.

to check them regularly, using lunar distances to get an absolute fix on the time. Cook used both methods on his two subsequent voyages, mapping coastlines from the Antarctic to Alaska to Hawaii, where in 1779 he was killed after another attempt to retrieve stolen goods by taking a chief hostage went horribly wrong.

It wasn't until the 1870s and 1880s, when transoceanic telegraph cables carried accurate daily time signals to all ports (which then relayed them to visiting ships by dropping a ball or firing a gun), that the lunar distance method was no longer required. In the early twentieth century, radio signals allowed sailors to check a ship's chronometer even while far out at sea. By that time, the territories and transoceanic routes discovered by Cook and other naval surveyors had enabled European nations to colonize and dominate the globe. As David Barrie, author of the 2014 book *Sextant*, puts it: "The new world order . . . depended on good charts."

Back on Tahiti, with the transit successfully tracked, it was time to celebrate. On June 5, 1769, once all of the observing parties had returned to Fort Venus, Cook held a feast to mark the king's birthday and invited the local chiefs, who toasted "Kihiargo" (the closest they could get to "King George"). The *Endeavour* stayed on Tahiti for several more weeks after that, during which time Cook surveyed the island's coastline. Then, when it was time to leave, he did something unprecedented, something that would ultimately help to reveal a very different view of the cosmos. He agreed to let Tupaia join the crew.

✵ ✵ ✵

Tupaia was a high priest of the ‘arioi, a secretive group who worshipped the war god ‘Oro. Originally based on Ra'iatea, an island 340 miles northwest of Tahiti, they combined a dark tradition of human sacrifice with flamboyant displays of music, comedy and erotic dance. When Tupaia was a young man, the ‘arioi, wearing scarlet costumes and black tattoos, had sailed in fleets from island to island, to awe and entertain the locals in return for feasts and gifts. In the late 1750s, after Ra'iatea was invaded by warriors from nearby Bora-Bora, Tupaia fled to Tahiti. There he became an adviser to Purea, and one of the island's most powerful men. But by the time the *Endeavour* arrived, Purea had been defeated and Tupaia's prospects were far from certain. As the strange visitors prepared to leave, he asked them to take him to England.

Banks, having enjoyed his conversations with the learned priest, imagined the entertainment value. "I do not know why I may not keep him as a curiosity," he wrote, "as well as some of my neighbours do lions and tygers at a larger expence than he will probably ever put me to." Meanwhile, Cook, though reluctant at first, accepted that Tupaia's knowledge of local islands could be useful. As the *Endeavour* eased out of Matavai Bay on the morning of July 13, Banks and Tupaia climbed to the masthead, where they stood for a long time, waving to the canoes that followed the ship before gradually falling away. Banks was unimpressed by the "clamorous weeping" of the islanders but thought he saw real emotion in Tupaia, who parted with "a few heartfelt tears, so I judge them to have been by the Efforts I saw him make use of to hide them."

Over the next month, Tupaia guided the *Endeavour* around the neighboring islands (which Cook named the Society Islands) including Ra'iatea, impressing the crew with his ability to navigate without the use of instruments or charts. He spent hours on deck—pointing out key stars at night; watching the patterns of swells during the day—and if asked could always point in the direction of Tahiti.

After this local tour, Tupaia urged Cook to sail farther west. There were more islands, he said, around twelve days' sail away. If Cook had followed Tupaia's instructions, he would have reached Tonga, Fiji and Samoa, more than fifteen hundred miles distant. But Cook had a second set of orders—"Additional and Secret Instructions" from the Admiralty—to follow once the transit was successfully observed. He was to search for the Southern Continent, a huge, fertile landmass that scholars back home were convinced must lie in the far reaches of the Pacific, to balance the weight of Europe, Asia and Russia in the north. On August 9, Cook set a course south.

He didn't find the fabled continent (on his second Pacific voyage he was able to show, after multiple crossings of the Antarctic Circle, that it doesn't exist). But on October 9, the ship reached New Zealand, where on the beach of Poverty Bay, Cook made a different discovery: that Tupaia was able to speak with the island's native chiefs. It's a meeting that has been described as an extraordinary moment in world history. Back home, the British couldn't make themselves understood across the English Channel, yet Tupaia and the Maori shared a common language over half an ocean.*

Over the course of his career, Cook was amazed to find that this shared heritage stretched even further. As one anthropologist puts it, Cook essentially "discovered Polynesia," realizing that all of the peoples on the islands in the ocean triangle bounded by Hawaii, Easter Island and New Zealand—specks in a vast six million square miles of sea—had the same language and culture: the same stone adzes, fishhooks, thatched houses and canoes. "It is extraordinary," Cook wrote when he visited remote Easter Island in 1774, "that the same Nation should have spread themselves over all the isles in this vast Ocean . . . which is almost a fourth of the circumference of the Globe."

Other European explorers had also been astounded by the presence of thriving communities on the tiny, scattered islands of the Pacific, separated by hundreds if not thousands of miles. Unlike Western navigators, Polynesian sailors had no instruments or charts; no telescopes, sextants or lunar tables. How could they possibly have reached the islands over such great stretches of open sea? Theories included

---

* As Cook and his colleagues surveyed the coast of New Zealand and recorded its culture and wildlife, Tupaia became their official spokesperson. The importance of Tupaia's role in helping the *Endeavour*'s crew to establish friendly relations with the locals was never fully acknowledged by Cook and has only been recognized by scholars in recent years.

that an ancient, inhabited continent had sunk, leaving only its mountaintops above water; or that the islanders were created in place by God. After getting to know Tupaia, though, Cook had a different idea.

In March 1770, shortly before leaving New Zealand for the journey home via Australia and Indonesia, Cook wrote down a long list of Pacific islands dictated by Tupaia, and mentioned a chart, "drawn by Tupia's [*sic*] own hands," that showed the locations of seventy-four of them. Although the original chart did not survive, Cook's copy of it is now kept in the British Library in London. The yellowed paper is splattered with island outlines, apparently stretching far beyond the Society Islands. Cook wondered whether the Polynesians really had voyaged across the Pacific, traveling from island to island, "with the Sun serving them for a compass by day and the Moon and Stars by Night."

Cook didn't suggest how they did this, though, and it seems that Tupaia was unable or unwilling to explain. In December, after more than a year on board the *Endeavour*, during which he survived hazards including scurvy and a near-catastrophic collision with the Great Barrier Reef, Tupaia died of fever in the disease-ridden port of Batavia (now Jakarta).* He would never bring his learning to London. But the map he left has fascinated scholars ever since.

✲ ✲ ✲

Archaeologists now agree that the inhabitants of the Pacific did migrate by sea as Cook suggested, moving eastward from Southeast Asia from the second millennium BC, and fanning out in several stages to Tonga, Hawaii and Easter Island before eventually reaching New Zealand, perhaps as late as AD 1200. But although stories brought home by Cook and other explorers initially inspired romantic ideas about

* A third of the *Endeavor's* crew succumbed to the diseases of Batavia, including the astronomer Charles Green.

the Polynesians being expert ocean navigators guided by the stars, by the 1950s this was replaced by skepticism.

Tupaia's chart looked all wrong, for example. Once the geography of the Pacific was understood, most of the islands he showed were unidentifiable or wildly out of place. Meanwhile, explorers' accounts of how priests like Tupaia navigated were sketchy, and in Polynesia itself, populations had crashed due to diseases introduced by the Europeans, with the loss of vast stores of cultural knowledge. Long-distance navigation skills were forgotten, and influential critics such as the New Zealand historian Andrew Sharp insisted they never existed at all. Polynesians couldn't possibly have found their way over long distances without astronomical instruments and charts, he argued. New islands must have been colonized by accident, when lost canoes were lucky enough to be swept onto distant shores.

Dissenters soon fought back, reconstructing traditional canoes and testing them over long journeys. In the 1970s, Ben Finney, an anthropologist at the University of Hawaii, helped to build a vessel named *Hokule'a* ("Star of Joy"). The canoe's twin 62-foot hulls were lashed together to support a deck, two masts with brown, crab-claw sails and a crew of sixteen. In 1976, guided by a Micronesian navigator named Mau Piailug (no Polynesians could be found with the necessary expertise), *Hokule'a* sailed two thousand nautical miles from Hawaii to Tahiti without the help of modern instruments. When she arrived after thirty-three days, she was greeted by an ecstatic crowd of more than seventeen thousand people. "None of us were prepared for that kind of cultural response," said Nainoa Thompson, now president of the Polynesian Voyaging Society. The voyage changed the identity of the Hawaiian people, he added later. "We went from being castaways . . . to being children of the world's greatest navigators."

More recent archaeological and genetic findings are revealing that the Polynesians once routinely sailed and traded between far-flung Pacific islands and even as far as the Americas, long before the arrival of Columbus. Meanwhile, researchers have been studying the few clues that are left to work out how they did it. In one of the most influential accounts, the sailor and Polynesian scholar David Lewis studied historical descriptions of Polynesian wayfinding and interviewed surviving Micronesian navigators to discover a system almost unimaginably different from the Western approach.

Polaris, so crucial for European sailors, is not visible in the southern hemisphere, and instead of measuring altitudes, Pacific sailors used a "star compass," taking their bearings from the rising and setting points of stars on the horizon.* When setting sail for a destination, navigators aimed their canoes toward the appropriate star. Each star is only visible on the horizon for a brief time, so keeping course through the night involved following a succession of stars, or "star path": an approach that required the sailors to memorize rising and setting patterns of hundreds of stars in all parts of the sky. This system was supplemented by observations of the Sun, Moon, wind direction and ocean swells, as well as signs of distant islands such as cloud formations and land-nesting birds. It's a process that involved navigators using all of their senses to detect and assimilate subtle cues: changes in the color of the water, even the taste of the sea. Lewis reported that if clouds or fog meant a navigator couldn't see the direction of waves approaching his canoe, he stood with legs apart to feel the swell patterns using the swing of his testicles.

Mastering these techniques involved long, rigorous training that began in early childhood. Navigators created complex memory maps using chants, stories and dances, entwined with visual metaphors—such as the diamond-shaped Southern Cross as a "great triggerfish"—as well as religious beliefs. One of the most important sources for Polynesian cosmology comes from Bora-Bora, where a Western missionary recorded an ancient chant recited to him in 1818 by an old woman named Rua-nui. It is a story about the creation of the cosmos,

* At any particular latitude, each star will always rise or set at the same point on the horizon.

and describes how when the world began, the stars sailed in their canoes to all corners of the sky, before creating the "kings of the chiefs of the earth." When Tupaia and his peers sailed by the stars, they weren't simply following compass bearings. They were retracing the sky voyages of their ancestors.

Rua-Nui's chant also describes how the sky was once supported by pillars that later became stars, including Antares, "the entrance pillar"; Aldebaran, "the pillar to tattoo by"; and Alphard, "the pillar to debate by." In 2010, researchers including the New Zealand surveyor and navigator Stan Lusby suggested that this passage might hint at a teaching schedule for priests like Tupaia, perhaps conducted within a traditional pillared house that represented the dome of the sky.

It might even contain a clue as to how Polynesians once found new islands in the first place. (In reenactments of ocean voyages, sailing directions are obtained from modern charts, then transferred to a Polynesian-style star compass.) Lusby suggests Rua-Nui's pillars refer not to single stars but pairs of stars. When the lower star is on the horizon, any star pair appears vertical only at a specific latitude. Lusby showed that pillars defined by consecutive stars in the ancient chant would have created a system of sea lanes running east-west across the Pacific. Canoes could have sailed north or south until a star pillar appeared vertical, then followed it east or west until hitting land. Perhaps early Polynesian priests echoed the celestial voyages described in the chant by sending out missions along the position lines defined by these stars. (Crucially, other crews could have followed and reached the same destination, even if the first boat didn't return.) Once the location of an island was known, star paths could have been derived to get there by a more direct route.

It's a speculative suggestion, but it shows that directed voyages of exploration and trade would have been possible with the tools that the Polynesians had available. And the image of a mythological pillar straightening on the horizon each night nicely demonstrates their holistic approach. Whereas Cook used accurate astronomical observations to calculate his position from tables and charts, his Polynesian counterpart would have relied instead on assimilating a complex web of sensory cues, memories, stories and beliefs. As French archaeologist Anne Di Piazza puts it, this is navigation not as a sum of knowledge, but as "a way of being and of conceiving the world."

It's a realization that brought Di Piazza, and her colleague Erik Pearthree, back to Tupaia and his map.

✡ ✡ ✡

Cook's chart of Tahiti, painstakingly plotted from astronomical observations he made around the island's coastline, is—like all maps—a fascinating blend of the real and the imaginary. Many aspects of his depiction resemble aerial views of the island: two unevenly sized lobes; a burst of mountains and rivers; a frill of coral reefs. But other features he included have no visible counterpart. The rippling ocean is replaced by a grid, numbered in degrees (counted west from Greenwich and south from the equator); a scale bar; and an arrow that points north. These conventions, developed over centuries, are now so natural that we barely notice them; yet they have come to define how we see the physical world.

The first people we know of to impose mathematical features onto maps were, not surprisingly, the Babylonians, who introduced both scale and orientation into sketches of local areas of land. The Greeks, though, were first to chart the entire globe. It's a task that was inseparable from efforts to map the sky. In the fifth century BC, the astronomer Eudoxus described a spherical cosmos centered on Earth, and to demonstrate, he built a celestial sphere: a map of the sky from the outside looking in. As well as stars, he marked significant features such as the celestial poles, the equator and the tropics (the path of the Sun through the sky on the equinoxes and solstices, respectively). When astronomers subsequently built terrestrial globes, they traced the same points and circles onto the surface of the Earth.

In the third century BC, the astronomer Eratosthenes compared shadows in distant cities to work out the length of the equator—in other words, the circumference of the Earth. It was then just a matter of geometry to calculate the length of any parallel circle around the planet, and to convert differences in latitude—calculated from astronomical observations—into distances. Humans finally understood the extent of the entire Earth, and could start to map onto it the places they knew (although as longitude generally had to be estimated from travelers' reports, the results weren't very accurate). Four hundred years later, Ptolemy created a state-of-the-art mapmakers' manual called *Geographia*, in which he explained different ways to project the Earth's curved surface onto a flat chart—techniques first developed by astronomers mapping the sky—and listed systematic coordinates for about eight thousand locations around the known world.

Ptolemy's work was forgotten in Europe during medieval times (though Arabic-speaking scholars knew about it). Instead, religious depictions of the cosmos called *mappae mundi* were popular. These were symbolic, not scientific: one well-known example from England's Hereford Cathedral shows the world as a single landmass surrounded by a circular ocean and centered on Jerusalem, its cities and towns mixed up with biblical scenes and overseen by Christ on a throne. But in 1406, *Geographia* was translated into Latin, and revolutionized westerners' view of the world. For the first time in over a thousand years, people realized that the world they knew (encompassing Europe, Asia and Africa) occupied only a small part of the Earth's surface. There was so much more to discover, and as explorers from Columbus to Cook sailed across the oceans, Ptolemy gave them the tools to record their discoveries.

In replacing myth and religion with longitude and latitude, it was finally possible to convert locations into trustworthy positions; to create an objective record of the physical world that was untainted by personal beliefs or experiences. It enabled Cook to plot a reliable course across his charts from his starting point to his destination, just as we do when we follow a map today. But in 2007, Di Piazza and Pearthree pointed out that this is not at all how a Polynesian navigator like Tupaia would have traveled. Generations of scholars with Western eyes and brains, they argued, had been looking at his chart all wrong.

Although many aspects of Polynesian navigation are still poorly understood, a key concept is that of *etak*, or "moving island" navigation, in which a sailor thinks of his canoe as being stationary throughout a journey, while the surrounding water and islands flow past.* When we follow a Western map, we imagine ourselves looking down on it with a bird's-eye view: the terrain stays still while we move across it. But a voyaging Polynesian remains absolutely at the center of his cosmos, following the stars as the ocean changes around him. He works out his position not by imagining distance traveled on a map, but by calculating the bearings to relevant islands relative to his current position, even when he can't see them. (This is analogous—on a much larger scale—to being able to point in the direction of a particular item of furniture in a familiar room, even if it's behind us.)

A training exercise from the tiny island of Puluwat Atoll, called "Reef Hole Probing," helps to demonstrate the approach. Trainee navigators imagine a parrotfish living in a deep hole in the reef of their island. They tell how poking a stick into the hole causes the fish to flee to the reef hole of another island, and as it does so they recite the star path that leads there. Then they mentally transfer themselves to that island and repeat the exercise, until all the neighboring islands have been visited. Clues like this caused Di Piazza and Pearthree to suggest that Tupaia's chart wasn't a map at all, at least not how we would understand it. Instead, it was a composite, or mosaic, of sailing directions with several different centers.

---

* This doesn't mean that Polynesians literally believed they were staying still; it's just how they thought about it for the purposes of navigation. In a similar way today, we often think of the Sun as rising and setting even though we know it's the Earth that is moving.

To test the idea, the researchers took the largest islands on the chart as a series of central points, and worked out the angles from each of those to its neighbors. Then they superimposed these onto Tupaia's chart and found that bearings centered on five departure locations, including Tahiti and Ra'iatea, matched the location of more than half of the islands on the chart, including the Australs, Cooks and Marquesas.* Unlike a Western map, on which islands "hold positions which are defined absolutely," they argue that the center of Tupaia's map is a "subjective coordinate," which depends on the position of the person reading it.

Di Piazza and Pearthree suggest that when Cook copied Tupaia's drawing, he added compass directions and a scale. His view of what a map should be was so ingrained that he—and countless scholars since—couldn't imagine it any other way. But once decoded, they say, Tupaia's once-discredited chart reveals knowledge of islands from Samoa in the west to the Tuamotus in the east, a region of ocean the size of the United States.

✡ ✡ ✡

By drawing lines of longitude and latitude around our planet, we changed our relationship with the space that we inhabit. The medieval *mappae mundi* were crammed with not just places but people, creatures and events, both real and mythical. Time and space were blended; the prominence given to each location depended on its perceived importance: scenes were painted as they would appear to the human eye. With the switch to Ptolemaic maps, this moral and historical framework was replaced by a mathematical one.

---

* If they're right, the chart only shows sailing directions from each center point, not distances. Cook and his colleagues recorded, however, that Tupaia also quoted sailing times to different islands (more relevant for Polynesian navigators than absolute distances because they incorporate information about currents and winds).

The new charts—compiled according to astronomical observations—were constructed to represent not a human viewpoint but a geometric projection, and were proportioned not according to myth or whim but a regular scale. In other words, they treated each location equally, as a simple pair of coordinates, regardless of its cultural significance. It's a change that seems natural and obvious today, but it had fundamental implications. Our subjective experience of the world was no longer the "truth." Reinforced by the discovery that Earth is not at the center of the cosmos, what these maps ultimately implied was the existence of a deeper, objective reality, a terrain that could only be accurately revealed once personal beliefs and impressions were stripped away.

In the seventeenth century, Descartes carried this concept to its conclusion when he described how to use numerical coordinates (now known as "Cartesian coordinates") to describe not just locations on a map but geometric shapes and lines: essentially leaving the physical universe behind and creating new realms of purely mathematical space. No longer defined by the physical places and events that fill it, space now stretches out regardless, according to a uniform, mathematical grid.

In other words, just as we have abstracted time, and God, we have also abstracted place. This Cartesian view, in which we move between fixed, objective points, underlies modern science and has led to breathtaking technological advances. The charts and instruments used by European explorers allowed their ships to conquer the Earth. We've since taken that approach to exquisite heights. We use ever more sophisticated technology to navigate not just over the ocean but across the solar system, while fleets of artificial satellites (fitted with atomic clocks) have replaced the stars, allowing us to track positions on Earth to within a few feet. With GPS information now routinely beamed to cars and phones, we can find our location without even looking out of the window, let alone up at the sky.

But there has been a price to pay. Psychologists and neuroscientists warn that when we rely on technology to perform tasks such as navigation for us, our awareness of our physical environment fades as we become immersed instead in an abstract, computerized world. Studies show that we tend to place too much faith in the accuracy of information from computer monitors, and to ignore or discount information from our own eyes and ears, an effect that has caused pilots to crash planes and GPS-following tourists to drive into the sea. A team led by the British neuroscientist Hugo Spiers found in 2017 that areas of the brain normally involved in navigation just don't engage when people use GPS. "When we have technology telling us which way to go," said Spiers, "these parts of the brain simply don't respond to the street network. In that sense our brain has switched off its interest in the streets around us."

Other studies have shown that people who regularly use GPS become less able to find their way without it, a phenomenon thought to be caused by structural changes in the brain as underused regions start to shrink. Just as sedentary lifestyles weaken us physically, overreliance on technologies to perform sensory or intellectual tasks appears to dull us mentally, and might even make us more prone to neurodegenerative conditions such as dementia. The more we rely on computers instead of our physical experience, the more we erode our own awareness and skills.

In one sense, then, those invisible lines of longitude and latitude have connected us to the universe in a way that early societies couldn't have imagined. Like moorings or guide ropes, they gave us a frame of reference, enabling breathtaking insights and abilities and allowing us to fix our position not just on the ocean but with respect to the planet, solar system and farther stars. But at the same time, the invention of abstract space was one more step in our journey from a subjective view of the universe to an objective one; from being inextricably entwined with—even creators of—the cosmos to becoming recorders and observers of an independently existing reality.

Tupaia's story throws into relief the choices we've made. Our view of space—as of time—now feels so self-evident, it's hard to see any alternative. It's easy to assume that a mathematical, objective approach is the best—if not the only—way to learn about the "real," physical world. Yet instead of discarding their experience of the cosmos, Polynesian navigators maximized its potential in order to explore millions of square miles of ocean. A mix of stories and songs, senses and instinct, enabled them to achieve—without technology—feats of navigation that as westerners we can barely imagine.

POWER

IN DECEMBER 1774, AS CAPTAIN COOK CONTINUED EXPLORING the Pacific on his second ship, the *Resolution*, a very different British vessel limped into the bustling young port of Philadelphia. Her passengers had set out across the Atlantic, looking for a new life in America. But during the three-month voyage, typhoid fever swept through the cabins. By the time the unlucky ship arrived, most of those on board were so ill they had to be carried onto the docks.

One of them was an argumentative but eloquent thirty-seven-year-old called Tom Pain. Like many who made the trip, he was out of options back home. The son of a corset-maker from Norfolk, he had tried and failed at several careers—tax collector, tobacconist, teacher—and eventually sold his last possessions to pay for passage to the New World. In the British colony of Philadelphia—less than eighty years old but growing fast—he found a vibrant mix of culture and wildness. The multistory brick houses, churches, libraries and perhaps the largest market in North America were riddled with pests, from mosquitoes to wild pigs, and in wet weather, carriages plowed the unpaved streets into mud.

Pain didn't see it for six weeks; close to death on arrival, he was taken in by a local doctor and confined to bed. But once he recovered, he started writing to his fellow colonists, and once he started, he didn't stop. He wrote about democracy and human rights, and about how society could be different; about the ridiculousness and corruption of the monarchy, and the natural-born right of every man to be free.

Under the new identity of "Paine," he became the most successful author the world had ever seen. One biographer has called him "possibly the most influential writer in modern human history," and it is no exaggeration to say that his arguments changed the world. Over the next few decades, he brought radical ideas to the masses, helping to bring about two revolutions and narrowly failing to trigger a third; events that saw him claimed a hero in America, chased from England and sentenced to death in France. George Washington praised him for "working a wonderful change in the minds of many men," while Napoleon Bonaparte said that "a statue of gold should be erected to him in every city in the universe." His final work, though, was deemed so heretical that his reputation was destroyed for two hundred years.

Where did Paine get his ideas? And how did this penniless drifter come to play such a central role in shaping the politics of the modern world? It's a story that shows how power and our views of the cosmos have always been entwined. It starts with a pirate ship, some astronomy lectures and a pair of globes.

✡ ✡ ✡

The notice in London's *Daily Advertiser* ran almost every day for two weeks: "To cruise against the French, the *Terrible* Privateer, Capt. William Death . . . All Gentlemen Sailors, and able-bodied Landmen, who are inclinable to try their Fortune . . . are desired to repair on board the said Ship." For nineteen-year-old Paine, the offer was irresistible.

He had grown up in the town of Thetford, in a thatched cottage next to the gallows, and left grammar school at fourteen to learn his father's trade. In 1756, he ran away to London and worked a few months as a corset-maker, but the low pay and fourteen-hour days didn't suit him, and he signed up to serve on the *Terrible*. Privateers were like legal pirates, licensed during wartime—in this case, the Seven Years' War—to capture enemy ships and split the proceeds. Paine's father, a pacifist Quaker, tracked him down and convinced him not to go—a lucky escape since the *Terrible* soon had a disastrous battle with a French ship in the Channel, which killed nearly everyone on board.

But ultimately the teenager couldn't resist choosing conflict and adventure over peace and quiet, as he would for the rest of his life. In January 1757, he joined the crew of another privateer, the *King of Prussia*. This voyage was a great success—they captured eight ships in six months—and he returned to London that summer with what must have seemed like a small fortune of thirty pounds. For the first time in his life, Paine had cash to spend. Biographer Craig Nelson suggests Paine would have treated himself to "the clothes of a proper urban gentleman": breeches, garters and stockings, silk-lined coat, felt hat. But he also had his eye on something more unusual. "As soon as I was able," Paine wrote later, "I purchased a pair of globes."

The London he returned to must have been an exciting place, in the midst of what has been called the most remarkable transformation of thought in European history: the Enlightenment. Sixty years before, Newton had published *Principia*, showing how the same equations govern the motion of everything from planets and comets to an apple falling from a tree. Now his ideas were sweeping through Europe, together with their exciting, empowering implications. Newton had literally revealed the workings of the universe, and he did it not by appealing to the authorities of the past—ancient scholars, or the church—but through his own observations and ingenuity. The cosmos was no longer a divine mystery, buffeted according to the whim of the gods. Newton showed that it ran according to regular, universal rules, which could be discovered and understood by anyone who cared to investigate. There was an explosion of activity and debate—an enterprise that was then known as natural philosophy, but is now called science. Reliance on traditional theories was replaced by an enthusiasm for modern observations and experiments, summed up by German philosopher Immanuel Kant's instruction, *sapere aude* ("dare to know").

At the same time, science became more accessible, with the latest ideas being discussed in popular books, lectures and clubs. One of the most prestigious was Britain's Royal Society (with the motto *Nullius in verba*, taken to mean "take no one's word for it"), but smaller societies sprang up in towns and cities all over Europe. The followers of the new science included not just aristocrats like Joseph Banks, who sailed with Captain Cook, but particularly "mechanics": artisans and craftsmen who taught themselves.

When Paine arrived in London with his earnings from the *King of Prussia*, he enthusiastically joined this community of Newtonians. Six months at sea would have given him plenty of time to ponder the stars, and he was fascinated by the workings and nature of the cosmos. As well as buying the globes—one for the Earth; one for the sky—he learned how to use an orrery (a mechanical device that demonstrates the movements of the solar system) and attended lectures by science popularizers such as the mathematician and lens-maker Benjamin Martin, and James Ferguson, a Scottish astronomer and author of a hugely successful book on Newton's ideas.

At coffeehouses and dining clubs, Paine and his new friends talked late into the night, about chemistry and comets, pendulums and prisms. But that wasn't all they discussed. The Enlightenment spirit encouraged its participants to question not just scientific dogma but all received ideas. "Their constant lectures and debates would over time evolve far from Newton's arithmetic," writes Nelson, "into a consideration of the most astonishing questions of the century, one of which would make Thomas Paine famous in every city across the globe: why should there be kings?"

✡ ✡ ✡

For all of human history, the events and relationships we see in the heavens have been central to the power structures that shape our lives on Earth. As described in chapter 1, elites may have been defined by knowledge of cosmic cycles as far back as Paleolithic times. As societies became more complex, ideas about the sky held them together. The shining orbs people watched on high provided the model for radiant emperors and kings, and civilizations arose in which the entire machinery of government was organized around the concept of living in line with the heavens: what historian of cosmology Nicholas Campion has called a "cosmic state."

Astronomy was helpful for predicting weather patterns and planning agricultural schedules, as well as stimulating scholarship that was useful in other areas, such as engineering and economics. But most of all, focusing on the sky provided the stability and discipline that allowed societies to succeed. Ensuring tight adherence to celestial events was really about power: using knowledge of the heavens to reinforce political ideology, justify rulers' status and keep the masses tightly controlled.

Pretty much all the examples we've discussed so far of societies looking to the heavens ultimately come down to attempts to safeguard the position of those in charge: such as the Babylonians, obsessed with using celestial omens to protect the king; the Egyptian pharaohs, who identified themselves with the Sun; or the early Jews and Christians, whose ideas of heaven reflected the structure of royal courts. As the French philosopher Bruno Latour put it: "No one has ever heard of a collective that did not mobilize heaven and Earth in its composition." It's hard to find a successful early civilization that did not organize itself around the sky.

Through most of history, this has worked in several related ways. Associating with the grandeur and perfection of the heavens helped to enhance a ruler's reputation. Being able to predict astronomical events in advance, which must have seemed magical to most people in ancient times, acted as proof that a ruler was either divine or had the approval of the gods. And seeing the hierarchy of a king and his court reflected in the sky made their existence seem natural and inevitable. It's much harder to question the power structure within a society if it's seen as an integral part of how the universe works.

In Imperial China, for example, the emperor was known as the "Son of Heaven," responsible for keeping life on Earth in harmony with the sky. For more than four thousand years, teams of Chinese royal astronomers monitored celestial phenomena, and the star maps and almanacs these scholars produced were closely guarded as state secrets. The emperor had to prove that he retained the divine right to rule by accurately predicting celestial events; if his astronomers failed to predict a major event like an eclipse, his rivals might take the opportunity to rebel.

Central American rulers, too, employed advisers to watch the sky, and affirmed their link to the heavens by impersonating celestial bodies. A tenth-century Mayan ruler from Yucatán, Mexico, who named himself after the rain god Chac, covered his palace with hundreds of Venus glyphs, as well as the numbers 5 and 8 (which represent Venus's synodic cycle). Archaeologist Ivan Šprajc, who studies astronomy in Mesoamerican culture, points out that the annual appearance and disappearance of this planet coincided with the arrival of the rainy and dry seasons. By identifying himself with Venus, says Šprajc, Chac took credit for bringing the rains that kept everyone alive.

Links with the sky were often built into the physical structure of buildings and monuments. The Mayans used steps and pyramids to create dramatic solstice displays—updated versions of sites such as Stonehenge. Equinoxes are marked by the domed roof of the grand Pantheon in Rome. The fifteenth-century Temple of Heaven in Beijing was designed to symbolize connections between heaven and Earth. One of the most impressive examples involves an entire city built to mirror the heavens. In the eighth century, Caliph al-Mansur, the ruler of the Islamic Empire, employed a hundred thousand workers to build a new capital at Baghdad.

Known as the Round City, it was laid out in a series of concentric circles with the caliph's golden palace at the center and streets radiating toward the outer walls. Historian Ibrahim Allawi argues that it was designed as a "grand, cosmic astrolabe." The circle was divided into quarters, with more streets in some quadrants than others, which Allawi believes represented the eccentricity of the Sun's orbit (the quadrants where the Sun moves faster in the sky were packed with more streets). Meanwhile, the circles may have represented the celestial equator and the tropics of Cancer and Capricorn, with the palace marking the celestial pole around which the whole sky turns. Al-Mansur wanted his new capital to be "the centre of a world empire of power and commerce," says Allawi, "and the navel of the entire cosmos."

Links between astronomy and power don't just belong to the ancient past, however; they are set deep into the foundations of our modern world. The pioneering sixteenth-century astronomer Copernicus, for example, borrowed the image of a powerful king to support his ideas about the Sun. In 1543, when he set out to reshape a solar system that had been accepted since ancient times, he was in a difficult position. There was no direct evidence for the counterintuitive idea that far from being the fixed core of the cosmos, Earth was in fact hurtling around its nearest star. So he drew on the prevailing political structure of the time to give his ideas credibility, writing about how the Sun "governs the family of planets revolving around it . . . as though seated on a royal throne."

As Copernicus's ideas became accepted, European kings started in turn to associate themselves with this central, royal Sun, with phrases reminiscent of those once used to describe Roman emperors like Constantine. In the seventeenth century, Philip IV of Spain was nicknamed the "Planet King" (after the Sun), while Louis XIV of France was universally known as the "Sun King." As late as the second half of the eighteenth century, British kings George II and III were described as "shining sovereigns" or "glittering princes" spreading their "superior rays." Despite their new arrangement, the heavens were still supporting the old structure of powerful, divinely inspired rulers, just as they had done for thousands of years. But not for long. Revolution was on the way.

✡ ✡ ✡

When Newton used a combination of observation and reasoning to show that the very same equations regulated the motion of everything in the cosmos—from apples to planets, from a grain of sand to the blazing Sun—he transformed how people thought, not just about the physical world but about who they were and how they lived. What Newton came up with wasn't just "gravity," or "action at a distance," says historian of science Mordechai Feingold. It was a new "system of the world."

The ground had already been prepared. In the early seventeenth century, the Italian astronomer Galileo's telescope observations—showing that the Sun had spots, the Moon had mountains, and planets had their own phases and moons—started to break down the perceived division between the corrupt Earth and divine, perfect heavens. Through his studies of motion, Galileo also came up with the fundamental idea that the universe is a mechanical system, ruled not by God's will but by physical principles. "Philosophy is written in this vast book, which lies continuously open before our eyes (I mean the universe)," he wrote. "It is written in the language of mathematics."

These ideas about the cosmos greatly influenced political thinkers such as the English philosopher Thomas Hobbes. Like Galileo, Hobbes believed in a mechanical, material cosmos, and he argued that all natural phenomena, even human thought, are simply the result of mechanical interactions between material bodies. In a work called *Leviathan*, published in 1651, Hobbes applied these principles to politics and society, rejecting the divine right of kings and arguing that people are born equal. But he was still far from democratic.

Without a strong power structure to hold them together, he saw individuals as analogous to chaotic particles, bouncing about in a violent, anarchic "state of nature" in which there would be no technology, industry, arts or science, and life would be "solitary, poor, nasty, brutish and short." To avoid this, he suggested, we must give up our individual rights in return for protection from a sovereign with absolute power.

In 1687, Newton replaced this aimless cosmos with a greater plan. Far from being chaotic, he showed, celestial bodies are ruled by the universal law of gravitation. Almost immediately, philosophers and political theorists started to think about what that might mean for society. If a single mathematical law governs the entire universe, they figured, surely similar universal principles must also apply to people.

One of the most important Enlightenment philosophers was John Locke, who, like Newton, thought that the best way to understand the world was through a combination of empirical evidence and reason. He, too, saw a material cosmos, defined by mechanical interactions and collisions, but he rejected Hobbes's aimless void. "The state of nature has a law of nature to govern it," he said in 1689. "And reason, which is that law, teaches . . . that being all equal and independent, no one ought to harm another in his life, health, liberty or possessions." Just as the same physical laws govern all bodies in the cosmos, from atoms to planets, the same moral laws should govern everyone, even the king. Locke still believed that we must exchange some of our natural rights for a ruler's protection. But in Locke's scheme, that ruler doesn't have absolute power. If he becomes tyrannical and infringes the rights of the people, he can be removed.

Along with Hobbes, Locke is seen as one of the founders of "liberal" political thought—including concepts of individual rights and limited government—on which today's Western political systems are based. Although he later corresponded with Newton, Locke wasn't necessarily directly influenced by the physicist's theories. He worked on his ideas for decades before Newton's work was published, and he was responding to Galileo and Hobbes among others, as well as his own experience of British politics. He grew up during the English Civil Wars, and when the Protestant prince and princess of Orange were later invited to the throne in the Glorious Revolution of 1688, Locke helped to draft the Bill of Rights that limited their power.

Nonetheless, the similarity of Locke and Newton's approaches—published just a couple of years apart—meant that people inevitably saw them as connected. Locke encouraged the link, referring to Newton as "incomparable" and to the *Principia* as "never enough to be admired." Meanwhile, Newton suggested in his preface to *Principia* that the "same kind of reasoning" he had used would lead to discovery of similar principles regulating other aspects of nature. As Feingold puts it, Newton and Locke became "emblems of a new era." Locke's work was seen as proof, he says, that building on Newton's scientific foundations would yield natural laws governing society, too.

Newtonian metaphors and principles soon seeped through politics. In a 1713 essay, the Irish philosopher George Berkeley compared social bonds to gravity that declines with distance; a few years later, the politician Lord Bolingbroke saw celestial mechanics in the evolving English constitution, arguing that the monarch "can move no longer in another orbit from his people, and, like some superior planet, attract, repel, influence and direct their motions by his own."

According to historian Richard Striner, one of the most powerful political concepts to come out of the response to Newton's work was "counterpoise": the idea of maintaining harmony by setting opposing forces against one another. Just as a combination of attractive gravity and repulsive centrifugal force keeps planets perfectly placed in orbit around the Sun, so different parts of society had to be finely balanced. The French political philosopher Charles de Secondat, Baron de Montesquieu, pioneered this concept in his 1748 treatise, *The Spirit of Laws*, one of the most influential works of political theory ever published. He suggested that to avoid abuse of power, political authority should be separated into legislative, executive and judicial powers, to make, implement and adjudicate laws, respectively.

Montesquieu was building on ancient theories of mixed government, in which elements of democracy, aristocracy and monarchy were combined. But he was also inspired by the idea that political forces, like celestial ones, had to be held in opposition, so the balance of power didn't either collapse toward central tyranny at one extreme, or hurtle away into anarchy at the other. He saw the limited monarchy of Great Britain, with its counterpoise of commons, lords and king, as close to an ideal system. "It is with this kind of government as with the system of the universe," he said, "in which there is a power that constantly repels all bodies from the centre, and a power of gravitation that attracts them to it."

These, then, were some of the ideas being enthusiastically discussed in the mid-eighteenth century, not just by elite politicians and philosophers but by young Paine and his friends in the coffeehouses of London, and their contemporaries across Europe. Newton's physics didn't provide any undisputed answers. Thinkers from different countries, religious backgrounds and political persuasions each used Newtonian metaphors to justify and validate their own ideas: some wanted to interfere as little as possible with society's natural cosmic rhythms; others argued for controls to emulate the opposing forces in the heavens. Nonetheless, there had been a deep and important change. There was a new source of natural order in the cosmos, and a new idea: that through laws and logic, people might glean the best way to govern themselves. The fundamental paradigm of a divinely inspired cosmic ruler had been smashed wide open, and the fight for what would replace it had begun.

✲ ✲ ✲

Paine ran out of money in six months. He left London and embarked on fifteen years of dead ends and misfortunes, during which he set up a corset-making business that collapsed; fell in love, then lost his wife in childbirth; and worked as a tax collector, tackling smugglers on the coast before being fired twice, once apparently for whistle-blowing and once after fighting for higher pay on behalf of his colleagues. His second marriage, to a tobacconist's daughter in Lewes, failed too, as did the business that the couple inherited.

Through it all, he kept returning to the lectures and debates of London, and in 1774, after selling his possessions and what remained of the tobacco store to pay off his debts, Paine moved to the capital once more, where he visited an acquaintance: the inventor, diplomat and fellow Newton enthusiast Benjamin Franklin. Try your luck in the American colonies, Franklin suggested. He gave his friend some letters of introduction to influential figures in the New World, and Paine spent the last of his cash on a first-class ticket to Philadelphia.

Once he had recovered from typhoid fever, he found that the new Enlightenment ideas about science and politics were being discussed in the clubs and coffeehouses of Philadelphia too. He slotted easily into this community, drinking and debating at a club called the Indian Queen and attending lectures hosted by the American Philosophical Society. And he soon found work, as the editor of a brand-new publication called *The Pennsylvania Magazine*. It covered subjects from beavers to Voltaire, but Paine's contribution was increasingly political, as he ran articles arguing against dueling, cruelty to animals and slavery, and even in support of women's rights.

Meanwhile, tensions between Britain and her American territories were escalating. During the seventeenth and eighteenth centuries, the colonies had been largely left to run themselves. But the Seven Years' War against France left the British government with huge debts, and in 1765 it started to impose heavy taxes on the American colonists: on paper, glass, paint, lead and tea. The colonists resisted, with widespread boycotts of taxed goods and the famous Boston Tea Party of December 1773, when hundreds of chests of tea were dumped into the harbor.

Britain responded with warships and troops on the streets, and Paine arrived just ahead of the first violent clashes in spring 1775. George Washington, a charismatic political leader from Virginia, was appointed commander of a new Continental Army, but despite the hostilities, there was little direct talk of independence. The majority of American colonists had no wish to be rebels or traitors. The fighting was about forcing better terms with Britain, not severing the link. For Paine, though, it was a turning point. "When the country, into which I had just set my foot, was set on fire about my ears, it was time to stir," he later wrote. He started working on an extended essay, a pamphlet full of ideas that he discussed with his Newtonian friends, including Franklin and the lawyer and landowner Thomas Jefferson. He called it *Common Sense*.

It was so controversial, he could only find one printer to publish it, and at first Paine didn't dare to put his name on it (the title page simply read "by an Englishman"). But it was an immediate sensation. The first print run of a thousand copies sold out within days; within three months it had sold 120,000 copies (out of a free population of just two million). Rival and bootleg editions, as well as handwritten copies, circulated not just in America but all over Europe, and were read aloud to rapt audiences.

In the pamphlet, Paine argued for American independence from British rule, with language full of Newtonian references: "In no instance hath nature made the satellite larger than its primary planet," he said, "and as England and America, with respect to each Other, reverse the common order of nature, it is evident that they belong to different systems." He attacked the very concept of monarchy—even Britain's limited one—arguing that as all men are created equal, it makes no sense to hand power to an inherited line of kings; in fact, "nature disapproves it, otherwise she would not so frequently turn it into ridicule by giving mankind an ass for a lion." He described the force that urges men to cooperate in society as a "gravitating power" and argued for the creation of a new kind of nation, with a democratic government and as many adult men as possible eligible to vote. "We have it in our power," he said, "to begin the world over again."

Influential figures such as Franklin, John Adams and Samuel Adams, had been discussing similar ideas in private. But Paine was the first to advocate independence in public. He used clear, persuasive language and images that ordinary Americans could relate to and understand. And he reframed the entire conflict. Far from being for traitors and rebels, independence was now honorable; in fact, it was mandated by nature's laws. Revolution wasn't about avoiding taxes but building a better world.

Historians agree it was an extraordinary moment. "One had to be a fool or a fanatic in January 1776 to advocate American independence," said historian Bernard Bailyn: everyone knew Britain was the most powerful nation on Earth. "Pain took an idea that Englishmen on both sides of the Atlantic cherished," adds Craig Nelson, "that theirs was the best government in the world, as it balanced the competing powers of monarch, gentry and commoner against one another . . . and detonated it." Or, as one Bostonian put it at the time: "Independence a year ago could not have been publickly mentioned with impunity . . . Nothing else is now talked of, and I know not what can be done by Great Britain to prevent it."

It was no longer a question of whether Americans should declare independence, but how and when. As in Europe, the political debate often incorporated images from Newtonian physics. The language of astronomy was already being used to describe the relationship between Britain and her colonies; in 1764 the colonial governor Thomas Pownall described the British Empire as a celestial system in which colonies orbited in their "proper sphere." For Pownall, the government, rather than the king, was now "the centre of attraction, to which these colonies . . . must tend." But Americans had to confront a much more fundamental shift. What would happen when the gravitational core of the system were removed?

One writer to the *Connecticut Courant* warned that if independence were declared too soon, the colonies might scatter like "so many balls in the air," but a correspondent to the *Pennsylvania Ledger* in April 1776 was among those who forged a new, more positive vision. The writer imagined a trip into space where travelers "found a republic amidst the stars." The Sun might seem to admiring mortals below as "the grand monarch of the heavenly bodies," yet zooming out to the wider cosmos reveals that in fact the sky is studded with such stars: constellations that are "united upon the principles of perfect equality." The power structure written in the skies was no longer a monarchy, but a republic.

Congress formally announced its independence that July, claiming "the separate and equal station to which the Laws of Nature and of Nature's God entitle them" in a declaration originally drafted by Paine's friend Jefferson. The most famous sentence reads: "We hold these truths to be self-evident, that all men are created equal, that they are endowed by their Creator with certain unalienable Rights, that among these are Life, Liberty and the pursuit of Happiness."

Scholars endlessly debate its sources; there are strong similarities with Paine's writings, and the ideas were clearly influenced by Locke and England's Bill of Rights. Many historians believe that the Declaration is also heavily indebted to Newton. The physicist isn't cited directly, but with his ideas so firmly fixed in the public mind, the references to "Laws of Nature" and "self-evident truths" couldn't have failed to remind readers of Newtonian principles. Jefferson himself owned both a bust and a death mask of Newton and declared him (along with Locke and the philosopher Francis Bacon*) as one of "the three greatest men that have ever lived."

Throughout the 1780s, Newton-inspired ideas and metaphors—including the concept of counterpoise—would continue to be key in drafting the U.S. Constitution. Delegates to Congress were essentially designing a new government and federal system from scratch. They had different ideas about how to do this, but all were concerned with

---

* Bacon's work in the seventeenth century was instrumental in the development of the scientific method.

designing a structure that would resist imbalances or abuses of power, and they often discussed the problem in celestial or mechanical terms.

Some of the delegates wanted as little regulation as possible—Delaware's John Dickinson argued that the states "ought to be left to move freely in their proper orbits"—whereas others wanted stronger ties; John Adams, for example, argued for a constitution that would control "those attractions and repulsions, by which the balance of nature is preserved." It's a debate that ultimately resulted in the checks and balances—the separation of powers and the bicameral house—that are still at the heart of U.S. government: a model of democracy, inspired by the natural forces of the heavens, that has since been exported around the world.

Before all that, though, came the matter of choosing the nation's flag. The design was set by a resolution of the Continental Congress on June 14, 1777, and there was of course only one conceivable way to represent the thirteen newly independent states. They were to be shown, the resolution said, as "thirteen stars, white in a blue field." There was a new constellation in the sky.

✡ ✡ ✡

After the revolution, Paine struggled to find a new purpose. "It did not appear to me that any object could afterwards arise great enough," he said, "to make me quit tranquility and feel as I had felt before." He returned to his love of science, conducting experiments on gases and candles with Franklin and Washington, and spent seven years designing a wide-span iron bridge.

In 1787, he traveled to Paris and then to London, introducing his bridge to European audiences. He was granted an English patent and had a 140-foot-long model built, but his interest in an engineering career soon faded. There was little enthusiasm from the public, and Thomas Jefferson (now the U.S. minister to France) was writing to him from Paris with news of dramatic events across the Channel.

Political mismanagement, crop failure and a huge national debt (from the Seven Years' War, and then the American Revolutionary War, in which French soldiers fought with the rebels) had plunged France into an economic crisis; as the aristocrats feasted, peasants across the country starved. In 1789, with parliament refusing to enact Louis XVI's tax reforms, the king summoned an archaic institution called the Estates-General, representing the country's three estates: the clergy, nobility and commoners. It didn't go as the king had planned. The commoners promptly formed their own new National Assembly, in which they held the balance of power, and invited the others to join. The French Revolution had begun.

That summer, the assembly adopted what's still seen as a crucial document in the history of human rights: the Declaration of the Rights of Man and the Citizen. Influenced by Enlightenment philosophers and drafted with Jefferson's help, it established equal rights in law for all citizens, as well as a free press. The assembly also voted to vastly reduce the king's powers, and to dismantle aristocratic society. From the start, though, events had a darker edge compared with America's revolution. Those perceived to represent the old regime were often brutally killed by rioters, including the July 14 storming of Bastille prison, after which the jubilant crowd raised the governor's head on a pike.

Paine was given honorary French citizenship for his role in the American Revolution, and he traveled to Paris several times to witness events firsthand. He was inspired by this new example of liberty and democracy, but in 1790, a hugely successful pamphlet by the philosopher Edmund Burke appeared in England. Burke argued that the uprising was a "contagion," in which an insolent and violent mob was destroying France's traditional society and values. Paine was incensed, and immediately began work on a blistering reply. The crowd that attacked the Bastille were far from being a mob, he said; they were heroic. "Never were more pains taken to instruct and enlighten mankind, and to make them see that their interest consisted in their virtue, and not in their revenge, than have been displayed in the Revolution of France." Whereas Burke justified his arguments through tradition and history, Paine argued that rights are exercised by the living, and that people are born equal and free.

Published in February 1791, *The Rights of Man* made Paine once more the fastest-selling author in history. But the Revolution was only just getting started. On the morning of June 21, Paine was woken before dawn by his Parisian host, telling him: "The birds are flown!" The king and his family had fled the palace. Paine rushed out but forgot his hat, with its republican tricolor ribbon, and was almost killed by the angry crowd.

He returned to London soon afterward, but the experience doesn't seem to have dampened his enthusiasm for the Revolution. In early 1792, he published the second part of *The Rights of Man*, which was more explicitly republican. He had realized that the principles of liberty and equality he fought for in America could be fully expressed in Europe too. If a traditional regime like France could change so dramatically, revolution could happen anywhere.

The book was so successful that the British government sent representatives to assess the loyalty of its troops, worried that the country could be on the brink of civil war. Booksellers were arrested, debating clubs were closed and Paine and his publisher were charged with sedition, an offense that carried the death penalty. On September 15, he fled the country, "surrounded by a hostile mob of chanting Dovermen." He arrived in France to cheers of "*Vive* Thomas Paine!" and a seat on its brand-new elected government, the National Convention.

It was a tense atmosphere, with constant talk of traitors and invasions, and a wave of state-sponsored executions. The convention was bitterly divided from the start, pulled between the extremes of the Jacobins on the left (led by the lawyer Maximilien Robespierre) and Paine's closest associates, the Girondins, on the right. In heated debates about the fate of the king, Paine argued for clemency, angering the increasingly powerful Robespierre. On January 21, 1793, in a move that horrified conservatives across Europe, King Louis XVI was sent to the guillotine.

In June, the Jacobins seized power in a parliamentary coup, arresting their Girondin rivals. Paine was refused entry to the convention building, and when he met a Jacobin colleague on the steps, he commented that a writer who had compared the French Republic to the Roman god Saturn, who ate his children, was surely correct. "Revolutions cannot be made with rosewater," the Frenchman replied.

And so began the Reign of Terror. The Jacobins arrested farmers and seized their crops, and stepped up political executions, including of the Girondins. In October, Paine was publicly called a traitor, but he couldn't return to England, where he faced the death penalty, or flee to America, because British ships were ready in the Channel to intercept him. So he once again started to write, as fast as he could. "I saw my life in continual danger," he said later. "My friends were falling as fast as the guillotine could cut their heads off, and as I every day expected the same fate, I resolved to begin my work."

Just after Christmas, the dreaded moment arrived: he was woken by security agents banging on his hotel room door. On the way to prison, Paine managed to give a copy of his precious manuscript to his publisher, and it was released in London in February 1794. As Paine waited desperately in a dirty, louse-ridden cell, his latest work would become a smash hit, outselling even his previous books. With monarchy seemingly on the way out, Paine used the new science of the cosmos to attack what he saw as the world's other great tyranny. Neither it—nor Paine's reputation—would ever fully recover.

✡ ✡ ✡

Ever since his London days, Paine had been considering what scientists' understanding of the cosmos meant for religious belief. "After I had made myself master of the use of the globes, and of the orrery," he wrote, "and obtained at least a general knowledge of what was called natural philosophy, I began to . . . confront the internal evidence those things afford with the Christian system of faith." He was convinced that rewriting politics wasn't enough. All national churches—whether Jewish, Christian or "Turkish"—were just human inventions, he concluded, "set up to terrify and enslave mankind, and monopolise power and profit." There would need to be a religious revolution too.

America's Founding Fathers had taken a big step in separating their new government from religious influence—as recommended by Paine in *Common Sense*, and before him by philosophers such as Locke and Montesquieu. The intention was to protect freedom of belief, so one religion wasn't privileged over another. But as Paine was writing in Paris, France's revolutionaries were going much, much further. They weren't just separating religion from the government, but wiping it out altogether.

The new regime had already seized land owned by the Catholic Church, dismantled religious orders and exiled or killed many priests. By autumn of 1793, Robespierre and his fellow Jacobins were systematically removing all religious symbols and beliefs from the country, replacing them with the secular values of the Enlightenment. Cathedrals and cemeteries were vandalized. Church bells were melted down to make guns. A new calendar was introduced, with ten-day weeks and three-week months, dated not from Christ's birth but from the founding of the Republic. Notre-Dame Cathedral was renamed the Temple of Reason.

It was the religious rethink that Paine had been waiting for. But he feared what kind of society might be left if organized religion were destroyed without anything to replace it. So in his last major work, *The Age of Reason*, he tried to imagine a church-free religion; to provide a new, more democratic spiritual framework that would stop people from falling into immorality and atheism. Events in France had "rendered a work of this kind exceedingly necessary," he said, "lest in the general wreck of superstition, of false systems of government and false theology, we lose sight of morality, of humanity and of the theology that is true."

To do it, he drew on ideas from the lectures he had attended in London so many years before. The teachers popularizing Newton's scientific work at that time often used the new understanding of the cosmos to prove or support belief in a Christian god, in a tradition called physicotheology.* The orrery-maker James Ferguson, for example, wrote that astronomy convinces us "of the existence, wisdom, power, goodness and superintendency of the SUPREME BEING! . . . An undevout Astronomer is mad." Mathematician Benjamin Martin argued that the scientific study of the heavens "creates an Idea every way worthy of . . . an infinitely wise, perfect and powerful Being."

Paine now adopted the astronomers' arguments, but with a twist. He took readers on a tour of the solar system, as his teachers had done, emphasizing the vastness and grandeur of the cosmos—"there is room for millions of worlds as large or larger than ours, and each of them millions of miles apart"—and arguing that a benevolent creator (the "Almighty lecturer") created the multiple planets so that humans might observe their movements and thus discover the laws of nature. Here is all the proof we need, he said, of the existence of an all-powerful God.

But then he plunged in the knife, arguing that the very nature of the universe he was describing proves Christian beliefs to be absurd. If God made even more worlds than stars, he asked, why should a unique savior have been born just on this particular one? How could we believe that "the Almighty, who had millions of worlds equally

* This tradition emerged from a series of lectures, endowed by the physicist Robert Boyle, to consider the relationship between Christianity and the new science. The first series, given in 1692 by the theologian Richard Bentley, was called "A Confutation of Atheism from the Origin and Frame of the World." Newton himself wrote to congratulate Bentley, saying that when he wrote *Principia*, he had wondered whether the principles he was describing might help to support belief in God: "Nothing can rejoice me more than to find it useful for that Purpose."

dependent on this protection, should quit the care of all the rest, and come to die in our world, because they say, one man and one woman had eaten an apple!" Far from strengthening Christian faith, he said, the existence of countless solar systems renders it "at once a little ridiculous; and scatters it in the mind like feathers in the air."

Paine used a similar argument against the concept of divine revelation, suggesting that a benevolent God, who loved all rational beings in the vast universe, wouldn't make himself available only to a few individuals in a particular time and place; and wouldn't restrict knowledge of his message only to those who happened to understand the language in which the scriptures were written. Instead, said Paine, he would speak in a way that everyone could experience and understand: through the language of the cosmos itself. "That which is now called natural philosophy, embracing the whole circle of science, of which astronomy occupies the chief place . . . is the true theology." God's word was not to be found in the Bible, but in Newton's laws of physics.

As with Paine's previous works, the ideas in *The Age of Reason* weren't completely original. He was following the eighteenth-century tradition of deism, inspired by Enlightenment thinkers such as Voltaire, Spinoza and David Hume.* Rather than believing in a personal god who constantly interferes in human affairs, deists saw the Creator more as a divine clockmaker, who winds up his mechanism then stands back and lets it run according to physical laws. Such beliefs weren't particularly controversial. Many of the Founding Fathers—Franklin, Jefferson, Madison, Washington—are thought to have been deists. But they were discreet about their views and respectful toward the church; many still felt that organized religion was necessary to keep the masses under control. Paine, on the other hand, launched a contemptuous and public attack.

* Some historians argue that Paine was also influenced by the plain-speaking, skeptical traditions of his Quaker upbringing.

The result, published in 1794, was explosive, one of the most controversial books in the history of religion. Although there was little reaction in France, *The Age of Reason* was particularly popular in Britain, despite the government's desperate efforts to suppress it, and even more so in the United States, where by 1796 it had gone through seventeen editions. This helped to fuel not only an upsurge in deism but a Christian backlash of epic proportions, including more than thirty published rebuttals, one of which was given to every student at Harvard, and a religious revival known as the Second Great Awakening.

Paine almost didn't live to see any of it. By June, the Reign of Terror was at its height, with tens of thousands of citizens in prison and a flow of executions every day. "To such a pitch of rage and suspicion were Robespierre and his committee arrived, that it seemed as if they feared to leave a man living," Paine remembered. "Scarcely a night passed in which ten, twenty, thirty, forty, fifty or more were not taken out of the prison, carried before a pretended tribunal in the morning, and guillotined before night." Then he once again became seriously ill with typhoid fever.

"I had then but little expectation of surviving, and those about me had less," he said, but the fever may have saved his life. On July 24, Paine was sentenced to death. There's a story that by this time Paine's temperature was so high, his three Belgian cellmates were given permission to leave the cell door open, to let in a cooling breeze. When the cell was later marked with a figure "4"—to show they were to be executed the next day—it was chalked onto the inside of the open door. That evening, the Belgians asked to close the door again, hiding the mark, and all four lives were spared.

A few days later, the Reign of Terror collapsed and Robespierre was himself executed. Paine survived his illness (though his health never fully recovered) and, in November, was finally freed. He lived in France another five years, and met the future emperor Napoleon Bonaparte, who told him he slept with a copy of *The Rights of Man* under his pillow.* Paine wrote two more parts to *The Age of Reason*, pointing out what he saw as absurdities, inconsistencies and immoralities in the Bible. And in 1802, he crossed the Atlantic once more, to live in the brand-new capital of Washington, D.C.

He didn't get the reception he hoped for. Though he had some loyal supporters, conservatives disliked his radical politics and many people were shocked by his fierce attacks on George Washington, whom Paine felt had abandoned him to die in the French prison. But mostly Paine was hated for *The Age of Reason*, which transformed him from a hero of the revolution into a "drunken atheist," "loathsome reptile" and even a "demi-human archbeast." Old friends stopped speaking to him. One time he was refused a seat in a stagecoach and pelted with stones. A local official refused to let him vote. He died in 1809 in New York, with only six mourners at his funeral.

"Poor Tom Paine! There he lies," runs the old children's rhyme. "Nobody laughs and nobody cries; Where he has gone or how he fares, Nobody knows and nobody cares."

✡ ✡ ✡

Centuries earlier, as discussed in chapter 4, major organized religions including Christianity split God from the cosmos, replacing the ancient deities in the sky with a separate, abstract creator. But people still generally saw the universe as infused with, and powered by, divine energy—what Plato called *pneuma*, or soul. That includes the scholars who built the idea of the universe as a rational machine. In the fourteenth century, when Dante described the cosmos as a mechanical clock, its cogs were driven by God's love. For Kepler, calculating elliptical orbits three hundred years later, the ordered movements of the Earth, Sun and planets were guided by their souls,

* The pair fell out later, with Paine criticizing Bonaparte after he became more dictatorial.

which radiated out across space. Descartes described our bodies, like the cosmos, as physical machines, but argued that what makes us human (as opposed to purposeless automatons) is the vital addition of an immaterial soul.

For many Enlightenment thinkers, too, the specific workings of the physical universe might have been predictable using mathematics, but its power and meaning ultimately came from God. Newton himself was a devout Christian, arguing that the beauty and regularity he saw in the heavens could only "proceed from the counsel and dominion of an intelligent and powerful Being." Others, like Paine, hoped that humanity would combine a rational, Newtonian cosmos with the religious wonder of deism.

That's not how it worked out. Paine's arguments certainly helped to puncture many of the claims and arguments of organized religion: "It was men like Paine, who, in face of persecution, caused the softening of dogma by which our age profits," said philosopher Bertrand Russell in 1934. But deism didn't replace Christianity, and *The Age of Reason* is now championed not as a spiritual text but as a vital step toward atheism. When Western philosophers and scientists recast the cosmos as a self-regulating machine, explained and guided by predictable, mathematical principles, they unwittingly removed the need for divine influence. The soul of the universe began to drain away.

The German sociologist Max Weber called this process "disenchantment"; the philosopher Friedrich Nietzsche described it as the death of God. After the Enlightenment, rulers drew their power not from divine will but from the consent and rationality of the people; Paine's story illustrates how the new cosmology helped to inspire ideas about democracy and human rights. But that wasn't all. The model of predictable physical systems that can be understood mathematically was enthusiastically applied to other areas of human life too, from money to the mind.

Philosophers such as David Hume (sometimes referred to as the "Newton of the moral sciences") showed that ethical frameworks for living could be derived without any need for religious teaching. The founder of modern free-market economics, Adam Smith, called Newton's theory of gravity "the greatest discovery that ever was made by man," and framed his own 1776 work, *Wealth of Nations*—in which he described markets regulated by invisible forces of supply and demand—as a new *Principia*.* From the eighteenth century on, proponents of the new field of psychology saw this, too, as a counterpart of Newton's physics, and they began to study mental phenomena using the tools of mathematics, leading to the publication of hundreds of psychological "laws."†

God wasn't dead, of course. But the balance of power in the cosmos had shifted. Physics was the new supreme ruler, on Earth as in the sky.

* Smith argued, for example, that governments shouldn't intervene in markets, so that goods and services could reach the "natural price . . . to which the prices of all commodities are continually gravitating."

† These psychological "laws" include Weber's law, published in 1834, quantifying the perception of change in a given stimulus, and Thorndike's law, published in 1905, which states that responses producing a satisfying effect in a particular situation become more likely to occur again. (Modern psychologists are more wary of "laws," preferring less ambitious terms such as "effect" or "principle.")

8

LIGHT

ON TUESDAY, MARCH 13, 1781, A MUSICIAN NAMED WILLIAM Herschel stood in the garden of his house in Bath, England, surveying the stars through a 7-foot-long telescope he had made himself. Just before midnight, he spotted an unfamiliar light sailing through the constellation Gemini. At first he thought it was a comet, but it had no tail, and it gradually became clear that the wandering beacon was in fact a new planet. He named it Georgium Sidus—"George's Star"—after King George III, but astronomers eventually settled on Uranus, for the ancient Greek god of the heavens.

It was a hugely significant find. Uranus was the first planet to be revealed since antiquity, and with an orbit far beyond that of Jupiter and Saturn, it instantly doubled the reach of the known solar system. But it was just one of a wave of discoveries that filled the sky during the late eighteenth and early nineteenth centuries. As revolutions reshaped the political landscape, increasingly powerful telescopes transformed astronomers' view of the heavens, allowing them to see farther and better than ever before. Herschel, working with his sister Caroline, was at the forefront of a shift from observing the movements of known celestial bodies increasingly accurately—mostly in the service of navigation—to surveying and exploring the entire sky.

Our solar system became busier as European astronomers discovered a host of visiting comets, as well as multitudes of small, dim objects that became known as asteroids, starting with Ceres in 1801 and soon followed by Pallas, Juno, Vesta and many, many more. Herschel discerned more moons of Saturn (Mimas and Enceladus) and Uranus (Titania and Oberon), and was the first to identify undulating ice caps on Mars. The consequences for the wider universe were even more profound. The stars had been seen as essentially fixed points on the surface of a relatively nearby dome or sphere. But as astronomers magnified further, ever more distant lights appeared, and they instead found themselves entering the giddying three-dimensional infinity of deep space.

The variety was stunning. Many so-called stars were in fact groups of two or three neighbors orbiting each other, around which, many speculated, there must in turn be other worlds. Meanwhile, astronomers including the Herschels catalogued thousands of mysterious and colorful objects that didn't look like stars at all—a panoply of clouds and clusters, spirals and spots that were dubbed nebulae (from the Latin for "cloud"). Although scientists chose names from classical mythology, the cosmos was no longer a stage for fantastic beasts or ancient gods. It was an endless astronomical garden, filled with natural, physical wonders to be collected and described.

As with all the best gardens, though, some knowledge was forbidden. The revolutionaries who hoped to rebuild society after the old authorities were swept away were also redefining how and what it was possible to know. French philosopher Auguste Comte, the influential founder of positivism, rejected moral and metaphysical arguments altogether, declaring in the 1830s that the Newton-style interplay of observation and reason was the only route to truth. He spent years defining the scope and methods of each of the sciences, stating what types of inquiry were (and weren't) legitimate, and for astronomy, he was clear. As earthbound observers, our knowledge of the distant stars—which we could see but never touch—would always be limited.

"We can imagine the possibility of determining the shapes of stars, their distances, their sizes and their movements," he wrote in 1835.

"But there is no means by which we will ever be able to examine their chemical composition." Britain's Astronomer Royal George Biddell Airy reiterated Comte's argument in 1857, when he warned that speculating about the appearance or nature of celestial bodies "possesses no proper astronomical interest." Scientists might observe the motions of the stars, but could never ask what they were made of, or how they worked.

Few rules have been broken quite so spectacularly. Just two years later, in 1859, two researchers were working late in their laboratory at the University of Heidelberg in southern Germany. As they looked out of the window, along the Neckar River toward the Rhine, they saw that the horizon was lit by a red glow; a huge fire was raging in the nearby port of Mannheim. What they saw in those flames, as one chronicler put it, "surprised them to their bones." And it proved Comte and Airy utterly wrong. What the Heidelberg pair discovered opened up a dramatic new window on the universe, and once again transformed our relationship with the sky.

✡ ✡ ✡

The German chemist Robert Bunsen is most famous today for the gas burner he helped to invent, its simple metal lines now a fixture of chemistry labs around the world. Teachers show their students how to adjust the amount of oxygen that mixes with the gas as it burns to achieve a hot, colorless flame. What they often don't explain is how that flame helped Bunsen to reach inside the stars.

Bunsen was by all accounts warm and charming, though generally splattered with chemical reagents (the wife of one of his colleagues said she'd like to kiss him, but would have to wash him first), and he had a cavalier attitude toward lab safety—his British colleague Henry Roscoe remembered seeing smoke rise from Bunsen's fingers and "the smell of burning Bunsen" as he removed the lid from a super-hot crucible with his bare hands.

Bunsen began his career studying deadly compounds of arsenic, with the help of a facemask attached to a long glass tube that allowed him to breathe fresh air from outside. In the 1830s, he discovered an antidote to arsenic poisoning, which later saved his life when a sample exploded (one of several lab explosions he survived). He also studied the workings of superheated geysers in Iceland, and invented a type of battery made from zinc and carbon, which he used to liberate pure metals from their ores in the recently discovered process of electrolysis.

One of those metals was magnesium. Bunsen was fascinated by the brilliant flash of energy it produces as it burns, and he turned his attention toward the chemistry of light. Chemists (and alchemists) had long known, for example, that if you sprinkle different substances onto a fire, they produce flames of different colors: magnesium flashes white, lithium burns rose-red and potassium turns the fire lilac. What did the colors mean?

In 1852, Bunsen moved to Heidelberg University. At first, the chemistry department was housed in an ancient monastery, with lab benches running under the cloisters, a fitting symbol, perhaps, of how a new type of knowledge had replaced the old. "Beneath our feet slept the dead monks and on their tombstones we threw our waste precipitates," Roscoe later remembered. "There was no gas in those days nor any town water supply." Heidelberg was hooked up to central gas pipes in 1853, and in a newly built lab a few years later, Bunsen developed his famous burner. Its transparent, soot-free flame allowed him to see the elements' colors more clearly than when burning them in a fire. Bunsen also used filters to isolate different hues, but it was still hard to make sense of the results: copper compounds produce bluish flames, for example, but so do arsenic and lead. Then his friend, physicist Gustav Kirchhoff, had an idea.

Kirchhoff, though just as brilliant in his own field, was in many ways the opposite of Bunsen. Thirteen years younger than the exuberant chemist, Kirchhoff was described by his colleagues as modest and shy. He was interested in how electricity behaves in different circuits and materials, and like Isaac Newton he had a talent for making meticulous observations of complex phenomena, then deriving elegant mathematical laws to describe them. Kirchhoff suggested that instead of comparing different colored flames by eye, Bunsen should pass their light through a glass prism, just as Newton had done in 1666.

When Newton allowed light from the Sun to shine through a small hole or slit and then through a prism, he famously found that though it seems white or transparent to us, sunlight is packed with color, bursting with all of the rays that we perceive as emerald, indigo, crimson or gold. (Scientists now understand that the different colors correspond to different wavelengths of light.) When this mix of light enters or leaves a prism, different hues are pushed off course by different amounts: the sunbeam unfurls into a rainbow. Using the same trick on the light from Bunsen's flames, said Kirchhoff, would spread the emitted colors into a spectrum, making it possible to measure them much more accurately.

The pair built an apparatus they called a spectroscope, which involved a set of lenses, a prism and a viewing telescope, all inside an enclosed box. They used it to investigate "flame reactions" for a range of substances—sodium, lithium, strontium, calcium, barium—and found something intriguing. For any particular element, it didn't matter what chemical form of it was burned, what temperature it was burned at or what gas it was burned in: the pattern of light that shone out from that element was always precisely the same.

Other prominent scientists had already used prisms to study the light produced in flames. They had identified characteristic bands of color emitted by certain substances, called spectral lines, such as the pair of intense yellow lines produced by sodium. But the combinations of lines produced by the multiple elements in different compounds were often confusingly complex. Kirchhoff's talent for making sense of these patterns, combined with Bunsen's clear-flame burner (which emitted no visible lines of its own) and ability to make extremely pure samples using electrolysis, meant that they were the first to fully realize what the results meant: each element has its own unique signature, hidden in the light.

The new method was objective. Instead of judging different shades by eye, even a color-blind person could identify elements from their flames, by measuring the precise spacing of the lines produced. And it was exquisitely sensitive. Bunsen and Kirchhoff carried out experiments in which one of them would heat a tiny sample in a corner of the lab, swinging a large, open umbrella to help the vapor diffuse into the air, while the other sat in the opposite corner, looking through the spectroscope. For detecting sodium and lithium in particular, Bunsen reported, the method "surpasses all others known in analytical chemistry as to definiteness and sensitivity," picking up as little as one part per million of salt in the air.

A couple of years later, they would use flame spectra to discover two completely new elements. Trace amounts of one were hidden in mineral water from Dürkheim; it produced two sky-blue emission lines, so they named it caesium, from the Latin *caesius*, "which the ancients used to designate the blue of the upper part of the firmament." The other was rubidium (identified by a dark red spectral line) in a rose-colored mineral called lepidolite. Other scientists soon embraced the new approach, triggering a growth spurt in the periodic table.

But that wasn't all. One evening in 1859, Bunsen and Kirchhoff saw the Mannheim fire through their lab window. They looked at the light with their spectroscope, and were surprised to detect characteristic lines of barium and strontium, even though the flames were at least ten miles away. A few days later, they were hiking in the wooded hills around Heidelberg, a route known at the university as "the philosopher's walk," when Bunsen had an idea. Perhaps their spectroscope could gaze not just across the Rhine plain but across the universe. It was an idea that Comte and his fellow positivists had declared off-limits to science; that was so outrageous, he knew people would think it was mad even to dream of it. Yet at the same time, it made perfect sense. "If we could determine the nature of the substances burning in Mannheim," Bunsen ventured to Kirchhoff, "why should we not do the same with regard to the Sun?"

☼☼☼

Almost sixty years before, in the summer of 1801, two buildings collapsed in a back street of Munich, Germany. There was only one survivor. After working for hours, rescuers pulled a boy from the ruins—a fourteen-year-old orphan named Joseph von Fraunhofer—and with him, the foundations of our scientific understanding of the Sun. When Bunsen and Kirchhoff turned their attention to the sky half a century later, it was Joseph's work that guided them.

After his parents died, Joseph Fraunhofer was apprenticed to a glass grinder, who according to biographers did not treat him well. He tried to learn math and optics from old textbooks, but his master discouraged his studies and his friends mocked him for wasting his time. The accident changed everything. The story of his miraculous rescue reached Maximilian Joseph, the prince (later king) of Bavaria, who gave him a generous grant of eighteen ducats. Fraunhofer used the money to escape his apprenticeship and buy his own glass-grinding machine. He continued to teach himself optics and, when he was twenty, joined a firm that produced astronomical and surveying instruments. He set to work on one of the most sought-after prizes in astronomy: to make a so-called achromatic lens.

Telescopes had been around since at least 1608, when a Dutch glassmaker tried to patent the idea of using a pair of lenses "for seeing things far away as if they were nearby." News of the invention quickly spread. In Italy, Galileo made his own telescope and turned it to the sky in 1609. Johannes Kepler built a better one in Prague in 1611. In the 1650s, Dutch astronomer Christiaan Huygens used a 12-foot-long version of Kepler's design to spot Saturn's moon Titan and sketch the Orion nebula.

But as astronomers tried to make lenses stronger, their images became increasingly distorted, with blurry fringes of color. In the 1660s, Newton realized this was because the lenses work like prisms, bending (or refracting) different colors of light by different amounts. To solve the problem, he built a "reflecting telescope" that instead of a refracting lens used a mirror, because all colors bounced off the mirror at the same angle, avoiding such distortion. But the available mirrors—made from an alloy of tin and copper called speculum—tarnished quickly. This in turn would eventually be solved in the 1850s, when longer-lasting mirrors were made by sandwiching a thin film of silver between two layers of glass. But from the mid-eighteenth century, opticians focused instead on building lenses that canceled out the distortion, combining layers of different kinds of glass to refract all colors by the same amount.

It was hard to make glass samples that were pure enough, and then to grind them accurately enough, but Fraunhofer was one of the best. He also made lenses for microscopes, and a new kind of instrument he called a heliometer, for accurately measuring positions in the sky. (In 1838, it was used to measure stellar parallax, giving the first ever estimate of a distance to a star.) To design the perfect lens, he needed to check exactly how much different types of glass bent various colors of light. So he made prisms out of his glass samples, and analyzed the resulting spectra through a small telescope. In 1814, when charting the path of colors in a narrow beam of sunlight, he noticed something strange.

Newton had reported that a prism smears sunlight into a continuous band of color. But when Fraunhofer looked more closely at the spectrum, he realized that this wasn't true. "I saw with the telescope an almost countless number of strong and weak vertical lines," he wrote. These mysterious lines were darker than the rest of the spectrum, and some were "almost perfectly black," as if those colors had been scratched out. He eventually recorded 574 examples, now known as Fraunhofer lines.*

* In 1802, a British chemist named William Hyde Wollaston noted seven dark lines in the Sun's spectrum, but didn't pursue the work. Fraunhofer seems not to have been aware of this research when he reported his own observation. He

As well as the Sun, Fraunhofer examined the spectra of the Moon, planets and a few stars. He concluded that light reflected from Venus is the same as that which shines from the Sun, but he saw different patterns in the stars. The light from Sirius, for example, was marked by three dark lines—one in the green part of the spectrum; two in the blue—that don't appear in sunlight. Different colors were missing from different types of celestial objects. But why?

In 1823, the glass-grinder's apprentice finally joined the establishment; Fraunhofer was made a professor of physics, and a member of the Royal Bavarian Academy of Science. The next year, he was raised to the nobility, adding the title "von" to his surname. In 1826, aged just thirty-nine, he died from tuberculosis, before he could find an explanation for the lines he found. But he had laid the first steps on which others could build. His grave in Munich reads *Approximavit Sidera*: "He brought the stars closer."

✫ ✫ ✫

Bunsen and Kirchhoff's challenge was to work out what their bright emission lines had to do with the dark lines in the Sun's spectrum seen by Fraunhofer. Other researchers (including William Herschel's son, John) had noticed that some of the lines seemed to coincide: for example, sunlight was interrupted by two prominent dark lines at the same position in the spectrum as the two bright yellow lines emitted by burning sodium.

In 1859, Kirchhoff carried out a crucial experiment. To check if these two Fraunhofer lines really did match the bright sodium lines, he superimposed them, allowing sunlight to shine through a flame in which he burned table salt (sodium chloride). Then he analyzed the resulting spectrum with his spectroscope. He expected the bright

went much further than Wollaston, using extremely pure glass samples and accurate measurements to record hundreds of lines.

light from the burning salt to fill in the dark gaps in the solar light. But he had a surprise. Instead of plugging the gaps, the bright emission lines disappeared, and the Fraunhofer lines became even blacker.

When he repeated the experiment with weaker sunlight, the bright sodium lines reappeared. Kirchhoff quickly worked out what was going on. Newton had argued that light is made up of corpuscles, or particles, but Kirchhoff saw it as a wave, and he knew that similar waves also transmit heat.* Starting from the principle that energy always flows from a hotter object to a cooler one, he proved theoretically that if a gas is hotter than its surroundings, its atoms emit a certain pattern of radiation. But if it's cooler, it absorbs that same pattern, creating dark lines instead of bright ones. This in turn explained how Fraunhofer's lines were produced in the first place. Kirchhoff realized that the outer layers of the Sun must be cooler than its glowing core. As the light passes through those cooler layers on its way to Earth, certain colors, or wavelengths, are absorbed. This meant that the dark solar lines revealed the presence of different elements in the Sun's atmosphere, in the same way that bright lines identified them in the lab. The pair of yellow lines missing from the Sun's spectrum suggested that one element in those cooler layers was sodium.

Kirchhoff and Bunsen realized they could probe what else was there too. The pair embarked on a feverish set of experiments, which Bunsen described in November 1859 as work that "doesn't let us sleep." They introduced everything they could think of into the burner flame, searching for colors that matched up with other Fraunhofer lines, and soon ticked off elements from calcium and chromium to magnesium, nickel and zinc.†

---

* In 1800, William Herschel found that when he placed a thermometer just past the red end of the spectrum, it heated up faster than one placed within the visible spectrum—this heat radiation is now known as infrared.

† In a memoir published in 1888, Kirchhoff recalled how researchers investigated whether there was gold, too. His banker apparently remarked: "What do I care for gold in the sun if I can not fetch it down here!" Shortly afterward,

Roscoe visited them in Heidelberg in 1860. "I shall never forget the surprise I experienced," he said, "when, in the back room of the old Physical Institute, I peered into the excellent spectroscope of Kirchhoff constructed there." In light that had been created in the heavens, 91 million miles away, he was transfixed to see the familiar signature of iron. "The conviction that our terrestrial iron also existed in the solar atmosphere took hold of me with overwhelming force."

Other scientists studying Fraunhofer lines had come close to some of these conclusions, but Kirchhoff was the first to explain precisely how emission and absorption lines were linked, and to suggest using them to hunt for elements in the Sun. The specific mechanism behind each signature pattern of lines wouldn't be explained until the 1920s, when scientists developed a model for the structure of atoms. Nonetheless, our view of the heavens was transformed. For all of human history, the Sun had been a beguiling yet ultimately mysterious celestial light. Newton finally made sense of its motions. But Kirchhoff and Bunsen turned it into a comprehensible physical object: a ball of glowing gas, cooler at its surface than inside, that behaved just like any other hot body and was made of just the same materials as found on Earth.

✡ ✡ ✡

News of Bunsen and Kirchhoff's discoveries soon reached all over Europe. In England, scientists including Roscoe gave public lectures explaining the revolutionary idea of using light to probe the physical ingredients of planets and stars. "[I]f we were to go to the Sun, and to bring away some portions of it and analyse them in our laboratories," explained the British astronomer Warren de la Rue at a lecture to

Kirchhoff was presented with a medal for his discovery, and a prize in gold sovereigns. Handing this over to his banker, he couldn't resist observing: "Look here, I have succeeded at last in fetching some gold from the sun."

London's Chemical Society* in 1861, "we could not examine them more accurately than we can by this new mode of spectrum analysis." One of those captivated by the news was an amateur astronomer named William Huggins, who had an observatory in his back garden.

Huggins came from a well-off family of silk merchants, and after a childhood attack of smallpox left him with delicate health, he was largely schooled at home. He had a talent for science early on (according to one story, he built an "electrical machine" when he was six, which caused him to run through the house excitedly, shouting, "I've had a shock!") and as a teenager bought his first telescope, gazing at the night sky through his window above the family shop in central London or standing on the roof among the smoky chimneys. He dreamed of studying at Cambridge University, but when his father's health failed, he instead stayed in London to look after the business. In 1854, when he was twenty-nine, he sold the shop and moved with his parents to the relatively clear air of Tulse Hill, five miles to the south.

There, Huggins built his observatory. Raised on iron pillars for an unobstructed view of the sky, it was connected to the upper floor of the main house by a wooden passageway, and topped by a rotating 12-foot dome. He bought a secondhand telescope—10 feet long, with an 8-inch lens—and pointed it toward the planets, sketching, among other things, Jupiter's clouds, swirls and spots. But he soon felt dissatisfied with such routine observations, and wondered if there was a better way to learn about the sky.†

In January 1862, he found it. Huggins attended an evening meeting at London's Pharmaceutical Society,‡ at which W. Allen Miller, a

* Now the Royal Society of Chemistry.

† He later wrote that soon after establishing the observatory, he became "a little dissatisfied with the routine character of ordinary astronomical work, and in a vague way sought about in my mind for the possibility of research upon the heavens in a new direction or by new methods."

‡ Now the Royal Pharmaceutical Society.

chemist at the University of London, gave a lecture on Kirchhoff and Bunsen's research. Hearing about "Kirchhoff's great discovery," Huggins later said in a colorful memoir, was like "coming upon a spring of water in a dry and thirsty land." Miller had explained how chemists could use flame spectra to analyze substances in the lab. But it was the possibility of transforming astronomy that would drive Huggins for the rest of his life. "I felt as if I had it now in my power to lift a veil which had never before been lifted," Huggins said, "as if a key had been put into my hands which would unlock a door which had been regarded as for ever closed to man—the veil and the door behind which lay the unknown mystery of the true nature of the heavenly bodies."*

Huggins and Miller were neighbors, and that night they walked home together. On the way, Huggins asked Miller to join him in applying Kirchhoff's methods to the stars. He then kitted out his observatory with the equipment of a chemistry lab, his astronomical instruments sharing the space with huge batteries "giving forth noxious gases," a large induction coil and shelves filled with Bunsen burners, chemicals and vacuum tubes.

Huggins and Miller set about comparing the dark lines in the star spectra with the bright lines from different earthly elements, lit by a giant spark from the induction coil. "The characteristic light-rays from earthly hydrogen shone side by side with the corresponding radiations from starry hydrogen, or else fell upon the dark lines due to the absorption of hydrogen in Sirius or in Vega," recalled Huggins.

* Huggins's comment is often quoted because it seems to sum up the excitement of astronomers at the time: the feeling that the study of the sky would never be the same. Historian Barbara J. Becker has questioned whether Huggins was really so prescient, arguing that his work started slowly, and pointing out that he wrote these comments several decades later, after the potential of spectroscopy had been demonstrated. Others at the time (such as Roscoe and de la Rue) were certainly grasping the significance of the new approach, however, while Huggins himself published his first paper on stellar spectroscopy within a year and never looked back.

"Iron from our mines was line-matched, light or dark, with stellar iron from opposite parts of the celestial sphere."

It was difficult, painstaking work. Starlight is desperately feeble—we receive around fifty billion times less light from the prominent star Vega than we do from the Sun, for example. The pair had to build their spectroscope from scratch, creating a slit just 1/300th of an inch wide (a clockwork mechanism slowly moved the hefty telescope to keep it perfectly lined up with the stars throughout the night), and a prism that was precisely designed so as not to waste a single scrap of starlight when it was spread into a spectrum. Even then, they couldn't see any lines at all except on the very clearest nights. And to match up the lines corresponding to different elements, they repeatedly had to look directly from the emission spectra—the brief, blinding flash of magnesium, say—to the barely discernible star lines, which pushed their eyesight to its limits.

Astronomers in the United States, Germany and Italy were also chasing star lines, but Huggins and Miller soon led the field. In 1864, they published descriptions of the spectra of fifty prominent stars, including pale red Aldebaran, orange-tinted Betelgeuse and brilliant white Sirius. They reported that star spectra are just as rich in lines as those of the Sun and that, similarly, many of the lines coincide with terrestrial elements, such as hydrogen, sodium, magnesium and iron. There were some differences between individual stars, but overall the message was clear: the building blocks of matter that we find on Earth don't just extend through our solar system, but across the universe.

As well as reporting this similar chemistry, the pair also noted that the elements they saw in stars were "some of those most closely connected with the constitution of the living organisms of our globe." Several writers—like Paine—had previously wondered whether other stars support living worlds like Earth. Huggins and Miller claimed their results provided the first scientific evidence for this dramatic conclusion: that the stars they observed "are, like our sun, upholding and energizing centres of systems of worlds adapted to be the abode of living beings."

After that, Miller returned to his original research and Huggins worked alone, turning his attention to those mysterious celestial objects called nebulae—diffuse wisps, patches and spirals that look very different from point-like stars. Astronomers had wondered about them for the past century, suggesting they could be anything from regions of "shining fluid" from which solar systems would eventually form to unimaginably distant galaxies or "island universes." First, Huggins targeted the Cat's-Eye Nebula, an intriguing blue-green disk in the northern constellation of Draco. He later recalled feeling a mix of suspense and awe as he put his eye to the spectroscope. "Was I not about to look into a secret place of creation?"

The Cat's-Eye Nebula belonged to a class of unexplained round shapes dubbed planetary nebulae. The latest, most powerful telescopes were revealing that many nebulae were in fact clusters of individual stars, supporting the distant galaxy idea, so Huggins expected to see thickets of dark lines, from all the starlight superimposed. What he actually saw, though, made him wonder if his equipment had malfunctioned. There was no spectrum at all; just one bright line of light. Then he realized that unlike any other extraterrestrial object he had pointed his equipment toward, this nebula was producing emission lines, just like a burning gas in the lab.

He soon picked out two other faint lines, and saw the same trio in the light from other planetary nebulae too.* They weren't aggregations of suns, he concluded, but enormous clouds of luminous gas.† Not all nebulae were the same, though. In other types, including several

* The bright blue-green line in these planetary nebulae didn't match any known earthly element, so Huggins speculated that it might belong to a "substance unknown," which came to be dubbed "nebulium." In 1927, it was shown to be an exotic form of oxygen and not a new element at all.

† Planetary nebulae are now understood as dying stars that have ejected a glowing shell of ionized gas into space.

spirals, Huggins did see faint spectra, and these were later confirmed to be galaxies like our own. After decades of confusion and speculation, the spectroscope could finally tell different types of these "wondrous objects" apart.

From his Tulse Hill garden, Huggins jumped from one scientific frontier to another. In May 1866, he made the first spectroscopic observation of a type of exploding star called a nova, after receiving a tip-off from an Irish correspondent that something strange was happening to the Northern Crown. Feeling skeptical, he scanned the sky as soon as dusk fell, and "to my great joy, there shone a bright, new star." The light faded over the next few days, but not before Huggins found that its spectrum contained a mixture of both dark and bright lines, and he concluded that in some "vast convulsion," a previously dim star had become engulfed by flames of burning hydrogen. Huggins's discovery captured the popular imagination, even triggering a sermon "from the pulpit of one of our cathedrals . . . that from afar astronomers had seen a world on fire go out in smoke and ashes."

He also analyzed the spectrum of a passing comet, in which he detected carbon. And he pioneered the use of the newly understood Doppler effect*—in which light from a source moving toward us is shunted toward the blue end of the spectrum, and light from a receding source becomes more red—to estimate stars' motion from their shifting spectral lines. The answers he got were wildly inaccurate, but it was a valiant first attempt.

Huggins combined all this with caring for his elderly mother. Grief after her death in September 1868 meant he lost the race to see the spectral lines of the red flames of the Sun's corona, in which the

* Austrian physicist Christian Doppler suggested in 1842 that if a star is moving, the wavelength of its light will decrease as it comes toward us—because each successive wave has less far to travel to reach us, so they arrive closer together—and increase as it moves away. Huggins wrote to the physicist James Clerk Maxwell, asking for his mathematical analysis of the effect, which he used to estimate the motion of stars, including Sirius.

astronomer Norman Lockyer identified a new element, helium; he also missed a personal invitation to Paris from Napoleon. After that, his main companion was a dog named Kepler.* But his life was about to change yet again. In Ireland, a girl was growing up who also loved the stars.

✡ ✡ ✡

Margaret Murray, the daughter of a well-to-do Dublin lawyer, lived in a Georgian town house by the sea. Her mother died when she was nine, and according to contemporary accounts, she learned astronomy from her grandfather. She was soon using a small telescope to make drawings of constellations and sunspots, and taught herself the new technique of photography. In 1867, when she was nineteen, she read a magazine article about Huggins's work† and was captivated. Following the instructions in the text, Margaret built her own spectroscope and detected the Sun's Fraunhofer lines.

It's not clear how Margaret and William met, but they may have been introduced by William's instrument-maker, Howard Grubb, who was based in Dublin. They married in 1875, when he was fifty-one and she was twenty-seven. Margaret gently steered her husband away from an interest in spiritualism, and nudged his fashion choices, encouraging him to grow his hair long and to wear velvet coats in the afternoons. She also devoted herself to William's research, and together they began

* Kepler could apparently answer simple mathematical problems by barking (several years before a German horse named Clever Hans gained fame for a similar trick), and was terrified of butchers' shops. Kepler's father and grandfather had the same fear—Huggins described this to the biologist Charles Darwin, who was so intrigued by this apparent example of "inherited antipathy" that he published an article about it in *Nature* in 1873. See Charles Darwin, "Inherited Instinct," *Nature* 7 (1873): 281.

† Historian Barbara Becker suggests this could have been "A True Story of the Atmosphere of a World on Fire" by Charles Pritchard, president of the Royal Astronomical Society, published in *Good Words* magazine in 1867.

one of the most successful husband-wife partnerships in the history of science.

From 1876 on, William's brief notebook entries were replaced by Margaret's detailed experiment reports. And she brought to the observatory her photography skills. For years, William had been sketching spectra by hand, but his competitors were starting to photograph them—in particular, the husband-and-wife team Henry and Anna Draper, privately wealthy astronomers from New York State, who obtained the first star "spectrogram" for Vega in 1872.* After Margaret arrived in Tulse Hill, she and William became the first astronomers to use the just-invented "dry plate" method,† and with this, plus a pair of powerful new telescopes (one refracting, one reflecting) donated by the Royal Society, they were soon back at the head of the field.

William did the developing, while Margaret—younger and more agile, and with better eyesight—took over much of the actual observing. Perching on a ladder for long periods was physically demanding, requiring highly sensitive eyes and hands, she said, not to mention "universal joints and vertebrae of india rubber." The glorious view helped to make up for it: "No imagination could fail to be struck with the wonders of the heavens sweeping round majestically in perfect peace within five miles of the greatest city and the greatest turmoil of the world." But she still sometimes wished that her husband had been a painter, not an astronomer. "He'd have been a happier man I think as a landscape painter. And I—ah! The artist is strong in me," she mused. "Nobody knows how wearing science is: and it does take faith to be as happy in straining one's eyes to see little patches of light or of

* Henry and Anna got married in 1867, and the next day, on what they later referred to as their "wedding trip," they selected the glass that would become the mirror for a 28-inch reflecting telescope. It took five years to build, and enabled them to photograph Vega's spectrum in August 1872.

† In which a glass plate is coated with a light-sensitive gelatin emulsion.

darkness . . . as in feasting them on the beauties of the fields and skies and woods."

Together, they photographed the spectra of planets including Uranus, Saturn and Mars, as well as stars, from the familiar Sirius and Vega to unusual, rare specimens such as the newly discovered Wolf-Rayet stars, which had towering emission lines instead of dark bands. They continued to explore the nature of nebulae and spent years trying (unsuccessfully) to photograph the spectral lines in the Sun's corona. Photography also enabled the couple to extend their spectra into the ultraviolet part of the spectrum, invisible to the human eye.

William was honored with mounting accolades and awards—appointed president of the Royal Society, knighted by Queen Victoria—while at home in Tulse Hill the couple received a string of distinguished visitors, including the emperor of Brazil. (A giant, bearded science enthusiast who had his own astronomical observatory, he impressed the pair rather more than another visitor, who responded to William's explanation of double stars by saying: "Yes, it is very interesting; but what has always puzzled me most is, how you astronomers ever managed to find out the names of all the stars.")

They were riding an extraordinary wave, a period of technical developments and scientific discoveries that regularly smashed the wildest dreams of a just few decades before. Victorian Britain had initially seen a resurgence of fundamentalist Christian belief, partly fueled by the horrors of the French Revolution and Napoleon's subsequent wars, which were widely blamed on the skepticism and rationalism of writers like Paine. But not for long. Scientific advances finally completed the job that the Enlightenment thinkers had started, with empirical methods breaking the Bible's authority as a source of physical knowledge.

Alongside the stunning results in astronomy, geologists were revealing the Earth's deep past, stretching further back than the few thousand years suggested in the Bible. The Babylonian tablets found by Rassam in Ashurbanipal's library showed that elements of biblical narratives such as the Great Flood were used by other cultures, centuries before the Bible was written. And of course, Charles Darwin's theory of evolution by natural selection, published in 1859, provided a powerful alternative explanation for how species—including humans—began.

For positivist followers of Comte (his ideas were made available in English in 1865), scientific facts were out there waiting to be discovered, and the knowledge they provided was absolute. Perhaps we'd never know everything about the universe, but as we ventured into the unknown, armed with the tools of science, we would march ever closer to the truth. As William put it in a presidential address to the Royal Society: "By each discovery the vision of the world has become more glorious, the wonder of it more amazing, while the chambers and palaces of Nature still unexplored remain the exhaustless heritage of all coming generations."

Toward the end of their joint career, in 1899, William and Margaret published a grand atlas of stellar spectra (beautifully illustrated by Margaret), which was described by the *Times* newspaper as "one of the greatest astronomical books of all time." It cemented their status as founders of the new science of astrophysics, and the universe it presented was not a static creation but an evolving entity. Their photographs of celestial spectra from novas to nebulae were accompanied by a discussion of how stars might evolve as they burn fuel through their lives, from young, hot, white stars such as Sirius and Vega toward reddish, fading Betelgeuse. They even speculated about how successive generations of stars might emerge from the exploded remains of the old. It was a vision of epic size and magnificence, yet all ultimately understandable from the secrets hidden in the light.

✡ ✡ ✡

In 1908, when he was eighty-four, William informed the Royal Society that he wished to return the two telescopes they had donated to him, as he could no longer make best use of them, and the decision was made to transport them to Cambridge University's brand-new Astrophysical Department. Howard Grubb, the Dublin instrument-maker, went to Tulse Hill to supervise their removal. He arrived to find the process of dismantling well underway, with telescope parts and attachments scattered over the observatory floor.

In the midst of the mess was William himself, wrapped in a large cape and sitting quietly on a packing case, while Margaret flitted about, monitoring the workers' every move. Finally, the huge 15-inch lens of the refracting telescope was placed safely in its box. Grubb signaled to Margaret. "She took Sir William by the hand and brought him across the room to have a last look at their very old friend, the object glass which had for so many years fulfilled its mission in bringing rays of light from a far distance to a focus," he recalled. "They gazed long and sadly before I closed the lid."

The field of study the pair helped to start has shaken human culture to its foundations, swiftly replacing the celestial tales of thousands of years with a new set of stories. The Pleiades aren't orphans, sisters or hunting dogs—or the mark of a celestial bull—but a cluster of young hot balls of gas, burning hydrogen into helium as they speed through clouds of cosmic dust. Aldebaran, the bull's flaming red eye, has become an aging giant, forging carbon, oxygen and nitrogen that will one day be catapulted into the depths of space. Venus—bright star of dawn and dusk, deity of love (or the rain)—is in fact an Earth-sized rock, hiding volcanoes and mountains beneath its clouds, and perhaps once a shallow ocean, though it's now fried by a runaway greenhouse effect, with a fiery surface hot enough to melt lead.

Spectroscopy has revealed that in a sense, Plato was right: we really do come from the stars. Astronomers now know that all of the heavier chemical elements from which our world is made—from carbon to uranium—were cooked inside stars eons ago, before being blasted into space in supernova explosions. Almost every atom in our bodies was once part of a star. Using information from spectra to pin down the life cycles of stars has also led to the discovery of wondrous new objects that ancient astronomers could never have imagined: supernovae, cataclysmic explosions in which some stars end their lives; neutron stars, like a gigantic atomic nucleus, only a few miles across but so dense that a teaspoonful weighs a billion tons; black holes, even denser, with such crushing gravity that even light can't escape.

The method has even yielded clues to the origin and fate of the entire universe. Measuring the Doppler shift in the frequency of starlight that reaches us—the method first tried by Huggins—tells us that in every direction we look, distant galaxies are speeding away from us, in other words, the universe is expanding. Projecting backward from this finding led to the theory of the Big Bang—the idea that 13.8 billion years ago, an infinitely small, dense point exploded into the universe, and has been growing ever since. Finally, we have a scientific creation story for the cosmos—the first to be based on technology and observation rather than human experience and myth.

Galileo is seen as a scientific hero for turning his telescope on the sky—bringing unimaginable cosmic riches into view for the very first time. The resulting detailed view of our planetary companions helped to show that they were physical objects just like Earth, and enabled their orbits to be tracked precisely enough for Kepler and Newton to describe them mathematically. But I think that the development of spectroscopy—from Fraunhofer to the Hugginses—was just as revolutionary, because it allowed astronomers to go beyond simply recording appearance and motion. These pioneers brought the entire cosmos to Earth, by showing that Comte was wrong: we don't need to touch distant objects, or hold physical pieces of them in our hands, to study what they're made of and how they work. We can read their secrets hidden in their light.

You could say they brought the stars closer, just as Fraunhofer's epitaph describes. But ironically, their work also created a profound separation in the way that, as a society, we experience and learn about the sky. Astronomers now build mind-bogglingly powerful telescopes that analyze not just visible light but every part of the electromagnetic spectrum. Perched on mountaintops or launched into space to avoid the distorting effects of the atmosphere, these exquisitely complex machines probe everything from the existence of water on Mars or the Moon to the structure of the universe's earliest galaxies. Telescopes attuned to high-frequency radiation detect the signatures of cataclysmic quasars or gamma-ray bursts, while instruments sensitive to the longer waves of infrared penetrate through dust clouds, revealing newborn stars.

But no one actually "looks" through them. Margaret Huggins lamented the shift from gazing at the heavens to squinting at tiny patches of light. Now we've gone much, much further. In today's astronomy, photons of light from the sky, along with the celestial secrets they contain, are picked up by electronic detectors, converted into digital data and crunched through impossibly complex equations by some of the most powerful computers on the planet. In 2016, bricklayer-turned-astronomer Gary Fildes described visiting Chile's Very Large Telescope (VLT) in his best-selling book *An Astronomer's Tale*. Incorporating four mirrors, each 27 feet wide, the VLT collects visible and infrared radiation and can distinguish points in the sky separated by less than a millionth of a degree. Here, at the forefront of today's attempts to understand the stars, Fildes was struck by the sight of scientists hard at work in control rooms, eyes glued not to their telescopes but to banks of screens: "They didn't look as if they had seen the real sky for days."

When scientists first split light with the spectroscope—turning its colors into numbers—they took one more step away from a subjective, qualitative view of the cosmos toward an objective, mathematical one: from an internal universe that we *experience* to an external one that we *calculate*. And with the development of electronic detectors, our sense of vision—how the cosmos looks to us—was finally erased from the picture altogether. In this sense, modern astronomy is radically different from any kind of cosmological inquiry or understanding that has gone before. It no longer requires us to turn our faces to the sky. Our dominant source of knowledge about the universe—what it is, how it was made, how it relates to life and to us—is now our instruments, not our eyes.

It's a shift that hasn't gone unchallenged, however. At the turn of the twentieth century, as scientists worked to remove personal experience from our ideas about the cosmos, a group of revolutionary artists was fighting to keep it center stage.

9

ART

REVIEWERS SAID IT MADE A MOCKERY OF THE THEATER. SOME critics complained that the show was wild and senseless yet boring all at once; others were so confused, they found themselves unable even to discuss what had taken place. *Victory over the Sun* was staged at the Luna Park theater in Saint Petersburg, Russia, over the two nights of December 3 and 4, 1913. Tickets cost a hefty nine rubles, but thanks to rumors that something scandalous was about to happen, they sold out within a day, and the packed crowd wasn't disappointed.

At nine p.m., the curtain was ripped in two, unleashing a riot of crazy costumes, abstract imagery and nonsense language, accompanied by clashing tones played on an old out-of-tune piano. The show's three young creators—composer Mikhail Matyushin, librettist Aleksei Kruchenykh and designer Kazimir Malevich—were pioneers of a Russian avant-garde art movement known as Cubo-Futurism, and *Victory over the Sun* was the world's first Cubo-Futurist opera.

The characters, played mostly by volunteer students after just two rehearsals, included two strongmen of the future, a time traveler, an evil-intender and a fretful traditionalist named "Nero and Caligula in the Same Person." Dressed in giant, geometric costumes made from cardboard and wire, and illuminated by roving spotlights, these figures spoke in a style that Kruchenykh called *zaum*, which was full of illogical associations, and featured pauses between syllables to emphasize their sound and emotional force above their rational meaning. Malevich's sets, too, abandoned any normal sense of illustration or decoration. Instead, he created geometric black-and-white panels (he had no money for colored paint) that played with perspective and created the illusion of depth.

There was little recognizable plot. Reconstructions suggest that the first act featured efforts by the Futurist strongmen to capture the Sun and imprison it in a concrete box. There was plenty of eerie violence: the evil-intender calmly shot the time traveler before attacking himself; a corpse dragged itself off by the hair. The last scene featured a speaker on the telephone: "What? They've seized the Sun? Thank you."

The second act jumped to a distant, sunless future, a new reality in which conventional human values had been destroyed. It was a place of "great lightness" with no memories, dreams or emotions; where humanity's slate is wiped clean: "You become like a clean mirror or a fish reservoir where in a clear grotto carefree golden fish wag their tails like thankful turks." A skull ran around on four legs. A bewildered fat man asked, "Where is the sunset?" and tried in vain to set his watch, before an airplane crashed into the stage.

No one had ever seen anything like it, not in Tsarist Russia, nor anywhere in the world. But there was method behind the madness. By deliberately breaking social norms, the trio wanted to upset the establishment, of course, but their real target was much bigger. They were declaring war on reason, the very backbone of Enlightenment thought. According to art historian Charlotte Douglas, who has studied the opera in detail, the barrage of absurdity was an attempt to transcend the rational, everyday world; the artists hoped to force baffled viewers to let go of logic and submit instead to an intuitive, emotional awareness of events. "It's an attempt to shock everyone," says Douglas, "even the Futurists themselves, into a new consciousness."

For all of the success of reason and science, Malevich and his friends rejected the central assumption that reality equates to the physical world we observe—all those objects and forces governed by logical equations and laws. "We picked the sun with its fresh roots," the victors sing after tearing the fiery globe from the sky. "They're fatty, smelled of arithmetic." They target the Sun as "the creator and symbol of visibility . . . of the illusion of reality," says Douglas. "It is Apollo, the god of rationality and clarity, the light of logic." In other words, it's the light from the Sun that fools us into believing that the things we see are real. For humanity to advance, the Sun must be destroyed.

That first production of *Victory over the Sun* is now infamous, and reconstructions of it are regularly staged around the world. (It still divides audiences: one critic called a 1999 version at London's Barbican Centre "a ludicrous waste of everybody's time.") But for the designer, Malevich, this anti-opera was just a first step in a greater quest to explore the nature of reality. It was a journey that took him deep into a cosmos filled with unseen vibrations and extra dimensions; from the vast reaches of the stars to the "infinite space of the human skull."

In this chapter, we'll follow the stories of Malevich and some of the other revolutionaries who transformed Western art in the first few years of the twentieth century. As scientists exploring the nature of matter began to question the reality of the solid, predictable cosmos they had been building since the Renaissance, painters and poets rebelled against the illusion of visible appearances, and the dominance of logical, rational thought. They started by shattering long-held artistic conventions. But ultimately, they were trying to remake the universe itself.

✡ ✡ ✡

Beliefs about the cosmos have always influenced art. Just look at how images of the sky have changed through history. Ancient depictions bustle with mythical creatures and gods, whereas after the Renaissance, painters—just like astronomers—replaced celestial beings with points of light. But the connection between cosmology and art goes much deeper than simply how we paint the stars. Changing ideas about the nature of the universe—about the type of space that we inhabit—inevitably shape how artists depict reality itself.

In the Paleolithic, as we've seen, the animals chasing around cave ceilings may have represented a cosmos that melded man and nature, earth and sky. The art of medieval Europe also represents a distinctive worldview, showing up in the Christian symbols and themes, of course, but also through the fundamental way in which objects and scenes are constructed. In common with maps of the period (artists didn't generally distinguish the two), there was no attempt to depict objects in three dimensions, or to incorporate any consistent point of view. Objects often float rather than being placed on a landscape; events occurring at different times are shown together; important figures or features are magnified in size. This reflected a culture that, as one critic put it, "did not equate the real with the material world." The "truth" or reality that artists wished to convey was contained in the overall experience or meaning of a scene, rather than how it would actually look.

All that changed when artists started constructing their compositions using mathematical rules. The big break came around 1413, when an architect called Filippo Brunelleschi, working in Florence, Italy, came up with the technique of "linear perspective." This involves plotting objects according to a horizon line and vanishing point, to create the illusion of depth and mimic how a scene would appear to the eye from a specific position. Adapted into a workable scheme for painters in 1435, the technique spread throughout Europe and soon became the only acceptable way to paint. From Leonardo da Vinci and Raphael's Renaissance masterpieces to J. M. W. Turner's expressionist landscapes, linear perspective would dominate Western art for the next five hundred years.

Intriguingly, the innovation happened just as cartographers adopted longitude and latitude lines. Ptolemy's *Geographia*, in which he divided Earth and sky into networks of meridians and parallels, was translated into Latin in the same city just five years earlier, in 1406. Historians argue about whether there was a direct connection. Ptolemy explained a geometrical projection very similar to Brunelleschi's; then again, Italian artists were already experimenting with perspective in a society that was increasingly interested in measurement, and drawing ever more accurate architectural plans. Either way, the revolution in both maps and art was the result of the same cosmological shift, from the malleable, discontinuous space of the Middle Ages to the uniform, logical grid of the Renaissance. For astronomers, mapmakers and painters alike, explorations of reality were now dependent on measuring and modeling the external, visible world.

This paradigm held fast in art until the nineteenth century, when realists started to portray the imperfections of real-life people and situations rather than creating idealized compositions, and Romantics aimed to depict how they experienced a scene instead of how it objectively appeared. The causes were complex: artists in an increasingly secular, industrial, urban society were moving away from biblical or classical themes and addressing the modern life they saw around them. Historians also cite the invention of photography, which suddenly enabled the creation of accurate images on demand; artists had to find something else to do. Both groups, though, were moving away from the construction of mathematical perfection to highlight the messy human act of seeing.

They were followed by the Impressionists, who by the 1860s had declared war on the centuries-old illusion of the canvas as an honest window onto the external world. Painters such as Monet captured instead their fleeting, subjective impressions, while allowing their brushstrokes to show. After them came the postimpressionists—the twisting patterns of Vincent van Gogh; bold outlines of Paul Gauguin; dots of Georges Seurat—rejecting realism even more vehemently with distorted shapes, vivid colors and thick daubs of paint.

At the end of the nineteenth century, the French painter Paul

Cézanne finally began to break the rules of linear perspective, merging foreground and background, and incorporating several perspective points at once; in 1907, Pablo Picasso, working in Paris, produced an 8-foot-square canvas called *Les Demoiselles d'Avignon*. It shows five naked women, entwined with blue-and-white drapes. Two of them glare at the viewer with huge, black eyes; the other three hide their faces behind African-style masks.

It was shocking, even to his closest friends. That's partly because of its shameless content and utter rejection of Western ideals of beauty. But it's mainly because he abandoned any illusion of three-dimensional reality. Instead of featuring realistic, receding curves, these women are all angry angles and flat, jagged sheets. With other like-minded artists, Picasso then went further—mixing viewpoints and slicing surfaces into discontinuous, geometric facets—the approach now known as cubism. This was more than finding new arrangements of paint on canvas. They were reimagining the very nature of space.

The birth of cubism is often portrayed as a natural progression of artistic ideas, with Picasso building on Cézanne's techniques as well as the simplified, geometric forms of traditional African and Iberian art, which Western artists were just discovering. Both were undoubtedly crucial influences. But there's another factor: a surge of scientific and technological developments that began in the 1890s and shook people's understanding of reality. Once again, the cosmos was becoming a very different place, and artists like Picasso couldn't help but respond.

A few decades before, the Scottish physicist James Clerk Maxwell had showed that electric and magnetic fields travel through space as waves of electromagnetic radiation, and concluded that light is just such a wave. But now other kinds of electromagnetic radiation were revealed, with powers that seemed almost magical. In 1894, engineers started using newly discovered radio waves to transmit unseen messages through the air. In 1895, physicists discovered X-rays—beams that could pass through solid matter and reveal buried features such as the bones inside our flesh. Meanwhile, atoms, thought for thousands of years to be the fundamental, indivisible building blocks of matter, were found in 1897 to contain even tinier constituents called electrons. That was swiftly followed in 1898 by the discovery of radioactivity—in which some elements produce invisible emissions, releasing energy and altering their chemical composition in the process.

Running alongside these developments were two other scientific themes, largely forgotten today but which captured the popular imagination just as modern art exploded into existence. One was the existence of a cosmic "ether," a delicate fluid—like jelly or whirling steam—that scientists believed must fill the entire universe, from subatomic pockets to the vast depths of space. In the 1880s, a major experiment failed to detect any sign of ether around the Earth, but scientists had no other way to explain how electromagnetic waves were transmitted through the vacuum of space, and the idea remained popular until after World War I. The ether "is as essential to us as the air we breathe," insisted physicist J. J. Thomson (who discovered the electron) in 1909; others even claimed it was the ultimate source of matter, with atoms themselves simply whirlpools in the ether sea.

The other idea that excited the public was the possibility of hidden extra dimensions of space. Since the ancient Greeks laid down the rules of geometry, no one had seriously questioned the commonsense idea that space has three dimensions. But in the nineteenth century, mathematicians showed that it is perfectly possible to construct other types of geometry with different rules. The French mathematician Henri Poincaré popularized the idea from the 1880s, arguing that the apparent three-dimensional nature of reality may exist not in the universe itself but in how we perceive it. As the Greek philosopher Plato articulated thousands of years earlier, perhaps the universe we perceive is just a limited shadow of what truly exists.

What all of this added up to, says art historian Linda Dalrymple Henderson, who studies connections between modern art and science, was an all-out attack on conventional ideas about the nature of reality.

Matter was no longer stable and constant. Space was no longer empty but full of hidden rays and waves. The new discoveries broke down long-held boundaries between visible and invisible; matter and energy; solid and intangible. And, crucially for artists, they raised big questions about the reliability and truthfulness of what we perceive with our senses, especially with our eyes. "The existence of invisible realms just beyond the reach of the human eye was no longer a matter of mystical or philosophical speculation," says Henderson. Science had proved it.

In the early 1900s, mathematicians such as Poincaré suggested imagining the fourth dimension by visualizing different perspectives of an object all at once. Picasso and his fellow cubists embraced the concept, with the poet Guillaume Apollinaire, one of Picasso's friends, later describing it as part of the "language of the modern studios." Other artists, such as Umberto Boccioni, took inspiration from the ether. One of the pioneers of Italian futurism, an art movement that grew out of cubism and celebrated the energy, violence and power of modern machines, Boccioni was fascinated by scientific revelations about the nature of matter, writing in 1911 that "solid bodies are only atmosphere condensed." One of his most famous paintings is an unsettling 1912 image of his mother, called *Materia* ("Matter"), in which the woman's form dissolves into a heaving aura of colors and shapes. The extent to which it merges object and environment was unprecedented, says Henderson, and was an attempt, she suggests, to show matter emerging from and dissolving into the ether.

Although their approaches were different, both Picasso and Boccioni were embracing a new paradigm of reality, in which matter and space are interchangeable and the solid, three-dimensional objects that we see and touch are not what they seem. It was this rule-breaking work that helped Malevich and his friends to create their past-destroying opera *Victory over the Sun*. But before they did, another Russian artist became fascinated by the idea of a cosmos filled with vibrations and waves. In exploring its possibilities, he would leave the advances of cubism and futurism far behind.

✡ ✡ ✡

Wassily Kandinsky was the kind of person who naturally looked beyond visible appearances. "Everything showed me its face, its innermost being, its secret soul," he once wrote, "not only the stars, moon, woods and flowers of which poets sing, but even a cigar butt lying in the ashtray, a patient white trouser-button looking up at you from a puddle on the street, a submissive piece of bark carried through the long grass in the ant's strong jaws . . ."

Born in 1866, Kandinsky grew up in Odessa in what's now Ukraine, in a genteel family (his father was a tea merchant from Siberia; one of his great-grandmothers apparently a Mongolian princess). He was always fascinated by colors; when he was young, he imagined they had lives of their own. In 1889, he traveled as an ethnographer to the remote, forested northern region of Vologda—where shamans still journeyed and the trees had souls—and was transfixed by the folk art he found there. The peasants in their traditional costumes were "like brightly-colored living pictures on two legs," he wrote in his diary, while the carved wooden houses were so richly decorated that going inside felt like entering a painting.

He graduated in law and economics from the University of Moscow and in 1896 was offered a professorship; he was on the verge of a successful teaching career. But he still felt pulled into another world. That year, he attended a Moscow art exhibition and for the first time saw paintings by Monet, including one from the series *Haystacks*. "That it was a haystack the catalogue informed me," Kandinsky later recalled. "I could not recognize it." He was pained at first that Monet hadn't been clearer, but then "I noticed with surprise and confusion that the picture not only gripped me, but impressed itself ineradicably on my memory." He turned down his academic future and took the train to Munich, Germany, to study art.

Kandinsky was tall and smartly dressed, with pince-nez glasses and his head held so high that he seemed, as one historian put it, "to look down at the universe." He quickly progressed through different art styles, from realism and impressionism to the bolder developments of expressionism and fauvism. But despite painting flatter scenes and barely recognizable figures, he wasn't satisfied. They didn't capture how he felt when he attended a performance of Richard Wagner's opera *Lohengrin* and saw colors dance in front of his eyes; or when the sun set over Moscow, melting all of the city down to a single spot that, "like a wild tuba, starts all of the heart and all of the soul vibrating." He later recalled that one evening around 1905, on entering his studio at dusk, he finally found what he had been searching for. Leaning against the wall was an incomprehensible yet "indescribably beautiful" painting, lit by an inner glow. He eventually realized that it was one of his own works, turned on its side. "Now I could see clearly that objects harmed my pictures."

In 1909, Kandinsky settled in Murnau, southern Bavaria, and began the transformation that would make him one of the great figures of modern art. The cubists and expressionists were moving ever further from the idea of art as a window onto the external world. But as U.S. art historian Donald Kuspit puts it, if they cracked the window, Kandinsky smashed it. He became one of the first modern Western artists to cut ties with the visual world completely.

Over the next few years he embarked on a series of works, often with music-themed titles such as *Improvisations* and *Compositions*. Some still contain veiled references to earthly objects but others fly free, unfurling into riotous explosions of colors and shapes that swirl and dance across the frame. For decades, historians thought Kandinsky's first purely nonrepresentative picture (and so possibly the first in modern Western art) was an untitled work that he dated to 1910, now commonly known as *First Abstract Watercolor*. Scholars now think it was made three years later, but even so, specialists such as Kuspit see in this picture a culmination of Kandinsky's ideas during this period.

It's mesmerizing. Forms from sharp, dark lines to watery smears, in colors from intense violet to ephemeral green, are thrown onto the canvas without apparent order or rhyme. The eye doesn't know where to look. The brain doesn't know what to think. There is no place or time, no orientation or scale, no cause or effect. We're trained to categorize everything we see into classes and compartments, yet, like *Victory over the Sun*, *First Abstract Watercolor* defies logical interpretation. Kuspit describes it as "an artistic leap into the unknown." It's not, strictly speaking, abstract at all, because it doesn't derive from any earthly figures or objects. Instead, Kandinsky is showing something completely new.

He wasn't alone on his journey. Others, like Czech painter František Kupka and Dutch artist Piet Mondrian, were also starting to create nonrepresentational art. Like the cubists, they were influenced by scientists' changing view of the universe. Kandinsky was particularly struck by the discovery that atoms—the building blocks of matter—weren't immutable and indivisible, as had long been thought. "The collapse of the atom was equated, in my soul, with the collapse of the whole world," he wrote in 1913. "Everything became uncertain, precarious and insubstantial. I would not have been surprised had a stone dissolved into thin air before my eyes."

He was also captivated by the implications of the ether. Boccioni had tried to capture it as a physical substance. But what excited Kandinsky was the potential for, as Henderson puts it, "a new kind of artistic communication." The recent invention of wireless telegraphy had made it possible to send invisible messages flying through the air. Several respected scientists, such as the British physicists William Crookes and Oliver Lodge and the French astronomer Camille Flammarion, predicted that vibrations in the ether might ultimately allow the transmission of thought itself.

Such ideas naturally inspired artists and writers. In 1905, the English philosopher and poet Edward Carpenter described nature as "an innumerable network and channel of intelligence and emotion"; in 1912, the writer Ezra Pound described poets as "on the watch for new emotions, new vibrations," telling them to write "in new wavelengths."

Many artists learned about the telepathic possibilities of the ether from the hugely popular Theosophical Society, cofounded in New York in 1875 by Russian émigré Helena Blavatsky. She drew on ancient philosophies and Eastern religions to argue that the universe emanates from a divine creator, and interpreted the ether as a kind of "world soul." At the beginning of the twentieth century, theosophists Annie Besant and Charles Leadbeater, as well as Rudolf Steiner in Germany, taught that we are all surrounded by vibrating, ethereal auras that represent our thoughts and emotional states.

Kandinsky joined the Theosophical Society in 1909 (Kupka and Mondrian were also members). He attended Steiner's lectures in Germany and owned theosophical books such as Besant and Leadbeater's 1901 *Thought Forms*, which includes illustrations showing the supposed appearance of mental states—colored clouds, cones, tentacles and starbursts—relating to situations such as "At a Street Accident" or "On Meeting a Friend." Kandinsky uses similar forms in some of his work. But he was doing more than painting pictures of thoughts.

He believed that everything in the universe has its own soul or life force, what he called its own "inner sound," whether "a thunderstorm, J. S. Bach, fear, a cosmic event." He wanted his art to communicate that inner sound directly, not to mimic its appearance but to cause it to chime in the soul of the viewer, like one vibrating string inducing a sympathetic tone in another, or a radio mast beaming messages to a transmitter across the sea. Eventually, he came up with his own language, dissecting the emotional consequences of colors and shapes, concluding that yellow is aggressive, for example, while blue is calm, and was convinced he was uncovering "cosmic laws."

It's clear from Kandinsky's 1912 book *On the Spiritual in Art* that he saw this endeavor as nothing less than a deadly serious mission to save humanity. He warned of the "nightmare of materialist ideas . . . which have made an evil, pointless joke out of the universe." Science had begun as a way to explore the objective, physical world. But with the decline of religion and the rise of positivism, it had extended its power over the whole of existence. External reality was redefined as the only reality. The objective knowledge of science was recast as the only knowledge there is. He quoted the famous German pathologist Rudolf Virchow, who quipped that despite dissecting thousands of corpses, he had never managed to see a soul. "In this era of the deification of matter," complained Kandinsky, "only the physical, that which can be seen by the physical 'eye,' is given recognition. The soul has been abolished as a matter of course." But the spirit, he insisted, "can only be recognized through feeling."

Like many of his fellow artists, Kandinsky couldn't accept the dry, mathematical universe left behind after the soul was banished. "The world sounds," he wrote. "It is a cosmos of spiritually affective beings. Thus dead matter is living spirit." By removing objects from his art, he was trying to prove that there is more in this universe than the physical, measurable world. Ultimately, he wanted to transform ideas about reality, about what it means to be alive; to show that the true nature of existence lies not in the external world but in the *experience* of it. "It was Kandinsky who made it clear that art is not a substitution for reality," says Kuspit, "but a reality in its own right."

The birth of abstraction was a pivotal moment in Western art. But back in Russia, Malevich, the Cubo-Futurist, was moving on from *Victory over the Sun*. Within a few years, he, too, would grapple with how to paint the true nature of the cosmos, and the answers he came up with made Kandinsky's work seem almost tame.

✡ ✡ ✡

Unlike Kandinsky, Kazimir Malevich grew up far from cultural centers and without the luxury of a good education; in fact, he was lucky to survive at all (five of his siblings died as children). He was born in 1878 to Polish-speaking parents in the Ukraine, where his father worked in sugar beet factories, moving the family from village to village. As a young man, Malevich traveled to Moscow to go to art school, though at first he didn't get in. In 1905, as Kandinsky contemplated the beauty of his sideways canvas, Malevich joined demonstrators in a violent Russia-wide uprising, the start of a chain of events that, just twelve years later, would topple the tsar.

By 1907, Malevich was a central member of the Russian avant-garde art scene. He was influenced by cubism and futurism, but concluded that these artists didn't go far enough in smashing the conventions of the past. *Victory over the Sun* in 1913 was his first big statement; then in February 1914, Malevich took part in a debate at the Moscow Polytechnic Museum. He and his friend, painter Aleksei Morgunov, both wore wooden spoons on their lapels. Malevich made a speech in which he publicly "rejected reason," while Morgunov insulted another speaker: "He bored me, the fool." The chairman quickly shut down the event to prevent a fight.

Malevich later explained that the aim was to destroy not just artistic traditions but the entire edifice of modern thought—reason, with its focus on logical cause and effect—which he felt had more to do with narrow-mindedness and cultural conventions than truth. After all, the laws produced by reason through history never lasted but kept being replaced. Staying within the rules was shutting out swaths of possible ideas, and stopping people from truly understanding the world. "Reason has shut art up in a four-walled box," he wrote. "Run before it's too late . . ."

But how to attack reason? It could only be done in absurdist, illogical ways: by breaking convention, attacking taboos. After his wooden-spoon announcement, Malevich developed an art style he called Alogism. He created nonsense poems, as well as paintings based on what Russian art historian Aleksandra Shatskikh calls clashing quotations: images and objects juxtaposed in ways that make no sense. This phase of his work culminated with *Cow and Violin*, painted on a broken bookshelf in early 1915, which any viewer at the time, says Shatskikh, "would have had to conclude that his reason was powerless to comprehend."

But something strange happened to his paintings. As Malevich said in a March 1914 letter: "Planes reveal themselves." His pictures were based on irrational arrangements of familiar objects (or tokens or symbols of those objects), yet flat shapes and planes of color started to intrude. "Marvelous geometric figures surfaced out of the depths," as Shatskikh puts it, and began to dominate his jumbled references to the visible world. One collage Malevich made that year features rectangles and triangles in white, black, blue and pink, partly covering a defaced image of the *Mona Lisa*. He described it as a "partial eclipse."

In spring 1915,* the eclipse became total. Malevich was working on a multicolored geometric canvas when he suddenly stopped, and rushed to paint over his work. Shatskikh, who has studied Malevich's career, concludes that he was overtaken by a sudden moment of inspiration, or "ecstatic illumination." Malevich recalled that as he worked, "fiery lightning bolts" were crossing the canvas in front of him. What's more, his fingerprints are visible in the paint; he was working so fast to cover the shapes, she says, that he "outstripped his own brush." Finally, teetering against a white background was . . . nothing. The entire image had been replaced by a mysterious, infinite, empty square of black.

One of Malevich's pupils recalled that the artist was so overcome by the significance of *Black Square* that he "could not eat, drink or sleep for a full week." He had let go not just of visible appearances, but of everything. He called it the Zero of Form, and for all of the experimentation and creation occurring in modern art at the time, no one else had ever produced anything like it. One recent critic called it "an

---

* Like Kandinsky, Malevich tried to backdate his breakthrough. He immediately sent a sketch of the square to Mikhail Matyushin, who was preparing to republish the original drawings and text from *Victory over the Sun*. Malevich pretended in his letter that he had drawn it for the production in 1913, but it was overlooked. Historians were fooled for decades, but now agree that *Black Square* was actually created in June 1915.

extreme act of art and philosophy." Malevich said that he "felt the night in myself and in it saw something new."

It was first exhibited in December 1915 in Petrograd,* in a show called the *Last Futurist Exhibition of Paintings 0.10*, now seen as one of the most important art events of the twentieth century. *Black Square* hung high in one corner, where religious icons are placed in Orthodox Christian homes. That prompted a sarcastic review from one angry critic: "Undoubtedly this is the very 'icon' which Messrs. Futurists have been proposing to replace Madonnas and brazen Venuses . . ." The critic saw in *Black Square* only "nothingness and destruction." But it did become an icon; one of the most powerful images in all of modern art.

Alogism was about attacking what had gone before. Now Malevich was ready to build something new, and he called it Suprematism. He started painting vivid, colored shapes on white backgrounds: a purposeful black cross; a joyful yellow trapezoid; a jaunty red square, pushing forward out of the white. Other more complex arrangements of colored bars and rectangles seemed to defy gravity, as if soaring in space.

At last, he was rejecting earthly appearances in order to create a new kind of reality: the transcendent future promised in *Victory over the Sun*. He described his empty backgrounds as the "white abyss"; smearings of white paint that created an unfathomable, contradictory sense of depth. Instead of the logical, three-dimensional space created by linear perspective, Malevich's fellow Suprematist El Lissitzky called it "the ultimate illusion of irrational space": flat yet infinite all at the same time. Meanwhile, Malevich's shapes sparked with tension and energy, as if balanced in a kind of cosmic counterpoise. He wondered whether "there are such bodies in the world, in space . . . It seems to me that they do exist, but we don't know them." In April 1916, he wrote about the process of creating a painting

* Saint Petersburg was renamed Petrograd in 1914.

called *Supremus No. 51* (which has since been lost): "I am overwhelmed by awe, I feel the touch of the cosmos . . ." Like Kandinsky, he felt that he was uncovering universal, preexisting forms and laws.

Shatskikh calls it intuition, a solo "leap," but Henderson argues that like his peers, Malevich was responding to new ideas about the nature of the universe and, in particular, a possible fourth dimension of space. Cubists visualized this by combining different points of view, but another method was sectioning, or slicing. In a popular 1913 book called *A Primer of Higher Space*, the U.S. architect Claude Bragdon published diagrams showing how three-dimensional cubes passing through a flat plane at various angles appear as different two-dimensional shapes. In the same way, he suggested, three-dimensional objects can be imagined as sections or slices through four-dimensional forms.

This was about more than just geometry. Many writers at the time, including Bragdon, gave the fourth dimension a philosophical or spiritual aspect, arguing that only by learning to visualize four-dimensional space can we perceive the true nature of reality. Bragdon published a parable about a race of people who perceived themselves only as two-dimensional sections, unaware of their true, higher existence as cubes. "A new horizon opens," wrote Charles Hinton, another hugely successful author. "[S]pace is not limited as we first think."

Henderson suggests that Malevich's Suprematist polygons were at least partly inspired by writers such as Hinton and Bragdon, as well as the Russian mathematician and mystic Peter Ouspensky, who called on artists to communicate higher dimensions to the masses: "The artist . . . must possess the power to make others see that which they do not themselves see, but which he does." That would explain some of the names that Malevich gave his paintings, such as his indomitable red square, entitled *Painterly Realism of a Peasant Woman in Two Dimensions*, or *Painterly Realism of a Football Player: Color Masses in the Fourth Dimension*, which features multiple overlapping shapes.

Like Thomas Paine rushing to finish *The Age of Reason* while waiting for arrest in Paris, Malevich worked fast, desperate to communicate his vision of a higher reality before it was too late. Russia had entered World War I in August 1914, with disastrous results; more than a million Russian soldiers were killed on the Eastern Front. Malevich, approaching his fortieth birthday, knew he could be drafted at any time: "When I lay these bones down," he wrote, "I don't want all to be instantly lost."

He was finally called up in July 1916, separated from his canvases while posted to Krivichi and then Smolensk. Then came the popular revolution he had been hoping for. Nicholas II was forced to abdicate in February 1917, and in October, Vladimir Lenin's Bolshevik party seized power, taking Russia out of the world war and plunging her instead into a civil war that ultimately cost many millions of lives. Malevich eagerly joined the new regime, and around this time his paintings changed. At the end of 1917, he showed canvases with planes of color that although vivid and crisp on one side, faded into nothingness at the opposite edge. Malevich said their stupendous speed was causing them to dissolve in space. Then he started painting barely discernible white shapes that melted into the infinite white abyss. "The cosmos is dissolution," he said. He was leaving even his new reality behind.

In mid-1918, he painted his second most famous picture: the irrational white space of the canvas broken only by a single, soaring white square. After that, perhaps the last step was inevitable. His first one-man show opened in Moscow in March 1920. It ended with *White on White*, and then—sensationally—a room of plain white canvases. Finally, every last trace of content and distinction had been removed. There have been many blank or monochrome canvases in modern and contemporary art (an all-white painting by Robert Ryman sold for $20 million in 2015) but Malevich got there first.

Critics at the time were so lost for words, they didn't even mention it in their reviews. Even his friends thought it was a joke. But Malevich was unrepentant. In a subsequent exhibition he accompanied a similar blank canvas with an equation: "Zero = All." And then he declared the end of painting. Shatskikh says he was trying to communicate the mental state of transcendence: the ecstatic feeling, championed by mystics through history, of being at one with the universe, when all separations and differences dissolve.

"Back in the summer, sitting in my room at my easel with bayonets hanging in my color brain, I thought about my grandeur in space, since I am alone inside it, after all," he wrote in November 1917. "I saw myself in space; hidden in the color points and stripes, I am there among them moving off into the abyss . . . I declared myself the chairman of space." Another time, he equated the entire universe "in which meteors, suns, comets and planets rush endlessly," with the human skull. You could say he was journeying through the cosmos in his mind, and trying to express in his art the infinite, immeasurable, irrational spaces that he found. "If all artists could see the crossroads of these celestial paths, if they could comprehend these monstrous runways and the weaving of our bodies with the clouds in the sky," he wrote, "then they would not paint chrysanthemums."

And with that he erased the split, so carefully drawn over centuries, between objective and subjective, between the external universe and our internal minds. In a sense, he returned to the mystical spirituality of the shaman. "I am the beginning of everything," he said, "for in my consciousness worlds are created." Kandinsky saw the life, or spirit, in everything around him. For Malevich, the true essence of reality wasn't out there at all. The entire cosmos is in us.

☼ ☼ ☼

Once he scaled the heights of Suprematism, Malevich didn't pursue his wooden-spoon attacks on reason. But other artists did subsequently build careers out of attacking humanity's norms and rules. One was Marcel Duchamp (famous for presenting a urinal as a work of art in 1917), who tirelessly targeted scientists' view of the cosmos, and said he wanted "to strain the laws of physics." In 1914, he cut wooden slats according to a 3-foot-long string that he dropped three times, creating "a new standard of measurement based on chance." In 1921, he shaved his hair into the shape of a comet, tracing his own erratic orbit on the streets of Paris.

Duchamp's work helped to inspire surrealism, one of the twentieth century's major art movements, which embraced the irrational and absurd. Positivism was stifling intellectual and moral advancement, complained poet André Breton, one of surrealism's founders, in 1924: "Experience . . . paces back and forth in a cage from which it is more and more difficult to make it emerge." Surrealists aimed to use the unconscious mind, as well as chance, to free the imagination and escape the restrictions and conventions of rational thought.

Even so, they couldn't help but respond to the latest advances in science. In the first literary expression of surrealism, Breton cowrote an extraordinary book called *Les Champs Magnétiques* (*The Magnetic Fields*). It aimed to capture the dreamy, unfiltered voice of the unconscious mind, and was packed with bizarre settings and clashing images. In one passage, "the smart set" journeys to find "buried suns" in a distant place where "on blood-stained shingly beaches one may hear the tender murmur of the stars." The section is called *Eclipses*, and it was written just as British astronomers traveled to Sobral in Brazil, and the island of Principe, West Africa, to observe a real-life solar eclipse predicted for May 29, 1919. Their mission was to measure the deflection of starlight during the eclipse, a problem that lead scientist Arthur Eddington described as "weighing light." No wonder surrealist Breton was fascinated. Now it was the scientists' turn to leave behind the comforts of common sense, and journey into territory that was strange beyond their wildest dreams.

Classical physics describes a universe in which objects exist and events occur within a fixed, absolute grid of time and space. It's an approach that was highly successful for describing everyday events on

Earth; by the end of the nineteenth century some physicists thought there was little left to discover. Then Albert Einstein, working in the Swiss patent office in Bern, asked himself what would happen if you could travel at light speed in such a universe, and concluded that for such an observer, light beams traveling in the same direction would appear to stop. Yet according to Maxwell's rock-solid equations describing electromagnetism, that was impossible; light always travels at the same speed.

So Einstein assumed instead that the speed of light must stay the same for everyone, regardless of how fast an observer moves, and reached a mind-boggling conclusion: that space and time change depending on your point of view. He published his theory of "special relativity" in 1905. For objects moving at significant fractions of the speed of light, his math described a weird terrain in which time flows at different speeds for different observers and space can expand and contract. The equations imply that space and time aren't separate but interwoven—time, not space, is the fourth dimension—while mass and energy are interchangeable, as described in Einstein's famous equation $E = mc^2$.

He later extended his ideas to incorporate gravity, concluding that gravity is equivalent to the force of acceleration and that massive bodies like the Sun pull objects toward them because they warp the fabric of space-time. He presented his theory of "general relativity" in November 1915, but at first it didn't spread beyond a narrow circle of specialists. The big test came from the eclipse chasers.

Einstein predicted that because the Sun bends space around it, any passing light must curve instead of shooting in a straight line. If so, the position of distant stars should appear to shift slightly when the Sun is in the same part of the sky, because their light veers off course. Stars close to the Sun aren't usually visible, but it would be possible to measure their position when the sunlight was blocked during an eclipse. Eddington announced the results in London in November 1919. The data was ambiguous, it turns out, but it didn't matter; the headline in the *Times* the next day read "Revolution in Science: New Theory of the Universe: Newtonian Ideas Overthrown."

With Einstein, physics tipped into a new era. Ether and the fourth spatial dimension (as conceived in the nineteenth century, at least) were swept away, as were absolute time and space. From that point on, the cosmos was relative. It was the first glimmer for physicists that there is no single truth "out there" to be discovered, that the universe we perceive depends very much on how we look.

There was more to come. Relativity deals with events across vast reaches of space. But physicists studying the very small—atomic and subatomic scales—had been hitting problems too. It had been accepted since the early nineteenth century that light travels as a wave. It spreads out after passing through a narrow slit, like a water wave does, and waves from nearby slits become superimposed, producing a characteristic interference pattern of peaks and troughs. But from 1900, physicists found that they could only make sense of basic observations, such as the pattern of radiation that hot objects emit, by assuming that the energy released from atoms isn't continuous, but comes in discrete packets, or quanta. Light was apparently not a wave after all, but composed of particles (which they called photons).

In 1913, the Danish physicist Niels Bohr applied the concept to the structure of atoms. The latest model had suggested that electrons orbit atomic nuclei like planets in a solar system; Bohr added that only certain orbits—corresponding to particular energies—are allowed. Electrons jumping between levels emit or absorb photons with different energies, corresponding to different frequencies of light. Bohr's atomic model revealed—at last—what Fraunhofer, Bunsen and Kirchhoff never explained: why gaseous forms of different elements emit and absorb characteristic patterns of spectral lines. But the idea of forbidden energy zones, with electrons disappearing in one state and instantaneously appearing in another, splintered classical models of reality.

The cracks kept getting bigger. By the 1920s, it was clear that not just light but all radiation and all subatomic particles showed contradictory characteristics of both particles and waves, depending on when and how they were observed. When it came to the minuscule quantum world, seemingly rock-solid concepts such as orbit, position and speed were falling apart. Physicists including Bohr, Werner Heisenberg and Max Born described these counterintuitive results mathematically and were led to an idea that would have been extreme even for the surrealists. They concluded that the particles we observe don't exist in any normal sense until they are pinned down by a measurement. The wave aspect consists not of physical matter or radiation, but *probability*; it describes not where a particle is, for example, but where it might appear at a point at which we decide to look.

The quantum theory they pioneered states that on subatomic scales, there are fundamental limits to what we can know (for example, we can't ever precisely measure both the position and the momentum of a particle at the same time). And we can't ever predict with certainty what the result of a particular measurement will be, just the likelihood of different outcomes. Classical physics—and common sense—say that if you rerun identical measurements over and over, you'll get the same result. In quantum physics, the answer can be different each time.

These ideas had huge philosophical implications. First, the discovery that the particles we measure can't be separated from the act of measuring them immediately triggered fierce debates among physicists about the status of external reality, and even whether there was a link to human consciousness, a debate that still hasn't been resolved. At one extreme, traditionalists like Einstein never gave up on the idea of an objective reality that exists independently from physicists and their experiments. At the other, Wolfgang Pauli and later Erwin Schrödinger wondered whether we create the reality we perceive, and argued that quantum physics would have to find a way to accommodate the human mind.

Second, the new ideas smashed the dogma that reality is ultimately knowable and predictable, following a logical chain of cause and effect. Scientists since Newton had believed that, in theory, if they could gather enough information about a system—even the whole universe—they'd be able to understand everything about it, and to deduce everything that would follow in the future. The birth of quantum physics made that belief appear obsolete. It taught instead that what we observe is a statistical sum of probabilities. On everyday scales, this creates the illusion of cause and effect—the cogs and wheels of a machine—but perhaps not everything in the universe is caused. At its root, it seemed, was fuzziness, a creative chance that can't be pinned down.

In this chapter, we have strayed far from our view of the starry sky. But these artists and physicists form a crucial part of our story nonetheless, because in exploring reality in other ways, they, too, were trying to work out what the cosmos is. They went beyond stargazing, and instead aimed to penetrate the truth behind those twinkling points of light. They rebelled against the certainty and clarity that had seemed so unshakeable just a few decades before, and by World War II, the nature of the universe had shifted.

As we'll see in later chapters, the vision of a separate, external reality, carefully constructed over centuries, would not be so easily quelled. Yet this burst of creativity contained seeds that couldn't entirely be put back, that still demand our attention today. Artists had used their emotion and intuition to deconstruct logical, three-dimensional space; attack the limits of our senses; and explore the subjective terrain of the human mind. Physicists took a very different approach, relying on math and reason, but they, too, left common sense behind, drawing back the curtain on a cosmos as unpredictable and contradictory as a Cubo-Futurist opera: at its largest and smallest scales interlaced with our own experience yet far beyond the reality that we perceive.

✡ ✡ ✡

Malevich, it seems, was always ahead of the curve. As the common-sense cosmos wavered, he moved on yet again. Desperate shortages of food and fuel caused by Russia's civil war—Lenin's Reds fighting the more conservative Whites—made life in Moscow increasingly difficult, and in November 1919, he moved to the Belarusian city of Vitebsk, to teach at an art school led by painter Marc Chagall.

Chagall was soon elbowed out; the school became a stronghold of Suprematism. And at this stage in his life, Malevich did become passionately immersed in astronomy: he always carried a pocket telescope, and spent hours memorizing the map of the night sky. Perhaps inspired by those journeys through his inner universe, now he dreamed of escaping the physical Earth. He was influenced by the pioneering Russian engineer Konstantin Tsiolkovsky, who had recently calculated how a rocket engine fueled by liquid oxygen and hydrogen might blast a craft into space. In 1920, Malevich predicted a cosmic future for humanity in which Suprematist machines would orbit between Earth and the Moon. He described them using the Russian word for "fellow traveler": *sputnik*.

Malevich wasn't painting, but he made and sketched geometric architectural models, or "architektons," including designs for orbiting space habitations that he called "planits." The enthusiasm for space travel spread to his colleagues and students. El Lissitzky published a children's picture book in which two cosmic squares visit Earth before soaring away. Ilya Chashnik's 1925 *Red Circle on Black Surface* is so vivid in its depiction of a lonely satellite orbiting a red sphere against the vast darkness of space that it's hard to believe it was painted decades before humanity ever left Earth.

The vision that Malevich helped to inspire would eventually lead his nation to beat the United States into orbit, with the launch in October 1957 of the world's first artificial satellite, Sputnik 1. But the new political climate was not so kind to his art. After the Soviet Union was officially formed in 1922, the Communist Party exerted increasing control over all areas of society, including artistic expression, with Lenin's successor Joseph Stalin eventually allowing only earnest, positive depictions of Soviet life (a style called socialist realism).

Malevich moved to Petrograd* in search of a freer climate, but in 1926, his institute there was closed after a Communist Party newspaper called it a "government-supported monastery" rife with "counter-revolutionary sermonizing and artist debauchery."

His artistic and teaching activities were restricted, many of his works were seized, and in 1930 he was arrested and imprisoned as an enemy of the state. Faced with destitution, the visionary was forced to return to realist painting, producing what he hoped were politically acceptable portraits of peasants and his family. He never gave up sneaking in Suprematist elements, though: his *Final Self-Portrait* of 1933, which shows him as a Renaissance artist, is signed with a black square.

Malevich died of cancer in May 1935, after his requests to return to the West for treatment were refused. His wish to have his grave topped with an architekton and a telescope for visitors to gaze at Jupiter wasn't fulfilled either. But the site was marked by a white cube, on which was painted a giant black square. I like to think of it as a portal to the cosmos he spent his life reaching for. Malevich had finally left Earth for the infinity beyond.

* Petrograd was renamed Leningrad in 1924. In 1991, its original name, Saint Petersburg, was restored.

10

LIFE

IN FEBRUARY 1954, A U.S. BIOLOGIST NAMED FRANK BROWN discovered something so remarkable, so inexplicable, that his peers essentially wrote it out of history.

Brown had dredged a batch of Atlantic oysters from the seabed off New Haven, Connecticut, and shipped them hundreds of miles inland to Northwestern University in Evanston, Illinois. Then he put them into pans of brine inside a sealed darkroom, shielded from any changes in temperature, pressure, water currents or light. Normally, these oysters feed with the tides. They open their shells to filter plankton and algae from the seawater, with rest periods in between when their shells are closed. Brown had already established that they are most active at high tide, which arrives roughly twice a day. He was interested in how the mollusks time this behavior, so he devised the experiment to test what they would do when kept far from the sea and deprived of any information about the tides. Would their normal feeding rhythm persist?

For the first two weeks, it did. Their feeding activity continued to peak fifty minutes later each day, in time with the tides on the oysters' home beach in New Haven. That in itself was an impressive result, suggesting that the shellfish could keep accurate time. But then something unexpected happened, which changed Brown's life forever.

The oysters gradually shifted their feeding times later and later. After two more weeks, a stable cycle reappeared, but it now lagged three hours behind the New Haven tides. Brown was mystified, until he checked an astronomical almanac. High tides occur each day when the Moon is highest in the sky or lowest below the horizon; that's when the lunar gravitational pull on ocean waters is strongest. Brown realized that the oysters had corrected their activity according to the local state of the Moon; instead of matching the East Coast swells, they were feeding when Evanston—if it had been by the sea—would experience high tide. He had isolated these organisms from every obvious environmental cue. And yet, somehow, they were following the Moon.

For a while, Brown's experiment became infamous, one of the most controversial results in biology. Scientists were just starting to appreciate that living processes vary according to environmental cycles such as the time of day, but every other major figure in the field was convinced these rhythms were ultimately driven by internal clocks; Brown's lone insistence that organisms are plugged into mysterious cosmic cues was viewed with disdain. The disagreement was about more than the nature of biological timekeeping. It reflected a deeper, philosophical split regarding the relationship that living creatures, including humans, have with our planet and the wider cosmos. Are we autonomous, self-running machines, or is life in constant, subtle communication with the Earth, Sun, Moon and even stars?

Brown ultimately lost the battle, and his decades of work were thrown out as flawed. Conventional scientific accounts barely mention him except as a cautionary tale: a warning about the dangers of straying too far from common sense. The field of "chronobiology" has since become hugely successful, with researchers uncovering intricate networks of molecular cogs and gears that keep time inside our cells, allowing virtually all creatures on this planet to anticipate the daily and seasonal movements of the Sun.

But there's still a fundamental mystery at the heart of biological clocks that has never been explained. And the stubborn trickle of evidence hinting that Brown might have been onto something is fast becoming a flood.

✩ ✩ ✩

That life on Earth moves in harmony with the Sun's daily path through the sky has been known for thousands of years. We wake in the morning and sleep at night. Flowers open and close depending on the time of day. Birdsong heralds the dawn. The earliest known written account of a daily biological cycle comes from Androsthenes, a naval commander under Alexander the Great, who described in the fourth century BC how the leaves of tamarind trees on Tylos (now Bahrain) opened at sunrise and closed at night. Scientific study began in 1729, when the French astronomer Jean-Jacques d'Ortous de Mairan noted that the daily opening of the leaves of a "sensitive" plant, probably *Mimosa pudica*, continued even when it was kept in the dark.

But these daily cycles were generally seen as passive responses to changing environmental signals such as temperature or light. Even de Mairan concluded that his plants must be able "to sense the Sun without ever seeing it." It wasn't until 1832 that a Swiss botanist named Augustin de Candolle first suggested that sleep-wake movements in plants such as *Mimosa* might result from an internal timer.

By the first half of the twentieth century, the possibility that simple organisms such as plants contain accurate clocks still hadn't been accepted; in fact, the very idea of studying it was controversial. That was part of a more general shift in attitudes about the possibilities of the natural world. We read in chapter 9 how the late nineteenth-century enthusiasm for an all-pervading ether, as well as discoveries such as radioactivity, had suggested the existence of a connected, unseen reality, beyond the commonsense events that we ordinarily perceive. Topics such as mind reading and clairvoyance were hugely popular with the public and fair game for scientific inquiry, taken seriously by some of the world's most respected physicists. But it wasn't long before this credulity began to melt away. Physicists concluded that there was no ether. And a string of embarrassing failures made scientists painfully aware of how easily human biases can lead us astray.

An exciting new type of radiation called N-rays, for example, discussed in hundreds of physics papers after it was first reported in 1903, was shown in 1904 to be imaginary. Fabulous phenomena from psychics to fairies were exposed as frauds. Decades of research into telepathy made no convincing discoveries. And in an episode that shaped the future study of animal behavior, a world-famous horse named Clever Hans, who could apparently answer questions and solve math sums by tapping his hoof, was shown in 1907 to be responding to unconscious cues from his owner. By the 1920s and '30s there was fierce skepticism, particularly in the United States, about the mysterious abilities and forces that had so fascinated scholars a few decades before. This included the idea that living creatures were somehow tracking time.

But this didn't stop Brown from becoming hooked on biological rhythms early in his career. Born in 1908, he grew up in the tiny seaside town of Machiasport, Maine. He spent his free time hunting, sailing and fishing, and became fascinated by the creatures clinging to Maine's rugged coastline, a liminal zone ruled by some of the largest tides in the world. In 1929, he began a PhD at Harvard University, studying color changes in Atlantic fiddler crabs.

These crabs blacken their skin around sunrise, to hide better from predators and protect their internal organs from the Sun's rays. At sunset, the color fades, until their bodies are pale, silvery gray. Brown was interested in what triggers these changes but his results varied wildly according to the time of day, and he began to suspect that even in the controlled conditions of the lab, the crabs were following a 24-hour cycle.

He became more intrigued by animals' timekeeping abilities during the final summer of his PhD, in 1934, which involved a six-week trip to an ocean research station in Bermuda. Brown spent the evenings sitting on the lab's wooden jetty, shining a flashlight into the water and watching the animals that swam into the light. He soon got to know the local inhabitants—from pipefish and jacks to baby turtles and eels—but one moonless night at about ten o'clock, the water suddenly filled with creatures he had never seen before: a rippling swarm of tiny, translucent shrimp.

They disappeared as quickly as they arrived but returned a month later, just ahead of the next new Moon. Brown caught a few and sent them to an expert at the British Museum in London, who identified them as *Anchistioides antiguensis*, an elusive species of which the museum had only one other specimen. Subsequent studies revealed that the shrimp appear in swarms only for an hour or two, just before midnight, near the time of the new Moon.* Back in the United States, Brown had been offered a research position at the University of Illinois, and now he knew what he wanted to study. What was driving these cycles in the shrimp and crabs?

With money from a summer teaching job and a loan from his brother, Brown bought a secondhand car and drove west for a new life in Illinois. But before he left, a senior colleague at Harvard warned him to avoid the dubious topic of biological rhythms if he valued his career. He heeded the advice, sticking to respectable topics such as color vision in fish, until in 1949 he was made a professor at Northwestern University, a lifelong position from which he couldn't be sacked. He could return to the "rhythm problem" at last.

Working again on fiddler crabs, Brown showed that even when kept in constant light and temperature, they would blanch and blacken each day in time with conditions on their home beach. If he

* Brown and his colleague identified two monthly *A. antiguensis* swarms: a large one three to four days before the new Moon, and a smaller one two days after the new Moon. Biologists have since found that the rest of the time they live inside sponges.

exposed the animals to light at different times, he could shift the phase, persuading them to cycle, say, six hours earlier, as if they were in the Baltic Sea, or twelve hours earlier, as in Singapore. But that wasn't all. He also studied the crabs' physical activity. Fiddler crabs spend most of their time hiding in burrows beneath the sand; they scuttle out in search of food only when the beach is exposed at low tide. Brown's captive crabs stuck resolutely to this 12.4-hour tidal rhythm despite being hundreds of miles from the sea, showing low-tide peaks in activity twice every lunar day. The animals were like miniature astronomical clocks, with the two cycles—one solar, one lunar—gradually sliding in and out of phase, just as they do in the sky.

By the early 1950s, a handful of other scientists were taking an interest in biological rhythms, in an impressive range of species. In Germany, eminent botanist Erwin Bünning recorded the "sleep movements" of bean-seedling leaves, while Jürgen Aschoff, a pipe-waving, argumentative physiologist, discovered a 24-hour rhythm in human body temperature by experimenting on himself. In the United States, the British-born biologist Colin Pittendrigh started investigating insect rhythms after noticing daily cycles in mosquito activity while tackling malaria in Trinidad, and Romanian-born Franz Halberg entered the field after his drug test results were ruined by daily fluctuations in the levels of white blood cells in mice.

Whereas Brown was captivated by the influence of the Moon, his rivals focused on 24-hour cycles. Whatever species they studied—bean sprouts, birds, flies, mice—they, too, found that in constant conditions, the rhythms continued. But in their experiments, the speed of the rhythms changed slightly in the absence of external cues, so that the peaks and troughs gradually drifted with respect to the solar day. Different individuals ended up with varying cycle lengths—each close to, but not exactly, 24 hours. The researchers concluded that the rhythms must be driven by private, internal timers within the organism's cells. Under normal conditions, the cycles were nudged by environmental cues such as light and temperature, to keep them precisely tuned to day and night. But they were perfectly capable of running on their own.

At first, Brown thought this too. But he started to doubt it was possible. In his lab, the crabs' lunar and solar cycles persisted accurately for months, even when apparently isolated from the surrounding environment; he couldn't imagine how an independent, internal clock could keep such good time. And then, in 1954, came the experiment with the time-shifting oysters. Despite being in a sealed darkroom, they had adjusted their activity according to the local movements of the Moon. Instead of relying on inner timers, he was convinced they were sensing signals from the sky.

These two approaches essentially paralleled the rival timekeeping methods used by navigators in the eighteenth century to calculate longitude. You can build a machine that keeps its own time, as clockmaker John Harrison did with his chronometer. Or you can take time references directly from the sky, as with the astronomer Nevil Maskelyne's method of lunar distances. Which solution did life choose?

Brown decided to investigate the most fundamental biological process he could think of: metabolism. He studied sprouting potatoes—in experiments that ran for years, tracking more than a million potato hours—as well as bean seeds, mealworm larvae, chick eggs and hamsters, shielding them all from changes in temperature, pressure and light. Although they were supposedly cut off from the outside world, he saw patterns in their metabolic rate that matched not just the movements of the Sun and Moon, but pressure and weather changes in the Earth's atmosphere. Even the potatoes "knew" not just the hour but the season of the year. It was as if life were pulsing in time with the planet.

Brown concluded that the organisms were sensitive to external geophysical factors, perhaps minute fluctuations in gravity, or even subtle forces that hadn't yet been discovered. In his rivals' experiments, supposedly proving the existence of independent clocks, Brown argued that the subjects weren't cut off from the environment after all. They were bathed in—and influenced by—subtle, rhythmic fields that varied as the Earth turned.

Such ideas did not go down well with his peers. Several of them, too, had fought to have their work on daily cycles taken seriously by other scientists, having been accused of everything from parapsychology to paranoia. Their professional respectability hinged on using rigorous, reproducible methods, and basing their theories on impeccable physical principles of cause and effect; Brown's claims of mysterious forces were dangerous nonsense that jeopardized the field. His measurements weren't accurate enough, they insisted, or he was seeing patterns in his highly complex data that simply weren't there. Yet Brown was charismatic and articulate, and he was swaying public opinion. Something had to be done.

The first major blow came in 1957, with an extraordinary paper in the leading U.S. scientific journal *Science*, in which a respected ecologist named LaMont Cole claimed that by juggling random numbers, he had "discovered the exogenous rhythm of the unicorn." The satire was aimed at Brown and his team, and its message was clear: their results were as imaginary as the unicorn itself. It was an unprecedented, personal attack and it "hit us very hard," Brown recalled later. "We were everywhere encountering innuendos from this article." In 1959, Halberg followed up by coining the term that now defines the field: "circadian." It's often said to refer to 24-hour cycles, but that isn't quite right—it comes from the Latin meaning "about a day," and Halberg chose it precisely to emphasize the key flaw in Brown's theory: that most free-running daily rhythms are *not* exactly 24 hours long. Tensions came to a head in June 1960, at a prestigious conference on biological clocks held at Cold Spring Harbor, near New York City.

This event is now seen as chronobiology's defining moment, at which Pittendrigh and the others set out their vision of circadian rhythms as internal and self-sustaining, controlled by oscillating biochemical mechanisms analogous to the cogs and gears of a clock. Everything was looking good for the young field, with its new terminology and robust theoretical framework. There was just one problem: Brown. He wasn't invited at first, but went anyway, the only speaker to argue for a core pacemaker driven instead by cosmic cues. He faced a largely hostile audience.

One of Brown's arguments came down to temperature. Everyone agreed that the timing of the biological rhythms was surprisingly resistant to even quite dramatic temperature changes. Crabs switch color and flies emerge from their pupal cases at the correct time regardless of how hot or cold they are. Dried seeds stored in constant conditions still showed an annual rhythm in their capacity to germinate, whether kept at 20 degrees below freezing or 50 degrees above. Yet the speeds of biochemical reactions vary hugely with temperature, Brown pointed out; as a general rule, rates double with every 10°C rise (and at -20°C, internal biochemical processes would barely be expected to function at all). His rivals could provide no explanation of how any biochemical mechanism could create a clock that was immune to such influences, whereas external, unvarying cues driven by the Sun and Moon would explain this property perfectly. By insisting that an internal timer existed, he warned, they risked "chasing a ghost." Pittendrigh retorted that it was Brown, with his mysterious, subtle influences, who was chasing ghosts.

After the meeting, Brown noticed that his papers were increasingly rejected, particularly by U.S. journals, and that others in the field, though they had always disagreed with him, now no longer cited his research at all. According to Brown, Halberg eventually admitted that at around this time, his rivals privately agreed to block, ignore or discredit Brown at every opportunity; for the sake of the field's development, they had to silence the rebel. Whether or not that's correct, they certainly stepped back from engaging with Brown and his ideas; from then on it was almost as if he didn't exist. Pittendrigh ignored

Brown's arguments completely at a major conference in 1964, for example, while in a seminal *Science* review published the next year, Aschoff refused to name his rival, simply noting that the "opposing hypothesis" had been criticized enough: "I therefore refrain from discussing it again."

Brown and his cosmic cues were cast out, and the study of biological rhythms became the study of circadian clocks.

✲ ✲ ✲

The resulting field has since transformed our understanding of how life works: not just animals and plants but humans too. Aschoff, for example, embarked on a pioneering series of experiments to investigate what happens when people are cut off from the Sun. After conducting pilot studies in an old World War II bunker, he built a dedicated isolation facility into a Bavarian hillside in 1964. Working with a physicist colleague named Rütger Wever, he shut students inside it for weeks at a time, tracking them with a battery of instruments including motion sensors and a rectal probe. The soundproof chamber was comfortable, with a sitting room, shower and small kitchen, but all clues to the time of day—such as a clock, radio or telephone—were banished (although there was a record player, which the more resourceful subjects used to time their boiled eggs). The volunteers' only contact with the outside world was by letter, while food, and locally brewed beer, was delivered at random times to a small space between the bunker's double doors.

Aschoff himself was the first volunteer, observed by Wever. During his ten-day stay, he guessed he was getting up slightly earlier each day, and was "highly surprised" to discover on release that his last waking time was actually three p.m. After that, more than three hundred volunteers—often students cramming for exams—"went underground" for three to four weeks each. Only four left early, and one, whose day length had spiraled to 33 hours, complained that he hadn't had enough time to finish studying and insisted on being let back inside.

Just as in other species, Aschoff found that the volunteers' daily rhythms continued even in constant conditions,* showing that humans have innate circadian clocks too. When deprived of information from the outside world, the sleep-wake cycle usually lagged slightly slower than the solar day, with a period averaging around 25 hours. Over the years, he and Wever showed that the cycles could be trained to follow signals such as bright light, temperature and social cues.

For a few of the volunteers, sleep patterns varied wildly, with day lengths reaching up to 50 hours, even though they didn't realize it. But their physiology—such as body temperature or excretion of metabolites—almost always continued to oscillate within a narrow band of 24 to 26 hours. This meant that their sleep-wake patterns fell out of step with their physiology, a phenomenon that Aschoff called "desynchronization." It was one of his most important discoveries—the first hint that there are multiple clocks in the body, which drive different functions, and that without appropriate external cues, they can become uncoupled. Volunteers reported feeling less well when this happened, leading Aschoff to warn that cutting ties with the Sun—for example, with regular shift work—might have damaging consequences for health.

The first clue to how the clocks actually work came across the Atlantic in 1971, from a Californian graduate student studying daily rhythms in fruit flies. Ronald Konopka isolated three mutant fly strains that lost the ability to keep time: one with a slowed-down rhythm of 29 hours, one with a too-short period of 19 hours and one with no cycles at all. All three, it turned out, had different errors in the same gene, which was subsequently identified by other researchers in

* Some of the experiments were done in constant darkness, some in constant light, and in some the subjects could choose when to switch the lights on and off.

1984. They named the gene "period," and found that the protein it encodes rises and falls in abundance every 24 hours. At last, they had a glimpse of the machinery inside the biological clock. Chronobiologists had found their ghost.

Since then, many other clock genes have been identified. They encode proteins that regulate each other in a complex network of feedback loops, ultimately creating what Brown had thought impossible: a steady cycle that pulses roughly once each day, in time with the Sun. Similar systems are found not just in fruit flies but in every type of life, from bacteria to people. These Sun clocks tell animals when to feed, when to run, when to sleep and when to digest. They allow plants to ration their starch reserves through the night, and to get their photosynthesis machinery up and running for dawn. They tell fungi when to form spores and insects when to emerge from their pupal cases; and signal thousands of species of ocean plankton to sink before dawn and rise to the surface each night—the largest movement of biomass on the planet. By tracking the shifting times of sunrise and sunset, the clocks can also drive seasonal changes, telling organisms precisely when to migrate, molt or reproduce.

Meanwhile, in humans, the study of circadian rhythms has become one of the hottest areas in medicine. Inner clocks regulate our sleep patterns, as well as body functions such as digestion, blood pressure, temperature, blood sugar levels, immune responses and even cell division. In fact, it's hard to find any organ or function that doesn't follow a daily cycle, and as Aschoff warned, we ignore these rhythms at our peril. In the two centuries since the first artificial lights switched on, our lifestyles have become increasingly detached from the 24-hour cycle of sunrise and sunset. Many of us stay up late, work varying shifts, hop between time zones. We work in gloomy offices during the day, and are exposed to light from computers, TVs and smartphones at night. That is a problem, because although our body clocks *can* run independently, if they aren't reinforced by external cues they can veer wildly off course.

The most important timing signal is sunlight, detected by cells in the retina of the eye. Information regarding the timing of sunrise and sunset is sent to a central pacemaker in the brain, which in turn regulates secondary clocks throughout the body. Not enough bright light during the day, or too much light in the evening, can dampen or disrupt these clocks, throwing the body's complex choreography into chaos. Rhythms weaken, or become disconnected from each other, causing health problems from insomnia and depression to obesity, cardiovascular disease and even cancer.* Leading chronobiology researcher Russell Foster has warned that despite the modern technologies that enable us to wake, sleep and work whenever we like, "human biology remains profoundly dependent upon the 24-hour revolution of the Earth upon its axis." Our increasing separation from geophysical cycles is a "ticking time bomb," he argues, the psychological and physical effects of which "might condemn whole sectors of our society to a dismal future."

Circadian rhythms are significant for health in other ways, too. Doctors are realizing that most medical conditions display daily fluctuations in their occurrence or symptoms, including heart attacks, asthma, bronchitis, cystic fibrosis, strokes, fever, pain, seizures and suicide, to name just a few. The time of day can determine how we'll respond to an infection or drug, or whether eating exactly the same meal will cause us to gain or lose weight. And even seasonal changes are important: the month in which babies are born affects their later risk of diseases such as dementia, multiple sclerosis and schizophrenia (with opposite patterns in the northern and southern hemispheres). Scientists don't understand exactly why (theories include early-life infection risk, nutrition and vitamin D levels) but it's clear that the position of the Earth relative to the Sun at the time you are born has health consequences that last for life.

* In 2007, for example, the World Health Organization classified shift work that disrupts human circadian rhythms as a probable human carcinogen.

In 2017, the field of chronobiology received the ultimate scientific recognition: a Nobel Prize, for the researchers who identified the period gene. "We on this planet are slaves to the Sun," commented the prominent biologist Paul Nurse. "The circadian clock is embedded in our mechanisms of working, our metabolism, it's embedded everywhere." Within just a few decades, the ability to anticipate the solar day has been transformed from a niche curiosity into a defining feature of life itself.

It's a wonderful success story. And yet, for all of these impressive discoveries, some of the questions raised by Brown have refused to go away. One idea that persists, despite skeptics' best efforts to banish it, is the all-pervading influence of the Moon.

✡ ✡ ✡

One of the earliest known pieces of art—a Paleolithic figure chiseled into the limestone of a French cave in Laussel, just a few miles from Lascaux—shows a woman with prominent breasts and one hand on her swollen belly. Originally sprinkled with red ocher, she holds a crescent-shaped horn carved with thirteen downward stripes, which most scholars think represent the number of days between crescent and full Moon.

Chronobiologists have tended to focus on daily rhythms, yet the belief that the Moon, too, has a profound power over life on Earth exists in traditional societies all around the planet. Whereas the Sun always looks the same, the Moon waxes and wanes in a perpetual cycle of birth, growth, death and—after three moonless nights—rebirth. In his influential survey of the world's myths and religions, historian Mircea Eliade argued that the Moon's phases allowed the earliest humans to measure the passage of time long before they began to track the more subtle seasonal movements of the Sun. (The word for Moon in languages from Sanskrit and Persian to Greek and Latin derives from the ancient root *me*, which means "I measure.")

Meanwhile, the lunar pattern of growth, death and rebirth fostered an overall view of a cyclical, ordered cosmos in which patterns of life on Earth reflected events in the sky.

Art such as the Laussel figure hints strongly that as far back as the Paleolithic, people saw links between the Moon and the human menstrual cycle (which averages 29.5 days, just like the lunar month) and length of pregnancy (nine lunar months). One expert has suggested that this synchrony was the initial observation "which gave birth in the mind of man to a mythology of one mystery informing earthly and celestial things." Another argues that it may have been women, calculating their cycles from Moon to Moon, who made the first reckoning of time. Eventually, according to Eliade, this lunar influence over fertility was extended until the Moon was seen as "*the* heavenly body above all others concerned with the rhythms of life."

Certainly, similar beliefs about the Moon's power over life existed by the time written records began. The Babylonians and ancient Greeks took it for granted that female fertility had a lunar source (the word "menstruation" is derived from the Greek *menes*, meaning Moon). In the first century AD, the Roman author Pliny the Elder wrote of the Moon's power "penetrating all things," from plants to animals to humans (and in particular shellfish, which he said grow and shrink with the Moon's phases); in the seventeenth century, the English philosopher Francis Bacon noted the belief that "brains in rabbits, woodcocks, calves, etc. are fullest in the full of the Moon." More recently, anthropologists have reported that in Greenland, young women were afraid to stare at the full Moon in case it made them pregnant, while in traditional societies from India to France, women drank water saturated with moonlight in order to conceive.

But is there any biological truth to the myths? Lunar cycles are of course important for nighttime activities such as feeding and foraging; nocturnal predators from owls, wolves and snakes to doodlebugs hunt by the light of the Moon. Perhaps more surprisingly, this effect can shape entire ecosystems. The mass daily sinking of oceanic plankton allows species from krill to crab larvae to hide from predators during the day, and it had been assumed that in Arctic midwinter, when the Sun doesn't rise, the migration stops. But in 2015, biologists found that across the Arctic Ocean, the movement instead switches from a 24-hour cycle to a 24.8-hour one, following the rising and setting Moon. The hunting behavior of mythical werewolves is driven by moonlight, the researchers noted: "Our data reveal a moonlight-driven reality."

The idea that the Moon directly governs biological processes such as fertility has been more controversial, but here, too, the evidence is growing. One of the first to investigate was a British zoologist named Harold Munro Fox, in the 1920s. He was intrigued that Pliny's remarks about shellfish being "full" at full Moon were still widely believed by fishermen around the Mediterranean and Red Seas, so he caught sea urchins with a hand net from a jetty in Suez, and found that the story was true. Throughout the breeding season, their gonads gradually filled with sperm or eggs each month, before spawning into the sea at full Moon.

Biologists now recognize that many aquatic species that reproduce by releasing sperm and eggs into the water use lunar phases (often in combination with daily and annual clocks) to time their spawning to a tight window, sometimes just a few minutes. Palolo worms, which live in Pacific coral reefs, shed their reproductive organs for just a few hours before midnight in the third quarter of the spring Moon (in October or November). At Australia's Great Barrier Reef, too, the sea comes alive each spring at precisely choreographed lunar moments. Hundreds of millions of corals over vast stretches of reef release trillions of eggs and sperm, creating an underwater blizzard of tiny bundles that dance in the current and turn the surface of the water pink with froth.

For creatures that exist on the border between land and sea, survival can hinge on using the Moon to predict tidal extremes. Twice each month (at full Moon and new Moon), the Earth, Moon and Sun are aligned, and the combined gravitational pull of the Moon and Sun creates extra-high and extra-low tides called spring tides. In between, the Moon and Sun are perpendicular in the sky, causing more moderate neap tides as the Sun's gravity helps to cancel out the Moon's pull. Spring tides are vital for the marine midge, which lives in algal mats just off rocky Atlantic shores. The seabed is exposed only at the lowest tide, at which point millions of midge larvae pupate into adult flies, mate, lay eggs and die with the rising waters, all within a few hours. Other species rely on spring high tides, such as Japanese land crabs, which climb down from the mountains where they live just in time to release their offspring into the sea.

Even on land, many creatures follow the Moon, either to coordinate their reproduction with other individuals or species, or give their offspring the best chance of survival. Nightjars such as the whippoorwill tend to hatch during a new Moon, so when the chicks' highest energy demand comes two weeks later, there's moonlight for the parents to hunt insects. Serengeti wildebeest use the Moon to time their conception to a tight window, so calves are born safely ahead of the species' mass migration in May or June. In 2015, scientists published the first example of a plant that reproduces by the Moon. *Ephedra foeminea*, a nonflowering relative of conifers and cycads, attracts insects during the full Moon by exuding drops of sugary liquid that "glitter like diamonds" in the nocturnal light.

Despite all these examples, though, lunar cycles in biology have been barely studied compared to circadian rhythms. Many scientists remain skeptical about their significance, particularly when it comes to reproduction, and researchers who do take them seriously complain that it can be difficult to get their work published. Until recently, much of our knowledge has come from isolated observations, often made by accident, from Brown watching shrimp in Bermuda to a Tunisian biologist who noticed a Peruvian apple cactus flowering in the moonlight outside his office window. Nonetheless, it's becoming clear that it isn't just humans who measure time using the Moon. From oceans to deserts to forests, lunar cycles are reflected in every corner of the planet, across all branches of life.

And that's not all. Although some organisms respond to changes in moonlight, many lunar behaviors continue even when it's cloudy, or in constant conditions in the lab. In the last few years, the first molecular studies of these phenomena—mostly in marine species such as coral, bristle worm, marine midge and rabbitfish—have found that lunar cycles are directly regulating biology. A 2017 study of *Acropora gemmifera* coral from Taiwan found that the activity of hundreds of its genes varied with the Moon. These include genes already known to be involved in circadian cycles, as well as in key functions such as cell signaling and cell division.* Other studies have identified photoreceptors sensitive to lunar light. In these species, running alongside their circadian clocks, there are inner Moon clocks too.

What about other species, even humans? The idea that we are influenced by the Moon has been particularly divisive, partly thanks to a U.S. psychiatrist named Arnold Lieber, who claimed in his 1978 book *The Lunar Effect* that violent events from murders to psychiatric admissions are more likely to occur at full Moon. Drawing on police and hospital data as well as ancient beliefs about the Moon and madness (the term "lunatic" comes from *luna*, Latin for Moon; the Old English equivalent is *monseoc*, "moon-sick"), Lieber argued that irrational and aggressive behavior is triggered by the Moon's gravity, pulling on water in our cells just as it pulls on the oceans. His ideas triggered popular beliefs about our susceptibility to the Moon that persist today, as well as a furious backlash from skeptics who insisted that

* Cell division is known to occur more often at night, at least in mammalian cells, probably to reduce exposure to the Sun's ultraviolet rays, which can cause errors when the cells' DNA is copied. But the authors of the coral study found that not all nights are the same; many genes involved in cell division were much more active at new Moon compared to full Moon, suggesting that in *A. gemmifera*, at least, the process may be timed to avoid moonlight as well as sunlight.

claims of lunar effects are not only statistically flawed but physically impossible. The Moon's gravitational pull on a person is tiny, they pointed out, equivalent to that of a drop of sweat on your arm. As recently as 2018, the very idea of lunar effects in humans was described in an academic review as "paranormal."

But some chronobiologists now suggest that although there's no evidence to support Lieber's theory of gravitational effects, the desperation to debunk this and other myths may have led to a blind spot about how the Moon really does influence our biology. After all, the range of genetically controlled lunar clocks across the animal kingdom, from fish to invertebrates, shows that they must date from very early in evolution, and it would be surprising if they didn't exist in some form in people too. And although the Moon's changing phases might seem irrelevant to us now, for early humans the monthly cycle between dark and bright nights would have driven opportunities for everything from sleep to hunting to sex.

In theory, such clocks might influence human fertility and reproduction. Trials investigating whether births and menstrual cycles vary with lunar phase have yielded mixed results. But there's no doubt that their average period is extremely close to that of the lunar month (there's even some evidence that men have monthly hormone cycles too). If there is a link, it's possible that our monthly cycles share a common origin with lunar clocks in other species but became decoupled from the Moon at some point in our evolutionary history. Or perhaps the lunar rhythm has weakened or disappeared in modern society simply because we're no longer exposed to the Moon's changing light* (in which case, along with other health problems, our artificially lit lifestyles may be disrupting fertility).

There may also be some truth to the link with mental health. Since 2013, several studies have found that sleep quality varies according

---

* And because the timing of many births is now controlled by artificial factors such as induction and cesarean section.

to Moon phase—with shorter, poorer sleep at full Moon—even in constant lab conditions, suggesting that a genetic lunar clock may play a role. Disrupted sleep, in turn, is known to trigger episodes such as seizures and epileptic fits, as well as exacerbate psychiatric conditions such as bipolar disorder and schizophrenia. Some evidence already exists linking seizures and epileptic fits with Moon phase. And in an intriguing study published in 2018, U.S. sleep researcher Thomas Wehr tracked patients with bipolar disorder and found that their abrupt switches between high and low mood were triggered by varying sleep patterns apparently linked to lunar cycles. One of the patients, who recorded his sleep times for seventeen years, went to sleep at the same time each day but woke up every 24.8 hours, with the Moon; Wehr suggests that instead of being guided by the Sun, his waking time had become erroneously linked with the lunar day.

It's an idea that needs further study. But Wehr's work highlights just how little we know about the influence of moonlight and lunar clocks, even on our own bodies, and about the consequences of separating ourselves from celestial cues. And either way, it is starting to look as though Brown was right about the ubiquitous influence of the Moon. Across all domains of life on Earth, behind the more obvious effects of the Sun, there's a hidden lunar pulse. The Moon allows organisms to anticipate changes in their local environment, such as water currents or nocturnal light. But it's also emerging as a global organizing force for life that, in combination with annual and daily clocks, enables different individuals and species to cooperate and coordinate their activities in time.

Over billions of years, then, organisms have evolved so that the rhythms of the solar system are literally built into their DNA. At the center of all living processes, inside every cell, there's a never-ending molecular dance, guided by the light that reaches us from the day and night sky, that mimics the patterns of the Earth, Moon and Sun.

Yet there is one more twist in the tale of biological clocks. Maybe light isn't the only celestial signal reaching us after all.

☼ ☼ ☼

What was the mysterious force that allowed Brown's oysters to track the heavens? Brown knew his rivals wouldn't take him seriously unless he could suggest a mechanism. So he spent the summer of 1959 carefully monitoring the creepings of 34,000 snails, collected from the mud flats of the New England coast. He was astounded to find that the snails could distinguish between different compass directions. Not only that; their preferred orientation varied over time, following both the solar and lunar day. He could influence or disrupt this behavior using magnets. At last, he believed he could explain how animals might detect local time even in a sealed lab: they were sensing daily changes in the Earth's magnetic field.

This field is mostly generated by molten iron that circulates within Earth's outer core. Overall, it's shaped as if the planet contains a huge bar magnet, with north at one pole and south at the other. But it is also influenced by external factors such as weather and magnetic storms—as well as the movements of the Sun and Moon. Radiation from the Sun ionizes atoms in the upper atmosphere, producing free electrons. Meanwhile, the Sun's heat causes atmospheric tidal winds that move these charged particles across the Earth's field lines.* The resulting electric current creates its own magnetism: a 24-hour ripple superimposed on the larger magnetic field of the Earth. A similar, smaller ripple occurs every lunar day due to the gravity of the Moon. These effects interact with each other, creating peaks and troughs during spring tides and neap tides. They're also dependent on the amount of sunlight falling on the upper atmosphere, so they vary with latitude and with the seasons.

The Earth's geomagnetic field is extremely weak, around a hundred

* The Sun's gravity has a small influence on the solar tide as well.

times smaller than that of a standard fridge magnet, and the undulations caused by the solar and lunar tides are even fainter.* Brown was one of the very first researchers to suggest that animals might have a magnetic sense (scientists had only just discovered that some fish are sensitive to electric fields), and he had no idea how his snails might detect such subtle changes. But he knew it could be clinching evidence for his theory of external cosmic cues.

He excitedly presented his results at the biological clocks symposium in 1960, telling the audience that living things are fantastically sensitive to very weak magnetic fields.† Although we can't see it, we're all immersed in an electromagnetic ocean, he insisted, with waves, tides and ripples that shift according to the relative positions of the Earth, Sun and Moon, keeping organisms in constant touch with the state of the solar system and the time of day. "We cannot yet explain these phenomena," he said, "but they emphasize to us, even more, our abysmal ignorance as to some of the forces affecting life."

The bombshell didn't convince his rivals, though; in fact, it hardened them against him. Just as they were setting out rigorous principles for the study of biochemical internal clocks, the notion of a subtle magnetic sense—reminiscent of long-debunked claims such as animal magnetism—was beyond the pale. And yet, though they shunned Brown in public, they didn't completely ignore his idea: quite the reverse. What's rarely mentioned today about Aschoff and Wever's famous bunker, built just a few years later, is that it contained not just one underground apartment, but two.

The parallel units were almost identical, with matching beds, kitchens and record players. But there was a very important difference: one of the apartments was completely enclosed within a hefty capsule of

* The Earth's magnetic field varies between around 30 and 60 μT (microteslas), and the daily solar variation is a swing of around 50 nT (nanoteslas).

† Brown showed later that other species, such as flat worms, could also sense magnetic (and electric) fields.

cork, coiled wire, glass wool (fiberglass) and steel, through which no electromagnetic radiation could pass: anyone living inside was completely cut off from the Earth's magnetic field. The aim was to show that the shielding made no difference to the volunteers' biological clocks, to prove, once and for all, that Brown was wrong.

Between 1964 and 1970, more than eighty volunteers stayed in the two units.* As Aschoff predicted, their circadian rhythms did continue. But there was a problem; the results in the two groups were not the same. In the unshielded bunker, isolated from clocks and sunlight but still exposed to Earth's magnetic fields, people's sleep and waking patterns departed from the solar day, reaching an average period of 24.8 hours. But when magnetic fields were also blocked, the volunteers' circadian cycles deteriorated further. Their day length slipped even more, growing longer. There was significantly more variation between individuals. And their different rhythms were much more likely to become uncoupled. As mentioned earlier, Aschoff championed desynchronization as one of his key discoveries. Yet over those six years, it *only ever occurred in the shielded bunker*, cut off from the Earth's magnetic field. Wever found that if he exposed the volunteers to a similar artificial field, all of these effects were reversed.

The results proved that we do have an inner clock that runs independently, regardless of any electromagnetic information from the outside world. And yet, that clearly wasn't the whole story. Even though the volunteers couldn't consciously perceive the Earth's vanishingly weak magnetism, the results suggested that their bodies could somehow sense it, and that this had a profound impact on the workings of their biological clocks. Wever published the data in a series of now-obscure papers in the 1970s; it was a "remarkable" result, he said, the first scientific evidence that humans are influenced by natural magnetic fields. But Aschoff didn't put his name to them, and he didn't even mention

* After that, one of the units was modified for temperature experiments, so they were no longer identical.

the shielding experiment or the apparent role of electromagnetic fields in his own papers on desynchronization. The existence of the second chamber was largely forgotten, and circadian rhythm research continued as if the magnetism experiment had never happened.

Just as these researchers were ignoring any links to magnetic fields, however, other biologists were being forced to address their effects, despite the dubious connotations. These scientists were studying the impressive ability of many animals to navigate across the planet, from turtles and salamanders to birds and bees. How did millions of monarch butterflies find their way thousands of miles from North America to a particular patch of fir groves in central Mexico each year? How did female loggerhead turtles, after growing up in the open ocean, return to lay their eggs at the very same beach where they hatched more than ten years earlier? How did racing pigeons fly straight home from distant places they had never previously visited?

Since the 1950s, biologists had been realizing that many species are expert in deciphering celestial signals. Butterflies track the Sun; moths follow the Moon. Starlings orient north from the celestial pole around which the stars turn. Dung beetles roll their dung balls in straight lines by orienting against the glowing streak of the Milky Way. Animals often combine these visual cues with information from their circadian clocks, allowing them to compensate for the time of day. Just like ancient human sailors, animals are clued in to their place in the wider cosmos, using the circling heavens not just to tell the time, but to navigate around the globe.

But this wasn't enough to explain the behavior of many species, some of which could still find their way even when the sky was overcast. It turned out that some animals detect patterns in the polarized light of the Sun and even the Moon allowing them to pinpoint the position of these celestial bodies even through clouds.* Then, in 1972, a

* Viking sailors are thought to have learned a similar trick, using polarizing crystals such as Iceland spar to locate the sun on cloudy days and navigate across Arctic seas.

German graduate student named Wolfgang Wiltschko showed that artificial magnetic fields similar in strength to Earth's could disrupt or alter the direction in which robins tried to migrate. It was the start of a flood of evidence that animals, from pigeons and sparrows to lobsters and newts, are sensitive to the magnetic field lines generated as the Earth spins in space. Wood mice and mole rats use them when siting their nests; cattle and deer orient their bodies along them while grazing; dogs—for unknown reasons—prefer to point themselves north or south when they relieve themselves. Other species, such as turtles, even appeared to have a magnetic map sense, telling them not just direction but position. Decades earlier, telepathy and clairvoyance were exposed as myths; Clever Hans was exposed as a fraud. Now organisms were revealing other abilities that were perhaps just as impressive. Life, it seems, really is plugged into the invisible electromagnetic world.

There was huge skepticism at first, of course. Natural magnetic fields were thought to be far too weak to influence biological tissue, so how could the signal possibly be detected? Well, life finds a way. Or, as it turns out, several ways. Fish have an electrical solution: they use internal networks of jelly-filled canals to measure the flow of current as they swim through a field (they need to be immersed in water to complete the circuit, so this method doesn't work on land). Another method involves physical forces: in 1975, researchers discovered "magnetotactic" bacteria that use chains of tiny magnetic crystals—made from a mineral called magnetite—as compass needles, to steer themselves down magnetic field lines and into the mud where they live and feed. Some researchers believe that in animals from birds to bees, similar crystals could enable magnetosensing by exerting mechanical pressure on neurons as they try to align with the Earth's field, although such receptors haven't yet been definitely identified.

In 1978, German biophysicist Klaus Schulten suggested a third possibility after studying a class of obscure chemical reactions influenced by quantum effects. Electrons have a quantum property that physicists call spin, and Schulten was investigating how light energy can trigger the formation of short-lived pairs of "radicals"—molecules with lone electrons, which can spin in either the same or opposite directions. These two spin states are chemically different, and the amount of time the electrons spend in each state can be influenced by magnetic fields. So even if a field is too weak to influence a chemical reaction directly, light creates an excited state in which it can then nudge the outcome one way or another. Imagine a fly unable to move a stone block by flying into it. If you balance the block on its edge, a fly striking at just the right position and moment might be enough to tip it and create a much larger effect.

It was an example of what everyone had thought impossible: a mechanism by which an extremely weak magnetic field can produce chemical cues large enough to be detected by the nervous system. "I thought, well, maybe that's the internal compass the biologists were looking for," said Schulten. But no receptors capable of forming radical pairs were known in living organisms, and when he submitted his theory to the journal *Science*, it was swiftly rejected. "A less bold scientist," said one reviewer, "would have designated this piece of work for the waste paper basket."

Two decades later, biologists studying fruit flies discovered proteins called cryptochromes, which form radical pairs when exposed to blue light. Cryptochromes are now known to be common in organisms; they're found in plants and fish, insect antennas, and the retinas of mammals and birds. And there's good evidence in several species that cryptochromes are indeed involved in magnetosensing and navigation. Researchers have suggested that they enable birds to "see" magnetic field lines, perhaps by perceiving a brighter image if they are facing in a certain direction.

Humans have cryptochromes too. Until recently, most scientists agreed that people can't sense magnetic fields. But in 2011, researchers put the human cryptochrome protein into fruit flies that lacked their own version and found that it restored the flies' magnetosensing ability perfectly. We're separated from fruit flies by 300 million years of evolution; it's highly unlikely that the protein function would be conserved so effectively if it wasn't still being used for the same thing in people. The finding hints that Wever was right: even if we don't consciously perceive it, our bodies are sensitive to magnetic fields. And here's the really interesting thing: Cryptochromes are actually best known not as magnetosensors but for quite a different reason. They are also crucial components of biological clocks.*

This discovery that biological clock "machinery" is sensitive to magnetism is still very new, and it's not yet clear precisely if or how magnetic fields are influencing our sense of time. One theory is that at least some species, such as honeybees, tell the time using daily tidal variations in Earth's field, just as Brown originally suggested. Others think the link may relate to how clocks resist temperature changes: a question posed by Brown and never fully answered. If you lose temperature compensation, you'd expect the body's rhythms to continue but to become less stable and more variable, and to start uncoupling, just as Wever found in the bunker shielded from magnetic fields. Perhaps it is an external cue after all—the Earth's magnetic field—that enables biological clocks to run regardless of temperature: not necessarily driving behavior directly, but providing the fundamental "tick" of the clock.

✡ ✡ ✡

When physicists banished the ether, it was the end of not just an outdated science theory but something much bigger. The centuries-old belief that a ghostly fluid pervades the universe, diffusing across space and through all matter, underpinned a view of reality in which the

* In some species, such as fruit flies, cryptochromes reset the circadian clock each day in response to pulses of daylight, while in vertebrates, they're the quintessential "gear" in the clockwork, helping to regulate its overall speed. (Some species, like monarch butterflies, have both types.) Cryptochromes are also among the genes found to cycle with phases of the Moon.

Earth, planets and stars are physically connected. This fluid, as the "world soul" of the Greek philosophers, was originally a divine influence radiating throughout the cosmos. It "suggested the existence of sympathies between God and man; between man and the stars," as one historian put it, and it meant that understanding the movements of the stars and planets was crucial for every aspect of existence, including human health. (The note "Rx" that still appears on medical prescriptions, often thought to be abbreviation for "recipe," is thought to be a corruption of the ancient symbol for the Roman god Jupiter. It's an appeal for healing to the cosmos itself.)

Modern science and medicine severed that link between life and the cosmos. Science moved from credulity to skepticism, demanding physical mechanisms rather than subtle influences. But more than that, it created a model of living organisms, particularly humans, as essentially isolated entities, separate from the workings of astronomy. We can see and measure the universe, but we are not physically linked to the Sun and Moon, let alone the planets or stars. This philosophy has fundamentally shaped the progress of science ever since, from positivist Auguste Comte's pronouncement that we could never know the workings of the stars to circadian clock researchers' decision to focus on inner mechanisms rather than cosmic cues.

It's a mind-set that has also come to define our modern way of life. We're more detached from the cycles of the solar system than ever before: urban life can be like existing in a temperature-controlled, artificially lit bubble, with flexible days and the seasons marked as much by Christmas trees and Easter eggs as celestial cues. In cities like Las Vegas or Singapore, you can walk for miles through underground casinos or shopping centers without ever seeing the sky. We live as if we are autonomous, separate beings, able to work, sleep or shop when we want, regardless of the Sun or Moon. That has given us unprecedented comfort and freedom, but as we've heard, we may be paying the price with our physical and mental health.

The blocking out of Earth's natural cycles is harming other species too. Nighttime light pollution now affects more than a fifth of Earth's land area, disrupting bird and turtle migration, bat feeding, amphibian breeding, plankton migration and plant growth, killing insects by the trillions and causing researchers to suggest in 2018 that it should be seen as a global threat on a par with climate change. The effects of noise from our electronic transmissions and devices are only just emerging, but scientists recently found that weak radio-frequency radiation, at levels previously thought to have no biological effects, disrupts clocks and compasses in insects, mice and birds (one research team found that the background radiation on their German university campus was enough to obliterate the magnetic sense of the robins they were trying to study).

Brown's vision was always different. Through the very nature of the atoms and electrons from which we're made, he saw living creatures as a continuous part of an electromagnetic cosmos. There's "no clear boundary," he said, between an organism's electromagnetic fields and those of its geophysical environment. "The organism and its physical environment appear to merge intimately for the timing of life." He never gave up on his ideas (though he acknowledged that external cues must be working together with some form of inner clock), and in later life, he became convinced that organisms use electromagnetic signals to sense each other as well as their wider environment. One chronobiologist later recalled how at a conference in November 1979, the audience, including Aschoff, reacted with embarrassed silence as Brown gave "a rambling talk . . . in which he seriously claimed that two plants kept in two separate environmental chambers in close proximity, influenced the circadian rhythms of each other."

Brown died by the sea in Woods Hole, Massachusetts, in 1983. As his rivals transformed biology with their study of circadian clocks, he was written into history as chronobiology's foolish old man. Yet since his death, many of Brown's findings have moved from ridicule toward scientific acceptance. In 1987, U.S. biologist Kenneth Lohmann, now a leading name in magnetosensing research, showed that sea slugs orient to the geomagnetic field in a way that varies with lunar phase—very similar to the effects Brown reported with mud snails and flat worms nearly three decades earlier. In 2012, Italian botanists found that chili plant seedlings can sense their neighbors and identify relatives even when chemical, light and temperature cues, as well as physical contact, are blocked. The researchers concluded that the seedlings were using previously unstudied mechanisms of communication, perhaps quantum magnetic or acoustic signals. Like all living organisms, the team pointed out, "plants have evolved in and adapted to an environment rich in naturally occurring and fluctuating geophysical waveforms." It makes sense that they would tap into them.

Just as quantum physicists realized in the early twentieth century that we can't separate ourselves from the universe that we inhabit, biologists are being forced toward the same conclusion. From bacteria and seedlings to turtles and humans, the movements of the Sun, Moon and even stars are not just distant events that we observe. They're directly influencing our immediate environment—and us. For our bodies and brains to work properly, we need to be in touch with these changing patterns of light, temperature and magnetic fields.

These regular fluctuations, caused by the motions of the solar system, may well have governed and structured biological processes since the earliest appearance of life on Earth. They now drive everything from individual shifts in mood or sleep to huge, planet-level patterns of migration, reproduction and predation. It's a complex web of responses that we are only just beginning to understand. "One of the most obvious lessons animals have taught us over the past few decades," said the eminent Princeton biologist James Gould, "is that much of what seems to be mere 'noise' is actually behavior too subtle for our current understanding or imagination." Meanwhile, some researchers aren't ruling out even longer-distance influences: from the idea that wobbles of our solar system within our galaxy have driven patterns of mass extinctions throughout evolution* to the possibility that tidal effects from the planets trigger harmful solar storms.

"The universe is one," Brown wrote in 1977, a few years before he died. "Cosmobiology is a field which must and will be explored." It's the kind of comment that saw him sidelined as a misguided eccentric. But I reckon he was onto something. He didn't get everything right, but he did, I think, see a fundamental truth about the nature of life and our relationship to the cosmos, the consequences of which we're only just beginning to grasp.

* These wobbles affect the number of mutation-causing cosmic rays that reach the Earth.

11

ALIENS

DECEMBER 27, 1984, WAS A WARM SUMMER DAY IN ANTARCTica's Far Western Icefield, with a lull in the normally biting winds and the temperature hovering as high as -20°C. The ice here forms a vast, flat plain that dazzles in the sunlight and is tinted slightly blue. Geologist Roberta Score and her colleagues spent the morning patrolling the area on snowmobiles, sweeping back and forth for hours in a search formation around a hundred feet apart. The monotonous routine could numb the eyes and brain, so just before noon, the team leader, John Schutt, called a break, and the group diverted to a nearby escarpment surrounded by ice pinnacles, carved by the wind, that look like giant frozen waves.

After enjoying the view from the top of the ridge, the geologists rode back toward their search area, taking care to avoid chasms in the ice and windswept piles of snow. And that's when Score saw it: a dark green spot against the glaring blue. She stopped her snowmobile and waved her arms to alert the others. Here, on this ancient, untouched stretch of ice, was what they had been looking for: a messenger from the depths of space.

Cosmic debris has been raining down on Earth ever since our planet formed. The ancient Egyptians saw the rocks as gifts from the gods, and made beads and daggers from meteoritic iron centuries before the metal could be smelted on Earth. The Greek philosopher Aristotle later assumed that a stone that "fell out of the air" must have been carried up by the wind. It was only in the nineteenth century that scientists accepted that stones really can fall from the sky: physical pieces of celestial objects, delivered directly to us, that contain clues to the solar system's past.

As well as tens of thousands of tons of dust, many thousands of pebbles and larger rocks are estimated to land on Earth every year, mostly fragments of the ancient asteroids that jostle and smash in a belt between Mars and Jupiter. Only a tiny fraction of falls are witnessed, and most of the material is lost in forests or oceans or worn to dust. But in some remote areas, such as the Antarctic icefields, these visitors can be preserved for thousands of years. In the 1970s, America's space agency, NASA, set up a mission to find them.

In 1978, Score answered a job ad from the new Antarctic Meteorite Laboratory at NASA's Johnson Space Center in Houston, Texas, and six years later, she made her first trip to Antarctica. After survival training at McMurdo Station on Ross Island, she and five other meteorite hunters were delivered by helicopter to the remote icefields of Allan Hills, 150 miles away. They spent the next six weeks living in two-person "Scott" tents: bright yellow pyramids that cling to the ice and can withstand gale-force winds. It was December, the height of summer, so the Sun never set, and they spent long, blinding-bright days out on the snowmobiles, eyes peeled for rocks, sustained by packed lunches of chocolate bars or raisins because anything else would freeze. On Christmas Eve, they all crammed into one tent for a stove-warmed feast of ham, lobster, shrimp and yams.

By the time Score made her find three days later, the team had picked up over a hundred meteorites between them. Straightaway, though, she thought this one was special. It was large—grapefruit-size—and it stood out as vivid green against the featureless blue ice. She and Schutt scooped it into a sterilized nylon bag, which they sealed with rounds of Teflon tape. In their field notes, after writing that the rock had a "shocked" appearance and was partly coated in a black crust (where the outer layer of rock had melted as it tumbled through the Earth's atmosphere), they couldn't resist adding the comment "Yowza-yowza." Wait until you see it, Score would tell her colleagues back home.

Back in Houston, it was Score's job to assign numbers to the team's finds, using the prefix ALH84 (for the location, Allan Hills, and the expedition year). She wanted her green mystery rock checked out first, so she put it at the top of the list. But once unpacked, she was disappointed to find that it looked ordinary after all, like a piece of gray cement. The color must have been caused by her tinted snow goggles, she figured, or a trick of the light. ALH84001 was later classified as an unremarkable chunk of asteroid and consigned to the laboratory stores.

It was another decade before this meteorite became a global sensation and household name. NASA researchers eventually realized that it was in fact a chunk of early Mars, far older than any other planetary rock ever discovered and containing a secret so explosive that it triggered a heartfelt speech from the U.S. president himself. The discovery caused a worldwide media storm, sparked conspiracy theories and science fiction plots, shifted NASA's future course and founded a new scientific field: astrobiology, the study of extraterrestrial life. Scientists are still arguing about what exactly is inside the rock that Score found on that Antarctic summer day. But our understanding of life—on Earth and in the wider cosmos—would never be the same.

✡ ✡ ✡

Are we alone in the universe? It's one of the biggest questions that humans have ever asked, deceptively simple to pose but for which either possibility—that we're the only vital spark in the entire barren eternity of existence; or that we're just one bloom in a far bigger, cosmic web of life—is almost too epic to comprehend. It feeds into other inquiries too, about who we are and what our existence means: What is life? Is humanity special? Why are we here at all?

The answers people give have ebbed and flowed through history, in step with their philosophical and religious beliefs. Ancient civilizations saw gods, spirits and souls in the sky. But the idea of physical aliens inhabiting other planets or solar systems goes back surprisingly far, too, at least to classical Greece. Followers of the "atomist" school of philosophy, who believed reality is made up of tiny, indivisible particles, argued that there are infinite atoms in the universe, and therefore infinite worlds. "To consider the Earth the only populated world in infinite space," wrote the philosopher Metrodorus of Chios in the fourth century BC, "is as absurd as to assert that in an entire field sown with millet only one grain will grow."

The cosmologies of Plato and Aristotle that came to dominate Western thought, though, had no room for other worlds or alien life: Plato argued that a unique creator implies a unique creation; Aristotle insisted that all elements find their natural place around a single center, the Earth. (Extraterrestrials were relegated to science fiction. In the second century AD, the Greek satirist Lucian* wrote the earliest-known story of interplanetary travel—intended as a dig at authors who told exaggerated travelers' tales—in which he was carried by whirlwind to the Moon, to discover men who rode on three-headed birds and fought the inhabitants of the Sun.)

Plato's and Aristotle's Earth-centered teachings were later enforced as law by the Christian church, and for centuries, at least in the West, it was forbidden even to speculate about the existence of life elsewhere. That began to soften in 1277, when the bishop of Paris declared it heresy to suggest that God couldn't make multiple worlds if he wanted to. But it was the emergence of modern astronomy that really opened the door. When Copernicus presented his heliocentric model for the solar system, says science historian Michael Crowe, he "changed our Earth into a planet and transformed stars into other suns." If ours was just one solar system among many, why shouldn't there be life in the others too?

In 1584, four decades after Copernicus's death, the Italian friar Gior-

* Lucian came from the Roman province of Syria, but he wrote in Greek.

dano Bruno described a universe of "innumerable" Suns and Earths, inhabited by living beings. And in subsequent decades, Galileo's telescope observations—lunar mountains, Jupiter's moons—confirmed other worlds do exist, at least in our own solar system. The idea of alien life was, if not accepted, then at least considered by scholars such as Descartes and Kepler (who wrote a novel about plants and serpentlike monsters on the Moon) and by Galileo himself. In 1698, astronomer Christiaan Huygens set out detailed theories about how organisms might be adapted to celestial environments such as dimmer, colder Mars. As Crowe puts it: "The era of the extraterrestrials had begun."

Enthusiasm for aliens reached a peak during the Enlightenment. As the huge extent of the cosmos became clearer, it seemed unimaginable to many that other life—even intelligent life—wouldn't exist somewhere. Prominent figures such as Immanuel Kant and Benjamin Franklin were in favor; legal scholar Montesquieu speculated about alien laws. In 1752, Voltaire satirized humanity's supposed intelligence in a short story that featured an extraterrestrial with more than a thousand senses.

Much of the debate swirled around a central question: did divine creation extend throughout the whole universe (with humanity just part of a vastly larger plan); or did God create the cosmos purely to put us in it? Thomas Paine insisted the former when he rejected Christianity for deism in *The Age of Reason*. Another popular argument was that God wouldn't have wasted his creative work by leaving vast areas of the cosmos uninhabited. During the nineteenth century, the Scottish astronomer Thomas Dick took this principle to the extreme when he estimated likely alien populations for celestial bodies according to their surface area, from four billion individuals on the Moon to many trillions on Saturn's rings.

Opposing figures included the philosopher and scientist William Whewell, who argued in 1853 that the universe being so perfect for our needs proves that God designed it just for us. Charles Darwin was scornful, privately describing Whewell's judgment that the solar system is adapted to us and not the other way around as an "instance of arrogance!!" But Darwin's cofounder of the theory of natural selection, biologist Alfred Russel Wallace, ultimately agreed with Whewell. "The supreme end and purpose of this vast universe," Wallace concluded in 1903, "was the production and development of the living soul in the perishable body of man."

For a while, it seemed (to enthusiasts at least) that improved telescopes were bringing these aliens within reach. As far back as the 1770s, William Herschel, pioneer of stellar astronomy, thought he saw forests and circular buildings on the Moon; historians now believe this conviction was what inspired him to build the superpowered telescopes for which he became famous. In 1822, Bavarian astronomer Franz von Paula Gruithuisen reported seeing a great walled lunar city. In 1834, New York's *Sun* newspaper ran a series of fictional yet widely believed articles claiming that the renowned astronomer John Herschel (William's son), who was in South Africa surveying the southern sky, had spied inhabitants on the Moon, from spherical amphibians and blue unicorns to yellow-faced, winged people.*

Herschel was mortified by the hoax, but he did believe that life abounds in the solar system. In the 1860s, he suggested "huge, phosphorescent fishes" as an explanation for a report of giant, leaf-shaped objects on the Sun's surface. "The Humanities of the heavens are no longer a myth," commented best-selling French astronomer Camille Flammarion a few years later. "Already the telescope brings us in touch with their countries; already the spectroscope enables us to analyse the air they breathe." At the turn of the century, U.S. businessman Percival Lowell, who was inspired by Flammarion's books to become an astronomer, built an observatory in Arizona (at which

* The stories were written by a young reporter named Richard Adams Locke, who appears to have intended them as a satirical comment on the wild claims being made by Dick, Gruithuisen and others. But readers believed every word. The fantastic finds more than doubled the *Sun*'s circulation and were widely covered in other publications: the *New York Times* described them as "probable and possible"; the *New Yorker* said they created "a new era in astronomy and science."

Pluto was later discovered) and used it to detail what he claimed were networks of artificial waterways on the surface of Mars.

Then the wave of optimism crashed. The alien sightings didn't stand up to scrutiny, and by the mid-twentieth century, scientists using improved methods of spectroscopy and infrared astronomy realized that the Moon was dead and conditions on Mars were far harsher than thought. The possibility of Martian plants (perhaps some kind of lichen) was still widely accepted. But images from probes sent to Mars in the 1960s showed a bleak, barren world, and hopes were finally dashed when NASA's *Viking* landers touched down in 1976. Ahead of the mission, the astronomer and science popularizer Carl Sagan excited the public with the idea of Martians perhaps as big as polar bears. "The possibility of life, even large forms of life, is by no means out of the question," he said. But no creatures wandered past the landers' cameras. Their instruments revealed that the atmosphere was vanishingly thin, and with no protective magnetic field, the planet's surface was battered by deadly levels of radiation from the Sun.

What's more, despite apparent positive results from experiments looking for signs of alien metabolism, other tests showed no detectable organics—molecules with a carbon backbone—in the soil. Since life on Earth is based on organic building blocks, it seemed the anomalous results must have a purely chemical explanation. Though not all of the researchers involved in the experiments agreed, NASA's conclusion was clear: there was no evidence of life on Mars.

Mars was seen as a test case, the most similar place in the solar system that we knew to Earth: if organisms existed anywhere else, it would surely be here. So the negative results severely knocked not just the chances of Martian life, but life in the universe in general. "Since Mars offered by far the most promising habitat for extraterrestrial life in the solar system," wrote Norman Horowitz, who ran one of the Viking experiments, in 1986, "it is now virtually certain that the earth is the only life-bearing planet in our region of the galaxy."

Meanwhile, biologists studying Earth life were realizing that the steps required for life to start seemed staggeringly unlikely and complex. Since the structure of life's heritable material, DNA, was worked out in the 1950s, they had been uncovering the intricate dance between DNA, RNA and proteins required to encode and transmit information from one generation to the next, and the multilayered mechanisms by which these genetic instructions are carried out within a cell. For many, the chances of such a system arising spontaneously from scratch seemed so unimaginably small that it surely couldn't have happened more than once, a view supported by genetic studies showing that all known life on Earth is descended from a single ancestor.

The crucial question regarding life elsewhere now came down not to God but to statistics: would life arise commonly given the opportunity, or is Earth an extraordinary fluke? The balance of evidence seemed to suggest the latter. In the 1980s, Francis Crick, codiscoverer of the structure of DNA, famously described the origin of life as "almost a miracle"*; geologist Euan Nisbet summed up the prevailing view of life as "an extraordinary accident on an extremely special planet"; Nobel-winning biologist Jacques Monod pronounced that "Man at last knows that he is alone." The search for extraterrestrial life might engage science fiction fans and conspiracy theorists, but was no longer a serious topic of scientific inquiry. As cosmologist and best-selling author Paul Davies puts it: "One might as well have professed an interest in looking for fairies."

But then ALH84001 emerged from the stores.

✲ ✲ ✲

David Mittlefehldt never intended to join the hunt for aliens. A geologist working at NASA's Johnson Space Center, he was interested in

* In fact, Crick suggested that life didn't begin on Earth at all, and was intentionally placed here by aliens (a view that was rejected by his peers).

asteroids and what they can tell us about conditions in the early solar system. In 1988, he asked Score's lab to give him some pieces of asteroid to study, and one of the samples they sent him was from ALH84001.

Mittlefehldt probed the composition of the rock samples by firing beams of electrons at them, which causes different elements to give off characteristic patterns of radiation. The results for ALH84001 confused him at first, because they didn't match the other asteroid samples he was studying. It was several years before he finally accepted the reason: this rock had been misidentified. In fact, its composition matched a tiny group of meteorites known as SNCs, named after three stones that fell in Shergotty, India; El Nakhla, Egypt; and Chassigny, France. Other researchers had shown just a few years earlier that the mixture of trapped gases they contain perfectly matches the Martian atmosphere as measured by the *Viking* landers. These rocks came not from asteroids but from Mars.

They must have been blasted from Mars's surface by asteroid or comet impacts. It seemed a vanishingly tiny probability that rocks thrown into the vastness of space from one planet might eventually land on another (and that we might then find them), but here was proof that over billions of years, with billions of chances, near miracles can become reality. Just nine meteorites were known from this select group. In October 1993, Mittlefehldt announced that he had found a tenth.

The news spread fast—Score was thrilled—and researchers around the world rushed to study pieces of her favorite rock. They soon realized that even among SNCs, ALH84001 was special, having formed from volcanic lava over four billion years ago. That made it incredibly ancient, almost as old as the solar system itself: more than three times as old as the next oldest-known Mars meteorite, and significantly predating any rock known on Earth. Studies of molecular signatures in the rock showed that it was ejected into space by an impact on Mars sixteen million years ago, and drifted through the solar system until thirteen thousand years ago, when it was captured by Earth's gravity and landed in Antarctica. There it became imprisoned deep in the ice, until the region's relentless winds exposed it again for Score to find.

And there was something else curious. The rock was full of tiny cracks, from an earlier impact that occurred when it was still in place on early Mars, and the surfaces of these fractures are dotted with tiny, flattened grains. Sometimes described as "moons" or "globs," they are only just visible to the naked eye, but under the microscope they appear as golden circles with black-and-white rims. They're made of carbonate, a type of mineral containing carbon and oxygen. On Earth, rocks containing carbonate (such as limestone or chalk) most commonly form in water—for example, when shells and skeletons from marine creatures accumulate and fossilize. Carbonates aren't usually found in meteorites, so Mittlefehldt asked Chris Romanek, an expert in carbonates working in the same building, to look at them.

Working with NASA geochemist Everett Gibson, who had helped probe the Moon rocks brought back by Apollo astronauts and was now studying Martian meteorites, Romanek blasted the carbonates with lasers to analyze the carbon and oxygen they contained. The results suggested that the carbonates had formed on Mars (rather than being contamination from Antarctica) and were deposited inside the rock after carbon dioxide, dissolved in liquid water, had flowed into the cracks. Today's Mars has an average temperature of -60°C, too cold for liquid water. But ALH84001 pointed to a far more hospitable past.

That wasn't all. The pair also saw microscopic shapes near the carbonate grains: worms and sausages that looked just like Earth bacteria, except much smaller. It was the beginning of a crazy idea. Could water flowing into the rock have carried not just carbonates, but tiny Martian microbes? In September 1994, Gibson and Romanek showed their images to David McKay, a senior NASA scientist who a few decades earlier had taught geology to the Apollo astronauts. They agreed to pursue the idea, in secret, along with one more person: a few days later, McKay approached Kathie Thomas-Keprta, a NASA chemist and specialist in electron microscopy.* "I kind of thought he was crazy," she said later, but she reluctantly agreed to help investigate the rock's strange features. "I thought I would join the group and straighten them out."

It wasn't long before she, too, was hooked.

✡ ✡ ✡

Across the Atlantic, at an observatory nestled among the lavender fields of southeast France, a young astronomer named Didier Queloz was wondering if *he* was crazy. He had spent the summer of 1994 analyzing results from a new instrument, a spectrograph that could measure the movements of distant stars more accurately than ever before. And the results made no sense at all.

Queloz was based at the University of Geneva, studying for his PhD with supervisor Michel Mayor. Together they were building a detector at the Haute-Provence Observatory that could measure Doppler shifts in light from stars. The idea was to measure the speed of a star moving toward or away from Earth by detecting how its spectral lines shifted toward red or blue, just as nineteenth-century astronomer William Huggins had tried to do with his telescope in Tulse Hill. Mayor and Queloz hoped they could do this accurately enough not just to track stars, but to spot distant worlds.

Astronomers couldn't look for planets in other solar systems directly, because any light from a planet would be swamped by the much brighter glare of the host star. But it had been suggested that a planet orbiting a star would exert a gravitational tug, causing the host star to wobble slightly, and that wobble might be detectable. Other

* Prior to 1996, Thomas-Keprta published under the name Kathie Thomas.

astronomers had tried, and seen nothing.* But Mayor and Queloz's new spectrograph was more sensitive than ever before, able to detect stellar variations in speed of just 10 meters per second.

In summer 1994, the spectrograph was finally ready, and Mayor went to Hawaii on sabbatical, leaving Queloz with a list of stars to observe. They were hoping to spot giant planets like Jupiter, which would exert the biggest possible tug. Jupiter takes about twelve years to orbit the Sun, so Queloz settled in for a long wait; to capture one complete orbit for a similar planet would take at least a decade. But almost immediately, he noticed something strange.

One star, called 51 Pegasi, wasn't behaving. Its speed was unstable, cycling in a pattern that repeated every four days. At first Queloz thought it was a bug in the software he had designed. Panicking, and too ashamed to tell Mayor that the instrument wasn't working properly, he spent the autumn testing everything he could think of and repeatedly observing the star. But the wobble refused to disappear.

Eventually, he allowed himself to consider whether the anomaly might be a gravitational pull after all. He calculated that to explain the wobble would take a giant: half the mass of Jupiter (or around 150 times the mass of Earth), yet so close to its star that instead of taking a year to orbit, like Earth, it completes the trip in just four days. Such a star-hugging monster was unlike anything in our solar system, against all accepted theories of planet formation, and it contradicted respected astronomers who had searched for planets and concluded there was nothing out there with a period of less than ten years. But Queloz couldn't see any other explanation. He sent a fax to Mayor in Hawaii: "I think I found a planet."

Mayor returned to Europe in March 1995, but by this time 51 Pegasi had dipped below the horizon. So the pair calculated predictions for

* Two planets were found orbiting a pulsar in 1992, but this discovery isn't generally seen as relevant to the search for solar systems like ours. Pulsar planets are rare, couldn't support life (as we know it, at least) and form in a very different way from those around main sequence stars.

the orbit of the putative planet, ready for when the star reappeared in the night sky in July. Then they traveled with their families to Haut-Provence to compare the behavior of the rogue star against their model. The first night, its speed matched their predictions perfectly. And on the second night, and the third, and the fourth. On the fifth night, recalls Queloz, "we said yes, it's a planet." Mayor later described the realization as "a spiritual moment." They were the first humans in history to know that other worlds really do exist beyond our solar system. Nearly fifty light-years away, in the constellation Pegasus, a giant gaseous planet was circling a star similar to our Sun. They named it 51 Pegasi b, and celebrated with local cake and sparkling wine.

Mayor and Queloz announced the news at a conference in Florence, Italy, in October 1995. The researchers in the audience were excited, but also skeptical. Previous supposed planets had all turned out to be false alarms. Geoffrey Marcy and Paul Butler, rival planet hunters in the United States, heard the news while working at an observatory in California. They spent the next four nights scrutinizing 51 Pegasi, trying—and failing—to debunk the Europeans' result. Then they emailed Mayor: "So your wonderful discovery is confirmed!"

That's when the press went wild. The astronomers had realized that 51 Pegasi b, as the first extrasolar planet, was a milestone for astrophysics; plus, its orbit was so strange that theories of planet formation would have to be rewritten.* But the public reaction took them completely by surprise, with front-page headlines around the world and so many calls and interview requests, says Queloz, that for the next six months it was impossible to work. "What we completely missed," he told me, "was the connection between planet and life."

51 Pegasi b itself is a searing-hot, inhospitable world. Locked close to its sun and glowing at 1,000°C, the chance for life there appears

* In October 2019, Queloz and Mayor were awarded half of the Nobel Prize in Physics for their discovery of 51 Pegasi b (the rest went to cosmologist James Peebles, for theoretical discoveries about the evolution of the universe).

to be approximately zero. But the planet's mere existence implied something far greater: if there was one planet out there, there must be more. "Other worlds are no longer the stuff of dreams and philosophic musings," commented veteran space reporter John Noble Wilford in the *New York Times*. "They are out there, beckoning, with the potential to change forever humanity's perspective on its place in the universe." In thousands of years of speculation about alien life, we'd never had a scrap of actual evidence to hint that we are anything but utterly alone. Now, at last, that had changed.

✡ ✡ ✡

Back in Houston, Thomas-Keprta's plan was at first to save McKay and the others from embarrassment: to prove to them that there was no evidence of Martian life in ALH84001. As she scanned the meteorite's microscopic terrain, with its orange moons embedded in the silver-gray rock, she became intrigued by tiny black grains in the rims of the carbonate globules, just nanometers across. She found that they were magnetic crystals, made of magnetite (iron oxide) and pyrrhotite (iron sulfide), just like the tiny compasses produced by magnetotactic bacteria on Earth. There are nonbiological ways to make these minerals, but this generally takes extreme conditions—high temperature and pH—so it was hard to explain how they had ended up in carbonates deposited at mild temperatures. Unless these crystals were made by bacteria too.

In January 1995, the team received a visit from UCLA's William Schopf, a world expert in fossil microbes, who had identified remains of the oldest-known life within 3.5-billion-year-old rocks on Earth. He was unimpressed by their evidence for alien bugs. You have no case unless you can show organic matter in the structures, he told the team. It seemed a long shot: after all, the Viking mission had found no organics on Mars. But Thomas-Keprta had already sent two flecks of ALH84001 to chemist Richard Zare at Stanford University. He had a powerful instrument called a laser mass spectrometer, which could identify even trace amounts of chemical molecules by vaporizing them. Not wanting to give away what she was working on, she had sent the samples under code names, Mickey and Minnie; Zare and his colleagues, unimpressed by the subterfuge, duly left them on the shelf. But after Schopf's visit, Thomas-Keprta finally persuaded them to take a look.

They found what the *Viking* landers had failed to: organics from Mars. Complex organic molecules called PAHs (polycyclic aromatic hydrocarbons) were concentrated within the carbonate moons. PAHs can form nonbiologically; they're found in everything from car exhaust to interstellar gas clouds. But on Earth, you also find them wherever life has been—for example, in petroleum or coal. The molecules that Zare and his colleagues found in ALH84001 were exactly what you would expect to see when bacterial cells decay.

For the team, the coming together of so many potential fingerprints of life had to be more than a coincidence. In early 1996, with excitement combined with "gut-fluttering dread," they submitted their paper to *Science*. After weeks of scrutiny by the journal's panel of reviewers (including an aging Carl Sagan), it was accepted for publication, setting in motion a sequence of events that reached the country's highest levels of power. In July, the team was called into the office of NASA chief Daniel Goldin, who grilled them for hours, then Goldin in turn was summoned to the White House to brief President Bill Clinton and Vice President Al Gore.

At a press conference on Wednesday, August 7, 1996, President Clinton addressed the world live from the White House Rose Garden. "Today, rock 84001 speaks to us across all those billions of years and millions of miles," he said. "It speaks of the possibility of life." If confirmed, he added, the implications "are as far-reaching and awe-inspiring as can be imagined." Then the TV networks switched to the packed main auditorium at NASA headquarters. McKay, Gibson, Zare, Thomas-Keprta and Schopf (to give a doubter's view) were lined up onstage. In front of them were a small chunk of ALH84001 in a glass case and a large crowd of eager reporters. Goldin addressed the cameras. The team's work had "brought us to a day that may well go down in history," he said. "We're now at the doorstep of the heavens. What a time to be alive."

One by one, the team described their lines of evidence: the carbonates, magnetite crystals and organics. None was conclusive on its own, but taken together, the team argued, they were evidence for primitive life on early Mars. They played an animation showing wormlike microbes on Mars, swimming in water and becoming trapped in the buildup of carbonate deposits inside cracks in the rock, before the rock was sent spinning into space, eventually landing in Antarctica. Finally, to an audible gasp from the audience, they showed their images of the putative "fossils" themselves.

Within days, almost a million people had seen the *Science* paper online. NASA staff counted more than a thousand stories broadcast on ALH84001 in the first week. News crews lined up at the meteorite lab in Houston for a glimpse of the original rock, while the scale of coverage in the world's newspapers and magazines eclipsed even that of the first Moon landing. *USA Today* called it "the headline man has been waiting for since the first human eyes looked into the heavens."

Not all the coverage was positive, though. A prominent Microsoft executive called ALH84001 "the biggest insult to the human species in almost 500 years." Zare's lab had to temporarily take down its website and contact details after religious fundamentalists complained that its research contradicted the Bible. And many scientists, too, were gearing up for a furious response.

✡ ✡ ✡

Over the coming months, other researchers vehemently attacked not just the ALH84001 paper, but its authors. "This is half-baked work that should not have been published," said one meteorite expert.

Another called the team "an inferior group of people"; a third complained of "distressingly biased" interpretations of "turd-like shapes." The critics challenged every line of evidence, arguing that the worm shapes were created in the lab; that the carbonates and organics were simply contamination from Antarctica; or that the carbonates, even if Martian, had formed at temperatures much too high for life.

The team's observations stood up to scrutiny, and there is now consensus that the features they reported did form on Mars, and that the carbonates were deposited in a watery environment at temperatures around 25–30°C. The debate shifted instead to the interpretation of those results. Whereas McKay and the others argued that adding several "maybes" together strengthened their case, others dismissed the carbonates, organics and worm-like shapes as all explainable in other ways. The magnetite crystals, though, were harder to ignore.

In 2000, Thomas-Keprta published a detailed study of the crystals. Though some could have formed inorganically, she reported that just over a quarter of them had a unique combination of properties* only ever seen in biological magnetite (and on Earth, considered trustworthy evidence for life). Meanwhile, another team studying ALH84001 reported seeing some of the crystals in linear chains, just as in terrestrial bacteria, surrounded by a halo that they interpreted as possibly the remains of a cellular membrane. The evidence for Martian life, said Thomas-Keprta, was now "compelling."

In 2003, the tables turned when critics proposed that magnetites like the ones Thomas-Keprta described could have formed from carbonate in the meteorite if the temperature spiked during a sudden "shock" event, for example the impact that blasted the rock into space. Researchers even used similar conditions to produce such crystals in the lab. In 2009, she hit back, pointing out that ALH84001 contains not just iron carbonate but also carbonates of calcium, mag-

* The properties are narrow size-range, chemically pure, free from structural defects and a rare "truncated hexa-octahedron" shape.

nesium and manganese. In a heat shock scenario, Thomas-Keprta argued, these minerals would decompose into oxides as well. The only way to explain the pure iron oxides seen in the meteorite, she insists, is through biology.

The NASA team (among others) has also been studying other Martian meteorites, including Nakhla and Shergotty, and found suggestive evidence there too. In 2014, for example, the researchers reported complex organic matter stuffed into veins and cracks of a Mars rock known as Yamato 000593, as well as tiny tubules that look just like tunnels found in some Earth rocks (ancient and modern) thought to be etched by microbes hunting for nutrients. David McKay passed away in 2013, but Everett Gibson and Kathie Thomas-Keprta are still working on the project: "We continue to support our original hypothesis," they told me in June 2019. But the debate has reached a frustrating stalemate. Though the NASA team is as convinced as ever that life is the most plausible explanation, critics still insist that they have failed to prove their case. What everyone does agree on, though, is that the team's work, and the surge of interest that followed, helped to transform the search for alien life.

In the early 1990s, NASA was struggling to find a purpose. With the collapse of the Soviet Union, the space race was over, and the agency's huge-budget missions were suffering from delays, cost overruns and major failures, including the loss of space shuttle *Challenger* in 1986 and the $1 billion Mars Observer orbiter in 1993. The White House was cutting billions from NASA funding; indeed much of the vitriol shown toward ALH84001 was because planetary scientists feared it would strengthen the argument that taxpayers' money was better spent elsewhere. "We're at the bottom of the pecking order in NASA's budget," meteorite expert Allan Treiman told *Newsweek* in 1997. "People are concerned that if this turns out to be as stupid as cold fusion we'll be out on the street."

In fact, the opposite happened. NASA chief Dan Goldin was already streamlining the agency by slashing bureaucracy, cutting jobs and pushing small, innovative missions (with the motto "faster, better, cheaper"). But that wasn't all. He also envisioned a new—scientific—mission for NASA: to answer big questions about the universe and our place within it. He had to convince the politicians, though, and ALH84001 came at just the right time. When President Clinton addressed the world to announce news of the meteorite, he described the research as a "vindication of America's space program and our continuing support for it, even in these tough financial times." He promised that NASA would "put its full intellectual power and technological prowess behind the search for further evidence of life on Mars."

As journalist Kathy Sawyer puts it in *The Rock from Mars*, her 2006 account of the ALH84001 story: "The allure of the extraterrestrial was back." Clinton halted the decline in NASA's funding, and the agency shifted resources toward planetary exploration. Mars missions were revived with a new generation of spacecraft, interested—for the first time since Viking—in seeking biomarkers and habitats relevant to life. In 1998, NASA founded its Astrobiology Institute, and a whole new field was born that now extends far beyond this one agency. The multipronged research into ALH84001, says Gibson, provided "the guiding idea." Rather than considering "aliens" in isolation, scientists from a broad range of disciplines—understanding how planets form; studying past life on Earth; detecting organics in clouds of interstellar gas—would now work together with one overarching aim: to understand the cosmos as it relates to life.

✡ ✡ ✡

The research that followed has reshaped our view of the cosmos within just a few decades. Take the hunt for planets outside our solar system, or "exoplanets." Within weeks of the announcement of 51 Pegasus b in 1995, researchers started finding more. But the field hit warp speed when NASA launched the Kepler space telescope, which scanned stars for a slight dimming of their light caused by a planet passing in front. It became the most prolific planet hunter ever, taking the count of known exoplanets to more than four thousand before it was decommissioned in 2018 (with thousands more candidates waiting to be confirmed). Kepler showed that planets are not only very common in our galaxy, but breathtakingly diverse, from hot Jupiters like 51 Pegasus b to mini-Neptunes perhaps covered by planetwide oceans. Many are tidally locked "eyeball worlds," with endless night on one side, searing day on the other, and a knife-edge of permanent twilight in between.

Researchers are now using spectroscopy to probe planets' chemical compositions, and here, too, they are finding an array of exotic specimens far beyond anything that exists in our own small corner of the galaxy. Examples include 55 Cancri e, a dense, scorching volcano world that orbits its star every eighteen hours; HAT-P-7b, a gas giant with clouds of vaporized corundum, the mineral that makes rubies and sapphires; and Kepler-7b, as light as polystyrene. Kepler-16b has a double sunset like *Star Wars*' Tatooine, while HD189733b sports a vibrant blue atmosphere and silicon clouds that rain glass.

Central to the effort is the hunt for planets that might harbor life, which (in the absence of a better description) astronomers define as planets similar in size to Earth and at the right distance from their star for liquid water to exist. These planets seem to be abundant too. A 2013 analysis of Kepler data concluded that around a fifth of all stars in the Milky Way are orbited by at least one such world. In 2016, a rocky planet was discovered in the habitable zone of our nearest star, Proxima Centauri; the next year, astronomers found a system with seven Earth-size planets circling the same sun (three in its habitable zone). In 2019, water vapor was detected in the atmosphere of a planet in the constellation Leo. Overall, astronomers estimate there could be around 8.8 billion potentially habitable Earth-size planets in our galaxy alone.

That's a huge shift in perspective, just as those commentators predicted after 51 Pegasus b was revealed back in 1995. A few decades ago, the idea of "other worlds" was pure fiction, as it had been for millennia. Now we're confronted with a diversity of planets beyond anything we could have imagined: a universe with more planets than stars. Even if life is vanishingly unlikely to arise on any particular planet, we know that in our galaxy alone, there are billions of chances for it to occur.

At the same time, biologists have been realizing that life itself is far more flexible and tenacious than they ever thought. The discovery of thriving ecosystems around deep-sea hydrothermal vents came as a shock in 1977. But since the study of "extremophiles" (organisms that favor extreme conditions) took off in the 1990s, it has become clear that on Earth, at least, wherever there is even a hint of liquid water, there is life. "What we previously thought of as insurmountable physical and chemical barriers to life," biologists commented in *Nature* in 2001, "we now see as yet another niche harboring extremophiles."

As recently as 2013, bacteria were found living in frigid, briny lakes beneath Antarctic ice and deep within the superheated rocks of the Earth's crust. Entire ecosystems glean their energy not from sunlight, as once thought universal, but from chemical energy deep inside the planet. Bacteria can thrive in conditions of high acid or salt, extreme gravity, crushing pressures or harsh radiation; there are strains that can feed on uranium, or breathe arsenic. Lichen grows in Martian conditions. Tiny but tough invertebrates called tardigrades have laid eggs in the near vacuum of space. Each new discovery has expanded our notion of life and, in turn, made it easier to imagine that it could have evolved elsewhere.

We're increasingly appreciating other potential habitats in the solar system, too, such as Europa and Enceladus (moons of Jupiter and Saturn, respectively), where astronomers believe vast water oceans lie under miles of ice; or Venus, which could have been habitable billions of years ago before it was scorched by a runaway greenhouse effect. Meanwhile, the fact that Earth and Mars have been exchanging asteroids like ALH84001 for billions of years has led several high-profile scientists, such as cosmologist Paul Davies and biochemist Steven Benner, to suggest that life that emerged on one planet may have hitched a lift to the other. More controversial figures, like astronomer Chandra Wickramasinghe, see the entire galaxy as a single, connected biosphere, an idea known as "panspermia," with each planet "selecting its genetic inheritance from a vast cosmic reservoir of genes." With the detection of organic precursors of life on comets and asteroids, and the discovery that some bacteria and fungal spores can survive deep space (particularly if shielded inside rocks), perhaps this doesn't seem as crazy as it once did.

Ideas about Mars itself, past and present, have also rebounded. In 1998, measurements made by NASA's orbiter *Mars Global Surveyor* showed that four billion years ago, Mars had a planetwide magnetic field that would have protected it from the damaging "wind" of charged particles from the Sun. There's now strong evidence from Martian rocks that this allowed the planet to retain a thicker atmosphere, rich in carbon dioxide, which kept the climate relatively warm, with liquid water in briny lakes, rivers and seas. In 2018, NASA's Curiosity rover sampled three-billion-year-old sedimentary rocks from the bottom of an ancient lake and found plenty of organic building blocks necessary for life. It all fits eerily well with the story told by McKay and the others back in 1996, of magnetotactic bacteria that once lived in salty, carbonated water.

Today the magnetic field is gone, leaving Mars beaten by harsh radiation and with an atmosphere so thin that liquid water soon boils away. Yet data from Mars missions has consistently painted a more life-friendly picture than expected. Scientists have seen plenty of frozen water, in polar ice caps and underground, as well as dark streaks on equatorial hillsides during summer that look like flowing water. In 2018, the European Space Agency's Mars Express orbiter used radar to show that just like in Antarctica, there's a large lake of liquid water, 12 miles across, deep below the ice at Mars's south pole.

Methane has also been repeatedly detected in the planet's atmosphere. Although this gas can be produced geologically, most terrestrial methane is made by organisms, including bacteria living in the Earth's crust, so some scientists believe it might signal the presence of similar microbes below Mars's surface. Upcoming missions from both NASA and the European Space Agency will drill into the ground to look, but the results are already casting new light on the 1976 Viking data. That mission's positive results were rejected because no organic molecules were detected in the soil, but it's now clear these are present after all. Viking biologists Gilbert Levin and Patricia Ann Straat insist Martian biology remains the most plausible explanation for their results.

Overall, the difficulties inherent in interpreting the ALH84001 and Viking results have forced scientists to think more broadly about how to know aliens when they see them. Just as planets can be staggeringly diverse, maybe living creatures can too. The life we know is carbon-based and reliant on liquid water as a solvent, but different chemistries have been proposed: perhaps life in the methane/ethane lakes of Saturn's moon Titan could use these hydrocarbons as a solvent instead of water. Perhaps different energy sources are possible: instead of sunlight or chemical energy as on Earth, maybe thermal energy, or kinetic energy. With so much potential variety, should we look for the ability to evolve, or to metabolize, or to encode information? Or perhaps any definition we come up with will be too limiting, and we should just look for something that doesn't fit.

One speculative possibility is that beings on a planet orbiting a neutron star might harvest energy from fluctuating magnetic fields, basing their genetic code on chains of magnets instead of chemical DNA. Others have suggested that a "shadow biosphere" of alien life, undetected by our conventional tests, might exist on Earth, "like the realm of fairies and elves just beyond the hedgerow." Whether or not we find such creatures, the very search is changing us, by stretching our ideas about what life itself is.

✡ ✡ ✡

An even bigger question than whether there is other life in the universe, of course, is whether there is other intelligence, other awareness. When we gaze at the sky, is anyone else out there looking back?

A quarter of a century before Didier Queloz was so flummoxed by the strange signals he saw in the sky, another young astronomer testing out a new instrument had her own brush with the unexpected. At Cambridge University in England, a PhD student named Jocelyn Bell (now Bell Burnell) was hoping to find more examples of recently discovered and extremely bright sources of radio waves called quasars. Bell's supervisor, Tony Hewish, had designed a large radio telescope. It covered the area of fifty-seven tennis courts and took the team two years to build, hammering over a thousand wooden posts into the ground and connecting them with many miles of wire.

The telescope was switched on in July 1967, with beams that swept a circle through the sky every 24 hours as the Earth turned. Any radio signals were recorded by pens that traced lines on rolls of paper: a hundred feet of it every day. Bell analyzed all of the data by hand, and after a few weeks she noticed the occasional appearance of a strange, flickering pattern on the charts. It was a mystery, as it didn't resemble either terrestrial interference or any known astronomical source. She called it "scruff."

Eventually Bell and Hewish decided to inspect the scruff with a faster recording, to capture the pattern in more detail. The signal came and went but finally, in November, she caught it, and was startled by what she saw. "As the chart flowed under the pen," she recalled later, "I could see the signal was a series of pulses . . . 1⅓ seconds apart." She had never seen anything like it. Such regular timing was surely artificial. But the signal only appeared when the beam pointed at the same small patch of distant sky.

The astronomers considered and eliminated a range of explanations: radar signals reflected off the Moon; satellites in strange orbits;

anomalous effects from a nearby metal shed. Eventually, after the signal was confirmed using another telescope (ruling out a fault in the equipment) and shown to originate far beyond our solar system, Bell was forced to consider an uncomfortable possibility: "Were these pulsations man-made, but made by man from another civilization?" The group jokingly dubbed her mysterious source the LGM star, short for "Little Green Men."

A few years earlier, physicists had proposed in a *Nature* paper that intelligent extraterrestrials wanting to communicate through the depths of space might send a radio signal, perhaps tuned to 1,420 MHz, the spectral emission line of hydrogen. Although that might seem like science fiction, they said, "The presence of interstellar signals is entirely consistent with all we now know, and . . . if signals are present the means of detecting them is now to hand." The U.S. astronomer Frank Drake soon undertook the first search: at the National Radio Astronomy Observatory in West Virginia in 1960, he spent four months pointing an 80-foot telescope toward two nearby stars, but heard nothing. Now it seemed that Bell might have stumbled upon just such a message.

As Christmas approached, the signal appeared increasingly alien-like. Though its timing never varied, the pulses varied in strength and sometimes stopped altogether. The signal only occurred in a narrow frequency band, more like a radar pulse than any known natural emission. And the astronomers calculated that its source was far too small to be a conventional star.

At this point, Hewish and the observatory head, Martin Ryle, started taking the alien possibility seriously. "As the days went by excitement rose," Hewish wrote in 1968. "Were the pulses some kind of message from an alien civilization?" They discussed how, if no other explanation were found, they might announce the news. Should they go to the Royal Society, or to the government? Ryle half-jokingly suggested that they burn the records and forget the whole thing. If news got out, people would surely want to send a reply, and there was plenty of evidence from the history of European exploration, he pointed out, that "the less contact you have with higher civilisations the better."

For Bell, though, the unexplained signal was mostly an annoyance. "Here was I trying to get a PhD out of a new technique, and some silly lot of little green men had to choose my aerial and my frequency to communicate with us." Then on December 21, she looked again at the charts and thought she saw more scruff, but this time in a different part of the sky. Late that night, she went out to the observatory for a more detailed look as soon as that part of the sky next passed into the telescope's view. At first, she couldn't get the equipment to work in the cold. "By flicking switches, swearing at it, breathing on it, I got it to work properly for five minutes," she recalled. She was just in time to catch the signal, and confirmed that it was another series of pulses, this time 1.2 seconds apart.

Bell went home for Christmas much happier, now satisfied that the scruff had a natural explanation: "It was very unlikely that two lots of little green men would both choose the same, improbable, frequency, and the same time, to try signaling to the same planet Earth." When she returned in January, she soon found two more similar sources. Meanwhile, Hewish had checked the original signal for a Doppler shift (if it were sent by aliens, they were presumably on a planet orbiting a star, so the speed of the source relative to Earth should vary) but found nothing. And another colleague suggested that straightforward fluctuations in the solar wind, which the radio waves passed through on their way to Earth, would explain why they varied in intensity.

With aliens discounted, Bell and her colleagues finally wrote up their paper, which was published in February 1968. They called the blinking sources pulsars. Astronomers now explain them as a highly magnetized type of neutron star (the collapsed core of a giant star, just 12 miles across) that spins extremely fast and emits a powerful flare of electromagnetic radiation that sweeps around like a lighthouse beam, appearing to anyone in its path as an intermittent pulse. Hewish and Ryle—but not their junior, female colleague Bell—shared a Nobel Prize for the discovery in 1974.

Even though little green men turned out not to be involved, the idea that we might one day hear a call through the depths of space has remained as tantalizing as ever. Listening efforts have mostly been privately funded and relatively small-scale (a NASA project started in 1992 was ridiculed by politicians and canceled a year later), and the skies have essentially remained silent. Perhaps the best candidate occurred in August 1977, when a volunteer astronomer at the Big Ear radio telescope in Ohio witnessed a surge of radio waves at 1,420 MHz in the constellation Sagittarius, so striking that he circled it on the printout and wrote "Wow!" in the margin. The signal was never detected again and never explained.*

In 2015, Russian billionaire Yuri Milner helped to push the Search for Extraterrestrial Intelligence (SETI) toward the mainstream when he donated $100 million to create a project called Breakthrough Listen, based at the University of California, Berkeley. That's enough money to develop new search technologies such as machine learning and buy thousands of hours on the world's top telescopes; enough to earn the support of some of astronomy's top names†; enough to target a million nearby stars. In August 2017, the project detected a set of repeated and unexplained radio bursts from a dwarf galaxy three billion light-years away, though astronomers say they most likely have a natural source.

---

* Astronomers have been sending signals, too. The Arecibo telescope in Puerto Rico transmitted one of the first messages intended for aliens, devised by Frank Drake and Carl Sagan, to a nearby star cluster in 1974. Martin Ryle, by this time the U.K.'s Astronomer Royal, was horrified such contact would be attempted without a proper public debate. He wrote to Drake, complaining that it was "very hazardous to reveal our existence and location to the Galaxy; for all we know, any creatures out there might be malevolent—or hungry." Arecibo also broadcast a reply to the Wow! signal, in 2012. Sponsored by National Geographic, this signal included around ten thousand Twitter messages, labeled with the hashtag #ChasingUFOs. In late 2017, a nonprofit group called METI International began its own efforts to message extraterrestrial intelligence by sending a music-based math primer to a nearby exoplanet.

† Including U.K. Astronomer Royal Martin Rees and (before his death in 2018) cosmologist Stephen Hawking.

Despite the unpromising results so far, the increasing popularity of the search has made alien intelligence a hot topic, with everyone from philosophers to computer scientists discussing what form it might take, and how we might communicate. "What was once an exotic small-scale scientific enterprise," says U.S. space policy analyst Michael Michaud, "has become a vast, multidisciplinary thought experiment about the nature and behavior of intelligence, both on and beyond the Earth."

For linguists, these questions have revived and extended an old debate about the nature of language itself: the extent to which it is based on innate, universal principles, versus being shaped by our physical nature and environment. Meanwhile, animal behavior experts say we should consider not just our own thought and language patterns but other types of intelligence found on Earth, from the collective memory of honeybees to the curiosity and problem-solving abilities of octopuses. Theologians ask how we might recognize alien species or cultures as deserving of moral status, even if they are very different from us: what is it that makes a life-form worthy of dignity and respect?

Or aliens might not even be biological at all. U.S. philosopher and cognitive scientist Susan Schneider suggests most sophisticated civilizations out there will be supercomputers—forms of artificial intelligence created by biological life-forms that have either died off or merged with their technology. She predicts these beings will likely be based on silicon, which allows much faster information processing than biological brains. Others question, though, whether knowledge and technological advancement should necessarily be seen as the defining force of cultural progress. NASA biologist Mark Lupisella has asked, for example, whether truly advanced aliens might not see intelligence as a tool rather than an end in itself. Can we envisage a civilization that pursues subjective values—such as fairness, empathy, diversity—instead of objective facts?

In other words, speculating about alien societies is really driving a discussion about humanity: what's important to us; who we are; what our future holds. Just as looking for life elsewhere makes us question how life relates to the cosmos and even the essence of what life is, the search for extraterrestrial intelligence has been forced to grow into something more profound. Again, we become part of something bigger. Human nature, rather than being the universal, inevitable state of intelligent beings, becomes one point within a vast ocean of possibility. I think that changes how we see humanity. Even if an alien signal never comes, the very concept gives us an "other" against which to locate ourselves, just as the Polynesian navigators visualized unseen islands to find their way against the stars.

✵ ✵ ✵

Despite all the discoveries since 1996, we still don't have a single proven example of life elsewhere in the universe. For the first time in human history we can use scientific methods to investigate empirically what's out there in the universe, but we can't yet answer the big question—Is there other life?—on Mars or anywhere else. What we have done very successfully, though, is break down the search into a series of smaller questions: Are other planets common? Is Mars habitable? Are organic molecules widespread? Can organisms survive in extreme conditions, even in space? And it's hard to ignore the fact that, so far, at least, the answers keep coming back "yes."

Wherever we look, the evidence seems to support the case that life in the cosmos is not the exception, but the rule. Over the past few centuries, science has demoted our existence from God's special creation to a chance aberration in an otherwise empty universe. Score's Antarctic stone and Queloz's whirling planet help to point toward a different possibility: life as a common occurrence. It's a return, if you like, to the concept of a living cosmos. Once again we see life in the sky.

And not just that. The idea of a bio-friendly universe forces us to reconsider the prospects for conscious, intelligent life. As astronomy historian Steven Dick told Vice President Gore at a White House summit meeting on ALH84001 back in 1996: "We are trying to determine whether the ultimate outcome of cosmic evolution is merely planets, stars and galaxies, or life, mind and intelligence."

Science is based on the idea of studying a purely physical, material reality. Subjective experience is stripped out so we can seek what's really out there rather than in our imaginations. That has led inexorably to a worldview in which the physical universe is all that exists. Yet scientific results are now nudging toward the possibility of not just other life in the cosmos, but other minds. Scientists are starting to consider the idea—which we'll explore further in the final chapter—that consciousness, *experience*, is not a one-off by-product of chemical evolution but a fundamental feature of the universe. We're glimpsing the possibility of a cosmos that's not just alive, but awake.

12

MIND

WHEN NASA ASTRONAUT CHRIS HADFIELD CLIMBED OUT OF the International Space Station for his first spacewalk in April 2001, it was the culmination of decades of training and preparation. A hard-headed, disciplined pilot, Hadfield had studied math, physics, engineering and robotics. He had flown over seventy different types of aircraft. He had spent fifty full days practicing spacewalks in the pool. "I was completely technically prepared for what was going to happen," he says. And yet, in a sense, he wasn't prepared at all.

When he first floated free in the vacuum of space, holding on to the spaceship with one hand, all thoughts of his mission—to prepare a 56-foot-long robotic arm for installation—temporarily left his head. Instead, he was "attacked by raw beauty." To his right was the velvet, bottomless bucket of the universe, stretching on forever and brimming with stars. And to his left, the whole world—an exploding kaleidoscope of color—poured by. "I found it stupefying," he said later. "It stops your thought."

Alone in his spacesuit, looking down on "six billion people and all of the history, all of the beauty and poetry and everything that is human," Hadfield says he learned something in that moment that a lifetime of books, lectures and calculations had been unable to teach: "the power of the presence of the world, as told to me by my ability to see it."

This book has followed how, in the Western world at least, we have inexorably removed personal experience from our understanding of the universe. People once inhabited an enchanted (if sometimes terrifying) cosmos of myths and gods and spirits, in which meaning shaped reality and celestial events flowed through human lives and beliefs. Over the centuries, we have used a sieve of mathematical laws and equations to strain ourselves out of the stars. There have always been those who fought against this trend, but overall the picture—the great achievement of science—is clear. We understand the universe as a separate external reality that exists regardless of how we perceive it: a physical realm that came into being billions of years before our ancestors first gazed at the sky and will continue for billions more after the last living creature is gone.

Our knowledge of this universe now comes from measurements made by detectors and processed by computers, rather than from any personal view of the sky. The approach has been breathtakingly successful. We can probe inside dust clouds; photograph impossibly distant galaxies; explore the afterglow of the Big Bang; detect minute ripples in the fabric of space-time. We can write an evidence-based account of how the universe exploded into being and make predictions about how it may end. It's a privileged view of space, that we are the first people ever to glimpse. By superseding our own eyes, we have achieved scientific insights and discoveries far beyond anything our mere senses could ever reach.

From a practical point of view too there's no longer any need to actually look to the sky. Ancient societies relied on the wheeling heavens to guide the intimate workings of their lives, but satellite navigation systems now locate us at the click of a button, and digital clocks tell the time far more accurately than the Sun ever could. Meanwhile, if we do happen to glance up, light pollution shrouds the view so effectively that we've forgotten what once was. Even the Milky Way, until a few decades ago a great river across the night sky, so bright that people saw stories in its shadows, is no longer visible at all to most people in

Europe or the United States. From a data-gathering perspective, this erasure is inconvenient but not catastrophic—astronomers can build telescopes on remote mountaintops away from artificial lights, or send them into space. But is the data all that matters? Was there nothing more than numbers in our view of the stars?

It turns out I'm not the only one asking that question. In the last decade or two, growing numbers of philosophers, psychologists, neuroscientists and even physicists have been trying to put conscious experience back into our understanding of the universe we find ourselves in—without sacrificing the hard-won methods and insights of science. Their work has implications for how we see ourselves, and how we relate to the world and the wider cosmos. For some, it's even calling into question the basic ingredients of reality itself.

✡ ✡ ✡

Writers through history give remarkably similar accounts of how they feel when faced with the night sky, regardless of their background or religious beliefs. In the first century AD, the astronomer Ptolemy said that searching out the wheeling stars made him forget his mortality: "My feet no longer touch the earth, but side by side with Zeus himself, I take my fill of ambrosia, the food of the gods." Nearly two thousand years later, the Swiss philosopher and poet Henri-Frédéric Amiel lay on a nineteenth-century beach, his back against the sand, looking up to the night sky and through the Milky Way. The sight induced a reverie that was "grand and spacious, immortal, cosmogonic . . . when one reaches to the stars, when one owns the infinite!"

Today, this is a view many of us rarely witness in full force, but when we do, it's just as mind-blowing. A few summers ago, on a feature assignment, I found myself in a tiny one-man tent, sheltering from a violent thunderstorm in the remote mountains of Mexico. When the rain finally stopped, I squeezed out into the night. I felt anxious and alone until I looked up, and was hit by a rush of adrenaline. Above me was a radiant, shimmering sea; an ocean of light that stretched not just from horizon to horizon but deep into forever. For a brief moment, I was lifted up, connected, home.

Thinking back, it isn't the individual constellations I remember, or the planets, or even the glittering ribbon of the Milky Way. It is simply the sheer awesome power of the sky. In London, where I live, the night sky is dull and dark with a neon orange glow, its emptiness broken by only a few struggling pinpricks of light. But here the veil was lifted, as if returning to me something that I hadn't even known was lost. On this moonless night, it seemed there was no blackness at all. There was only silver. Only stars.

Until a few years ago, scientists ignored this more human side of stargazing, preferring to leave such musings for artists and poets to explore. Now, though, they are fast realizing that far from being merely aesthetic, direct contact with the cosmos can have profound, practical effects for our mental health, and how we choose to live. One of the pioneers is Dacher Keltner, a psychologist at the University of California, Berkeley. After years spent studying negative emotions such as anger and fear, he wanted to investigate a positive aspect of human experience, something that could trigger powerful and long-lasting changes. He chose "awe." In 2003, Keltner copublished scientists' first working definition of this emotion, describing it as the feeling we get when confronted with something vast that transcends our normal frame of reference and that we struggle to understand.

It's an emotion that combines amazement with an edge of fear, in which the force we confront is so huge it dwarfs us, or even threatens to consume us altogether. Whereas wonder, much loved by scientists, is more cognitive, often involving attempts to solve a puzzle, awe seems to block rational thought. It's the moment when we are forced to give in to mystery; when we acknowledge how much there is beyond us that we do not understand. According to Keltner, the vastness can be physical or conceptual, and there are many potential sources, such as a powerful leader, an act of supreme sacrifice or virtuosity, or an epic natural landscape: a forest, desert, ocean or gorge. Researchers around the world have since been eliciting awe in volunteers using everything from dinosaur skeletons to tall trees. But there's nothing bigger than the cosmos. One of the most reliable and commonly used methods to inspire awe is to show people photos or videos of the starry sky.

The results have been surprising. It turns out that even mild awe, as triggered in lab experiments, can significantly change our mood and behavior. First, looking at awe-inspiring images seems to break habitual patterns of thinking, making people more creative, and more interested in the world. Psychologists in Arizona found that after experiencing awe, volunteers' memories of a short story improved because they were less biased by prior assumptions, so they could focus more accurately on what was actually happening.

In another study, people who felt awe came up with more original examples in tests, found greater interest in abstract paintings and persisted longer on difficult puzzles. There are also lasting effects on health and quality of life. Keltner's team has found that after experiencing awe, people feel happier and less stressed, even weeks later. This extends to physical effects: one recent study found that awe cuts levels of cytokines, which promote damaging inflammation, and activates the parasympathetic nervous system, which calms the fight-or-flight response.

What's even more interesting, though, is that awe seems to make us nicer people. After feeling awe, volunteers become less worried about personal concerns and goals. Instead, in study after study, they describe themselves as feeling more connected to other people and the world. They make more ethical decisions, are more generous and are more likely to make sacrifices to help others. They care less about money but more about the environment. They feel as if they have more time.

Researchers think that by expanding our attention to encompass the bigger picture, awe literally shrinks our sense of self. In a 2017 study, Keltner found that after feeling awe, people signed their names smaller and drew themselves smaller (but with no drop in their sense of status or self-esteem). In 2019, neuroscientists in the Netherlands reported that watching awe-inspiring videos quiets activity in the brain's "default mode network," which includes parts of the frontal lobes and cerebral cortex, and is thought to relate to our sense of self. "Awe produces a vanishing self," Keltner told me. "The voice in your head, self-interest, self-confidence, disappears." As a consequence we feel more connected to a greater whole: society, Earth, even the universe.

Researchers debate how and why awe evolved. Keltner sees powerful rulers as a primordial source; others have suggested that this potent combination of social bonding and creative thinking was first triggered by natural forces such as thunderstorms. I think we shouldn't underestimate the influence of the night sky: the spectacular ocean of light that has illuminated every society since the first glimmerings of human thought.

✡ ✡ ✡

"Beautiful, so beautiful!" exclaimed the Soviet cosmonaut Yuri Gagarin, a few minutes after he was blasted into orbit in April 1961, becoming the first human being in space. He was talking not about the stars or the cosmos, but about our own planet. He later wrote and signed a message to the rest of us: "People of the world, let us safeguard and enhance this beauty—not destroy it!"

Even more potent than looking up at the night sky, it seems, is looking back down. Throughout human history, mystics and shamans—from cave dwellers to meditators to artists—have traveled through the cosmos in their minds. But in the latest blink of human progress, a few hundred people have finally made the trip into space for real. Spacefaring nations wanted to demonstrate technological prowess and gain scientific understanding. Yet the most powerful message that astronauts have brought back isn't factual, or scientific. It can't be crunched by a computer; sometimes they struggle even to express it in words. The knowledge these travelers most want to share is utterly dependent on their conscious experience of space; it is not measured or calculated but sensed, and felt.

Above all, they tell of the beauty, vibrancy and fragility of Earth, and the powerful recognition that it is precious and needs to be protected. In 1971, Apollo astronaut James Irwin glanced home from the surface of the Moon: "That beautiful, warm, living object looked so fragile, so delicate, that if you touched it with a finger it would crumble and fall apart." NASA astronaut Ron Garan, who orbited Earth in 2008, also saw our planet as "like a living, breathing organism," with only a paper-thin atmosphere to shield everything on it "from death . . . from the harshness of space."

They often return determined to protect the environment. Russian cosmonaut Yuri Artyushkin felt "a strong sense of compassion and concern for the state of our planet and the effect humans are having on it . . . You are standing guard over the whole of our Earth." After millionaire games designer Richard Garriott became the world's sixth space tourist in 2008, he sold his SUVs, installed solar panels and began investing in green energy and electric cars.

Others talk about the triviality of national boundaries and political conflicts, while astronauts of all nationalities emphasize that we are inhabitants of the same planet. As Apollo 14's Edgar Mitchell put it: "From out there on the moon, international politics looks so petty. You want to grab a politician by the scruff of the neck, drag him a quarter of a million miles out, and say, 'Look at that, you son of a bitch.'" Looking at Earth from space, "you don't see the barriers of color and religion and politics that divide this world," said Apollo 17 astronaut Gene Cernan. What struck him instead was "humanity, love, feeling, and thought."

This "Overview Effect" was named in the 1980s, but has recently gained growing attention. In 2008, a group of astronauts, scientists and space experts created the Overview Institute to share astronauts' insights more widely and "move humanity closer to world peace." The 2012 documentary film *Overview* and the 2018 National Geographic channel series One Strange Rock brought the phenomenon to mass audiences, while advances in virtual reality and space tourism mean that many more people could soon experience the effect directly. And in 2016, psychologists analyzed the effect, describing it as a powerful example of awe. To see Earth from outside, they suggest, is the ultimate shift in perspective, forcing astronauts' focus away from themselves and toward what matters for the planet as a whole.

Far from being terrifying, the effect typically involves feelings of euphoria and unity. Skylab 4 science pilot Ed Gibson experienced "inner peace" in orbit, while astronaut Jeff Hoffman called it "a state of grace." Astronauts often describe feeling at one with humanity, the Earth or even the entire cosmos. Mae Jemison, who spent a week on the space shuttle *Endeavour* in 1992, felt connected "with the rest of the universe." Hadfield was struck by a "sense that there is something so much bigger than you, so much more deep than you are, ancient, [with] a natural importance that dwarfs your own."

Irwin "felt the power of God," while Edgar Mitchell, on his way home from the Moon, felt that the entire universe was "in some way conscious." "I was overwhelmed with the sensation of physically and mentally extending out into the cosmos," he said. "There was a startling recognition that the nature of the universe was not as I had been taught." It's as if his sense of self shrank so far that the boundary between him and the rest of space simply dissolved.

✡ ✡ ✡

You don't have to blast into orbit to feel at one with the universe, of course, or to lose your sense of self. This effect is often associated with religious experiences, induced by rituals such as prayer or meditation. At the turn of the twentieth century, the psychologist and philosopher William James famously collected examples from a variety of faiths, from the higher awareness of enlightened Buddhists to the meditation of Christians such as Saint John of the Cross, who talked about entering "an immense and boundless desert, an abyss of wisdom," and Saint Teresa, who described her soul becoming united with God.

The experiences that James cites aren't all overtly religious. Poet Alfred Tennyson used to repeat his own name to himself to induce a waking trance in which "individuality itself seemed to dissolve and fade away into boundless being." But all share the experience of merging with a greater awareness. The German writer and idealist Malwida von Meysenbug was kneeling on the seashore when she experienced a "return from the solitude of individuation into the consciousness of unity with all that is . . . Earth, heaven, and sea resounded as in one vast world-encircling harmony." Canadian psychiatrist Richard Maurice Bucke was in a hansom cab after an evening discussing philosophy with friends when he became "wrapped in a flame-colored cloud." Directly afterward, "I did not merely come to believe, but I *saw* that the universe is not composed of dead matter, but is on the contrary, a living Presence."

Bucke named such experiences "cosmic consciousness," and said they confer on a person "a sense of immortality . . . not a conviction that he shall have this, but the consciousness that he has it already." James called them "mystical," and concluded that although people interpret the details differently depending on their previous beliefs, they are all describing essentially the same thing: the sense of unity with a conscious cosmos, characterized by vastness, timelessness, safety and rest (and what many of his sources described as "love"). A few decades later, the British writer Aldous Huxley suggested that this common core of insight forms "an immemorial and universal substrate that underlies all religious and spiritual paths."

The irony is that astronauts, blasted into space with the power of science and technology, seem to be returning with the same core message about the universe that spiritual explorers have relayed for millennia. So can these experiences tell us anything useful about the universe, or ourselves?

They all appear to involve a state of consciousness that psychologists now call transcendence, in which the self doesn't just shrink but melts completely away. Such extreme events are rare and have been hard to study directly; all James could do was collect personal accounts. But scientists have recently been granted a powerful new approach for researching new dimensions of consciousness: psychedelic drugs. These drugs triggered an initial burst of excitement when they were discovered in the 1950s—LSD was created in a Swiss lab; mushrooms containing the hallucinogen psilocybin were brought back from shamanic cultures in Mexico—and found to trigger transcendent states by binding to serotonin receptors in the brain.

Tens of thousands of studies on psychedelics were conducted in the 1960s and 1970s, using these compounds to treat alcoholism, drug addiction and depression. But the studies weren't always rigorous, and some overly enthusiastic researchers handed out the drugs like sweets. Fearful of the safety risks, not to mention the antiestablishment sentiment that the drugs seemed to fuel, authorities in the United States and Europe banned them, shutting down the research. Just in the last decade or so, though, a few scientists have once again been granted permission to investigate the effects of psychedelics, in strictly controlled conditions. The first study, led by pharmacologist Roland Griffiths at Johns Hopkins University, was published in 2006. Participants described experiences such as merging with an ultimate, conscious reality, which some interpreted as God. One talked about being "in the void," where time and space didn't exist, only love. Afterward, they reported a stronger belief that in some sense, our existence continues after death.

Three-quarters of the volunteers experienced what Griffiths and his colleagues rated as a "complete" mystical experience, suggesting, the scientists said, that rather than being a freak distortion of consciousness, such states are "biologically normal." Around two-thirds of the participants rated the trip as among the most meaningful experiences of their lives. Since then, evidence for the benefits of psilocybin—taken in highly controlled settings—has been growing. People who take psilocybin rather than placebo feel happier and more altruistic afterward, and report higher well-being and life satisfaction more than a year later.

As in the 1960s, there's a focus on using drugs like psilocybin to treat psychiatric disorders such as depression and addiction, as well as anxiety in cancer patients facing a terminal diagnosis. A 2016 trial of cancer patients led by Griffiths reported that after taking psilocybin, participants had reduced depression and anxiety as well as increased well-being, sense of meaning in life, optimism and acceptance of death, benefits that were sustained six months later. One of them wrote in his journal that he had visited a place of "pure joy," which taught him that "earthly matters" such as food, music, architecture—even cancer—were trivial, and that our existence continues without end.

"I now have an understanding," he wrote, "of an awareness that goes beyond intellect . . . that my life, that every life, and all that is the universe, equals one thing . . . love." Another patient, who took part in a study in California, had been so anxious about dying that she couldn't enjoy the time she had left. But as soon as the psilocybin started working, "I knew I had nothing to be afraid of," she said. "It connected me with the universe."

Meanwhile, neuroscientists at Imperial College London led by Robin Carhart-Harris have been scanning the brains of volunteers high on psilocybin and LSD to investigate what these drugs do to the brain. They found that psychedelics reduce activity in the default mode network, just as awe does, an effect that correlates with a sense of boundlessness, and loss of self. "My feeling is that it's the same thing," Carhart-Harris told me. "Psychedelics are hijacking a natural system and fast-tracking people to these experiences of awe."

The results of these studies, then, help to explain what's happening in the brain not just when people take psychedelic drugs, but when they gaze at the stars, connect with nature, meditate, or travel into space. As well as reduced sense of self, the boundaries between normally segregated bits of the brain break down, boosting creativity and flexible thinking. This may be why such states can trigger long-lasting changes in attitudes and personality—something that's unusual for adults, in whom these traits are generally seen as fixed. In fact, Carhart-Harris and others suggest that such states may help to reverse rigid patterns of thinking that we develop throughout our lives. Young children are highly flexible and adaptable, with a fluid sense of self. But as we mature into adulthood, our identity hardens, and our thoughts and behaviors congeal into well-worn paths.

Prescribed patterns allow us to be more efficient, so we don't have to work everything out from scratch. But they reduce the capacity for new ideas and can be harmful, as in depression, when people become trapped in ruts of negative thought. Carhart-Harris suggests that awe, working through the neurotransmitter serotonin, can loosen those chains. If we're confronted by a challenge that overwhelms our existing ways of thinking, it allows our brains to adapt.

The traditional scientific view, which underpins our modern society, is that our rational, waking consciousness gives the most accurate and useful view of reality, whether that's our immediate surroundings or the cosmos as a whole. In areas of life from business and politics to medicine and education, we tend to trust and therefore prioritize rational thought. Meanwhile, we dismiss awe and wonder as childish, and discount transcendent states as a meaningless if not downright suspect distortion, a messy artifact of how our brains are wired. But this picture is turning out to be flawed.

The research described so far in this chapter suggests that instead of thinking about reality versus hallucination, we might see different conscious states as more like a dial, in which our perception is highly filtered at one end, versus a broadband stream at the other. Pushing too far in either direction leads to mental illness and an inability to function in the world. But both represent reality, and the benefits being shown in studies of awe and psychedelic states alike suggest that for a healthy life—and society—we need to balance the two.

Having narrow filters in place gives us a strong sense of self. It allows us to focus on details, think logically and engage efficiently with the physical world. But this perception is necessarily limited, isolating and shaped by preexisting biases and beliefs. The expanded awareness of awe and transcendence, however, injects flexibility, creativity and connection. It allows us to grasp a bigger reality, to look beyond narrow daily concerns and make decisions that not only make us happier as individuals, but also sustain the planet and work for the benefit of humanity as a whole.

Keltner, for one, worries about the disappearance of awe in society. A focus on smartphones and screens rather than the broader horizons of nature means we're rarely forced to confront the fear of the vast unknown. "We believe that awe deprivation has had a hand in a broad societal shift that has been widely observed over the past 50 years," he and a colleague wrote in the *New York Times* in 2015. "People have become more individualistic, more self-focused, more materialistic and less connected to others."

There are different ways to balance the dial—apart from drug treatment, experts suggest everything from engaging with nature and the arts to mindfulness meditation and immersing ourselves in virtual reality. Above all, I think we should fight to preserve the biggest, most mind-blowing experience out there, central to human existence for millennia but now fading fast: the starry sky. Contemplating the stars doesn't necessarily teach us any new facts. It is of even greater value, because it allows us to think differently about what we know.

But that's not quite the end of this path. One key characteristic of extreme transcendent states, shared by many stargazers, mystics, astronauts and drug trippers alike, is the insistence that the great consciousness they experience is utterly real; that their own awareness is just one infinitesimal drop within a vast ocean that underlies all of existence. For those who experience it, this knowledge is often unassailable, a conviction so powerful that it can change how people live their lives and erase their fear of death. Is this, too, a glimpse of how the universe truly is?

William James quoted the nineteenth-century poet John Symonds, who said he often asked himself "with anguish, on waking from that formless state of denuded, keenly sentient being. Which is the unreality?" James didn't experience mystical states himself: "My own constitution shuts me out from their enjoyment almost entirely." But an experience with nitrous oxide intoxication left him with a lifelong conviction that our normal waking consciousness "is but one special type of consciousness whilst all about it, parted from it by the filmiest of screens, there lie potential forms of consciousness entirely different." He insisted that "[n]o account of the universe in its totality can be final which leaves these other forms of consciousness quite disregarded."

During the later twentieth century, any such hint of "cosmic consciousness" became firmly associated with pseudoscience; to suggest it in respected scientific circles would destroy a career. The accepted answer to Symonds's question is that the feeling of merging with a greater awareness is an illusion: part of the noise that results when our brain's perspective is widened too far. Most scientists would agree with journalist Michael Pollan, who popularized the research on ecstatic states in his 2018 book *How to Change Your Mind*, and concluded after all his experiences: "I still tend to think that consciousness *must* be confined to brains."

In the last few years, though, as researchers have begun to take transcendent states more seriously, opinion is starting to shift on this fundamental question too. Rebel philosophers and physicists are exploring once again the idea that mind might hold together the very threads of reality.

✡ ✡ ✡

This new argument over how our minds relate to the cosmos—how the mental relates to the physical—is just the latest battle in what has been a fierce and long-running philosophical war. As we've seen, René Descartes split soul from matter in the seventeenth century, framing the universe as a machine that runs according to physical rules. This model was strengthened by Galileo, who insisted that the book of nature is written in mathematics, and Newton, with his hugely successful physical laws. But not everyone agreed that the universe's objective and subjective sides could be separated so easily.

Newton's contemporaries Gottfried Leibniz and Baruch de Spinoza, for example, saw the physical and mental worlds as deriving from the same substance; both advocated versions of panpsychism, which sees mind as a fundamental attribute of physical matter. The eighteenth century saw the rise of different forms of idealism, which goes further, arguing that the physical world in fact derives from mind. The Enlightenment philosopher Immanuel Kant, reacting against the idea that we can completely understand the physical world through reason, argued that the reality we perceive—even down to the structure of space and time—is inevitably a function of how our minds work, and can't tell us anything definite about what lies beyond. Irish bishop and philosopher George Berkeley rejected that physical objects can even exist independently of our minds, arguing that "[t]o be is to be perceived."

By the nineteenth and early twentieth centuries, as James explored consciousness and modern artists fought to escape logic and realism, it was seen as a fairly obvious point that physics, which studies objective, physical properties, leaves something of nature untouched. James was influenced by the French philosopher Henri Bergson, who emphasized the limits of math and logic. We hit problems, said Bergson, when we start to believe that the abstract rules and laws of physics are somehow more accurate, more real, than the experiences they were derived from in the first place. We turn life inside out. We conclude that our perception, our existence, is a limited version of the underlying mathematical truth. As Bergson put it: "We give a mechanical explanation of a fact and then substitute the explanation for the fact itself." When in fact, he argued, it's the equations and graphs that fail to capture the full, vibrant will and complexity of the actual universe in which we live.

Even philosopher Bertrand Russell—champion of logic and science—pointed out in the 1920s that physics can only reveal the behavior of matter, not its intrinsic nature (including whether this involves consciousness). "Physics is mathematical not because we know so much about the physical world, but because we know so little," he said, "it is only its mathematical properties that we can discover." Arthur Eddington, whose 1919 eclipse observations confirmed Einstein's theory of relativity, built on Russell's argument in 1928. In fact, he pointed out, there's one case where we do know matter from the inside: our own brains, which are, of course, aware. Isn't the simplest assumption that the rest of matter is of a similar nature? It seems "rather silly," he argued, to insist that physical matter must be inherently nonexperiential, and then wonder where experience comes from.

At the same time, quantum mechanics was itself casting doubt on what really goes on at the bottom level of existence. As physicists continue to drill down, to study matter at the very smallest scales, objective reality seems to slip through their fingers. Instead of finding particles with definite properties, existing at definite locations, their experiments and hard-won equations yield only probabilities. All possible realities seem to remain in play until they make a measurement, at which point that cloud of possibility abruptly collapses into the single reality that is observed. Hence the famous Copenhagen Interpretation: that it makes no sense to speak about an objective reality other than what we observe. This led to the idea that rather than physicists recording what's already out there—the dogma since Newton—the very act of looking somehow calls the outcome into being.

From the start, some physicists were horrified by the notion that our conscious minds might determine when and where particles appear. The Moon doesn't exist only when we look at it, Einstein famously insisted. His colleague Max Planck saw the attack on objective reality in science as a "moment of crisis" that threatened civilization itself. But other quantum pioneers embraced a role for the mind, hoping that their work might help to unify science with mysticism. Einstein's conviction that reality exists independently of the mind was "philosophical prejudice," said Wolfgang Pauli. Erwin Schrödinger argued that "the material universe and consciousness are made out of the same stuff," and suggested that science needed "a bit of blood transfusion from Eastern thought."

After World War II, this heated debate faded. The role of the conscious observer was never resolved, but through the 1950s, physicists came up with other interpretations for their strange results—perhaps particles are guided by a hidden pilot wave, unreachable by measurement; or reality splits into multiple parallel universes (an idea known as "Many Worlds") every time an observation is made. These approaches, even if they couldn't be proven in experiments, allowed the concept of objective reality to be preserved. And quantum theory itself turned out to be exquisitely accurate in terms of predicting the behavior of the physical world. Scientists dialed down the philosophical arguments and got on with studying that world.

Throughout the second half of the twentieth century, this approach to understanding nature went from strength to strength. Physicists and astronomers armed with general relativity and quantum theory improved their understanding of the universe, building a detailed picture of its history right back to an instant after the Big Bang. Meanwhile, biologists gained unprecedented power to explain life's mysteries. The 1953 discovery of the structure of DNA, when combined with the theory of natural selection, was a great leap forward in understanding how traits evolve and are inherited.

Even seemingly subjective human attributes—our emotions, perceptions, morals—could be objectively explained as behavioral dispositions, selected for their survival value. And different conscious states were increasingly shown to correlate with physical states and mechanisms in the brain. Our awareness "can be bisected with a knife, altered by chemicals, started or stopped by electricity, and extinguished by a sharp blow or by insufficient oxygen," pointed out Steven Pinker in his 1997 book *How the Mind Works*. So much, he said, for the supposedly immaterial soul.

Science, it seemed, proved that consciousness is not a fundamental or necessary ingredient in the cosmos. Instead it's a side effect or byproduct of evolution, entirely caused by and dependent on the physical activity of our neurons. The smell of coffee or the prick of a needle, the all-consuming force of a mother's love or our transcendent awe at the glittering stars: all of these experiences are decorative. They have no causal role, because everything that happens in the universe can ultimately be reduced to particles and forces, and is decided by physical laws. We might feel like we're actively thinking up ideas or making choices, but our awareness is simply the output of neurochemical pathways in the brain.

There were grumbling questions. If life is a random accident, why does the universe seem so perfectly fine-tuned for our existence, with physical attributes from the speed of light to the properties of the carbon atom set at just the right level for living, thinking creatures to emerge? Then there's the simplicity, predictability and even beauty found deep within the equations that describe the structure of the cosmos, where we might have expected messy chaos. In response, physicists appealed to the concept of the multiverse—one version of which involves an array of infinite parallel universes, continually bubbling out of each other and all with different physical laws. Just to be able to ask such questions, we'd have to be in a universe that can support advanced forms of life, even if there are countless others that can't. No purpose or design needed, just blind chance.

Another mystery was why consciousness should have evolved at all, if zombies with the same neuronal activity but no inner experience would function just as well as humans. Or how the rich, qualitative nature of awareness could emerge simply by rearranging dead, dumb atoms. This long-standing question was highlighted in 1994 by philosopher David Chalmers when he called on his peers to address the "hard problem": how can pain, curiosity, the color red, ever be fully described by the equations of physics?* Some responded by simply eliminating the question. A champion of this approach is philosopher Daniel Dennett, who argues that the idea that anything beyond physics exists—what something *feels* like; a special, extra subjectivity that distinguishes us from zombies—is "an illusion." Beyond the objective properties that scientists study—the physical activity of neurons, and the measurable behavior that results—there is nothing about consciousness to explain.

The claim, ultimately, is that science's ability to make sense of the world is so powerful, it can explain everything, making not just religion but philosophy obsolete. With it, we transcend the human perspective, the human cosmos, and can see things as they really are. Don't be misled by our passions and poetry: humans are "robot vehicles blindly programmed to preserve the selfish molecules known as genes" (Richard Dawkins); "nothing but a pack of neurons" (Francis Crick); "just a chemical scum on a moderate-sized planet" (Stephen Hawking). It's a view that's not great for our pride, perhaps, but has been undeniably successful in terms of predicting and manipulating the physical world. "The reductionist worldview *is* chilling and impersonal," admitted cosmologist Steven Weinberg in his 1992 book *Dreams of a Final Theory*. "It has to be accepted as it is, not because we like it, but because that is the way the world works."

The latest generation of celebrity physicists has adopted a more

* He was building on a classic 1974 essay by Thomas Nagel, "What Is It Like to Be a Bat?," which argued that however sophisticated science gets, it can never tell us about the inner experience of other beings who are sufficiently different from us.

conciliatory tone. Maybe we are just accidental passengers in an otherwise barren, pointless universe, but we can still value our own brief, unique window of intelligence and self-awareness. Particle physicist Brian Cox suggests we celebrate ourselves as "the mechanism by which meaning entered the universe." In his 2017 book *The Big Picture*, cosmologist Sean Carroll advocates "poetic naturalism," emphasizing that we're "thinking, feeling people" with many ways of talking about the world. String theorist Brian Greene followed in 2020 with *Until the End of Time*, devoting several chapters to how religion, literature and art contribute to the "nobility of being."

At its root, though, their view of humanity is as hard-line as ever. Carroll insists there is "no special mental realm of existence." Whatever stories we might tell ourselves in daily life, he says, our feelings are simply "sets of words" we use that map onto physical states of neurons in our brains. There is nothing else. As we make and celebrate our own meaning, we must accept that our inner lives—our sensations, desires, values, feelings, choices, beliefs—have no existence or significance in the physical world.

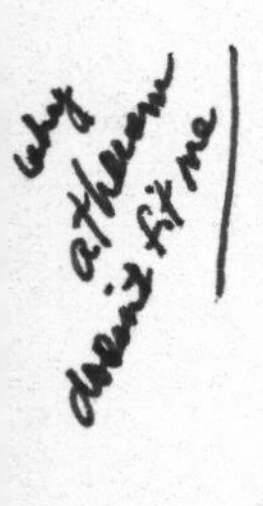

Greene, too, anticipates that consciousness will turn out to be fully explained by conventional physics. "The math *does* rule," he writes. "We are physical beings made of large collections of particles governed by nature's laws . . . We feel that we are the ultimate authors of our choices, decisions and actions, but the reductionist story makes clear that we are not. Neither our thoughts nor our behaviors can break free of the grip of physical law."

For today's scientific mainstream, this is the final step in our understanding of reality, the ultimate destination of the journey traced in this book. There is no mental realm that physical measurements can't reach, and even if science hasn't yet filled in all the details, its approach and methods can ultimately tell us everything we need to know. Sure, we can each find our own meaning in life. But as far as the universe is concerned, "you" (as in your mind, your experience, your self) are either a side effect of these blind interactions or don't really exist at all.

This worldview has been pushed hard as the only rational alternative to belief in supernatural gods or souls. But as we fly further into the twenty-first century, the terrain is starting to shift. Respected figures are increasingly arguing that even without God, science is missing something big. In 2012, the atheist philosopher Thomas Nagel complained in his book *Mind and Cosmos* that the conventional mix of materialism and Darwinism "is incapable of providing an adequate account . . . of our universe." He was widely criticized for it—Steven Pinker tweeted that the book exposed "the shoddy reasoning of a once-great thinker," while Dennett said it was "not worth a damn"—but Nagel's not alone. High-profile scientists, too, such as physicist Paul Davies and biologist Stuart Kauffman, reject the idea of a supernatural God but have questioned whether cosmic puzzles such as fine-tuning and consciousness can really be dismissed as random accidents.

Davies, Kauffman and others have suggested that the laws of physics as we know them might not explain everything. Maybe there's some extra principle, yet to be discovered, that has been nudging the universe toward complexity, so that life and consciousness could arise. And a small but growing minority of philosophers is responding in a different way, by reviving an idea that even a few years ago would have been laughed out of town: panpsychism. Maybe, they argue, we have our fundamental understanding of reality upside down. Perhaps consciousness—rather than being an illusion, or a late, accidental addition to the universe—is everywhere after all.

One of the pioneers of this movement is the British philosopher Galen Strawson. He says he never expected to end up as a panpsychist; he describes himself as a staunch materialist, who believes that everything in the universe is physical. But he also insists that consciousness is real: the fact that we are conscious, he says, is "literally the most certain thing we know." And he rejects the idea that consciousness emerged from nothing. If atoms could leap into such a radically new nature just by rearranging themselves, he argues, it would be a miracle every time it happened, a yawning explanatory chasm on a scale found nowhere else in science. "That's what forces me to think that there must be experience or consciousness at the bottom of things." That it has been here the whole time.

Science has vastly overreached itself by insisting that matter can't be aware, he says. Russell and Eddington's argument is watertight; physics simply cannot tell us about the intrinsic nature of the particles it studies. And if consciousness is simply an extra aspect of regular, physical matter (albeit one that we can't measure with scientific instruments), there's no jump to explain. Just as evolution shaped rudimentary matter into our complex physical bodies, suggests Strawson, "evolution also found consciousness and shaped that too" into our senses and minds.

The vision of a conscious cosmos didn't come easily. When the implications of the idea really hit him, "it blew my mind for about a fortnight," he says. And when he published a paper advocating panpsychism in 2006, "I was mocked." But since then, the idea of consciousness existing beyond the mammalian brain has become increasingly acceptable. Biologists are considering the possibility that consciousness might extend further down the animal kingdom at least, to octopuses or bees. Some are arguing, controversially, that even plants and slime molds are not just intelligent, but aware.

Neuroscientists are contemplating what consciousness might look like in computers and aliens, while a popular new approach called integrated information theory (IIT) suggests that awareness arises in any physical system that processes information in a particular way. And in philosophy, there's what Chalmers describes as "a new generation of philosophers who think we need to revise our view of the physical world to accommodate consciousness." Some are drawn to IIT, but following Strawson's work, there has been a resurgence of more radical versions of panpsychism. Young academics who previously kept panpsychist views to themselves are now building careers on it, including conferences, magazine articles and popular books, such as *Galileo's Error*, published in 2019 by British philosopher Philip Goff.

Contrary to popular belief, modern panpsychists don't suggest that chairs, or rocks, or spoons are conscious. Some panpsychists propose that fundamental particles such as quarks or electrons have some form of simple awareness, but these don't sum up to any larger consciousness, except in special cases such as brains. More relevant for the research on transcendent states described earlier in this chapter, others talk instead about one vast field of awareness, a view known as "cosmopsychism."

Most physicists today describe the universe as a set of interwoven fields, with fundamental particles as energetic oscillations within those fields. There's an electromagnetic field, a gravitational field, as well as fields for the various subatomic particles such as quarks, neutrinos and electrons. We see objects, from photons to planets, as having their own identity, but they are also part of a greater whole. What if a fundamental property of these fields is consciousness? Freya Mathews, a cosmopsychist philosopher of nature based in Australia, uses the metaphor of a great ocean coursed by currents and waves. In this view, our minds are like intricate whirlpools or vortices within that ocean: a continuous part of the overall field, yet also with a discrete, inner experience that's inaccessible from outside.

Whereas Strawson says he was led to panpsychism by rational arguments, others have been influenced by their experiences. Mathews, for example, has described a visit to Hamilton Downs, an old cattle station in rural Australia, during which she felt "as if one were riding something alive out there, a dragon, a great serpent, a current of energy." Another is Israeli philosopher of consciousness Itay Shani. Early in his career, Shani became dissatisfied with how conventional theories ignore first-person experiences. Then a trip to the rocky coves of Nova Scotia, Canada, transformed his perspective.

Staying with friends among forests and ocean waters, he began to feel unusually relaxed and connected to his surroundings. After a few days, he visited a site called Peggy's Cove, where the granite has been carved and scoured by ancient glaciers, and was overwhelmed by its beauty. "The rocks were shining with light," he says. Time faded, and he felt as if he had left his body and was crossing the universe toward distant stars. Then lights were pouring through him, he says, with a deep sense of peace, elation and deep unity, as if beneath all of reality there is "one beating heart."

For Shani, the idea of a conscious cosmos not only makes better sense from a purely rational perspective, it also explains the spiritual experiences that people have reported for millennia. Our normal, waking consciousness filters out all but a narrow spectrum of experience; perhaps transcendent states let us in on another aspect of reality, allowing contact with a field of awareness beyond. This is an idea found deep in many Eastern philosophies, such as in the timeless, unchanging being or "mother" of Taoism. More recently, it has been popularized in the West by figures from alternative medicine advocate Deepak Chopra to parapsychology researcher Rupert Sheldrake, often mixed up with unsupported claims such as telepathy and faith healing, and generally ridiculed by scientists. But Shani argues that we shouldn't dismiss the central idea of a larger field of awareness. We know there's a built-in limit to what science can tell us, he told me. "It's what James and his generation were emphasizing again and again . . . We have to find a way to take experience seriously. And that includes claims about transpersonal experience as well."

Like the psychologists studying awe and transcendence, many of the philosophers advocating some form of cosmic awareness believe that the approach has practical implications. "There's not a day since then that I don't think about it," Shani says of his experience at Peggy's Cove. "It becomes a bearing for whatever you do." He thinks panpsychism helps us to understand the universal human yearning for connection, for "something more." "Here's a way of trying to explain that yes, you are an incomplete part of something," he says. The research shows that transcendent experiences can benefit well-being by reducing our focus on ourselves. Living a meaningful life, Shani suggests, might involve "being better attuned to the centre of things which is beyond."

Mathews, meanwhile, sees panpsychism as vital for our continued existence on this planet. Western materialism, she argues, frames the world as an inert backdrop to human presence, and nature as a commodity, a source of raw material for us to manipulate and dominate. And now the biosphere is unraveling. To heal the damage, she wants us to recognize the Earth as an active subject, a presence in its own right, that deserves care and respect, and (through the insights we gain from transcendent experiences) might even impart wisdom of its own.

✡ ✡ ✡

Panpsychism isn't the only suggestion today for how to incorporate consciousness into the universe. Other ideas are coming from within science itself, including quantum physics. Although the field shifted toward realism after World War II, one giant of twentieth-century physics held on to the Copenhagen interpretation. John Wheeler was a younger contemporary of Einstein and Bohr; he later popularized the term "black hole" and coined others such as "wormhole" and "quantum foam." For Wheeler, the idea that the universe started with the Big Bang billions of years ago and later gave rise to life, and humanity, is all wrong. He suggested instead that the cosmos is "constantly emerging from a haze of possibility." And we're not just passive observers—we help to create it.

Wheeler rejected the idea of the universe as a machine running according to predetermined rules. There are no physical laws or quantities sitting magically out there regardless of us, with this or that numerical value. He used the motto "it from bit," which "symbolizes the idea that every item of the physical world has at bottom—at a very deep bottom, in most instances—an immaterial source as explanation; that what we call reality arises in the last analysis from the posing of yes-no questions and the registering of equipment-evoked responses." He was emphasizing that everything we think of as physical reality—every particle, field, even space-time itself—ultimately derives from *information*, from our knowledge of the results of the measurements that physicists make.

In the 1970s, Wheeler suggested a thought experiment to help demonstrate his point. It's a variant of the famous double-slit experiment, which reveals some of the weirdness of the quantum world. The original involves shooting photons through a screen with two slits. If you check which path each photon takes, it passes through one slit or the other. But if you don't check, it seems to go through both at once. Wheeler argued that you could delay the decision whether to check until the end of the experiment—after the photon has already passed through the screen—and get the same result. In other words, the observation would determine the past path taken by the photon, even though it wasn't observed until *after* its journey had already taken place.

It wasn't possible to check this when Wheeler made the prediction, but it was confirmed in a lab in Maryland in 1984. The photons' paths through the experiment weren't fixed until physicists made their measurements, strengthening the case that there is no hidden reality already out there; even past events are determined by the observer. This result has since been repeated in a series of variants, including with entire atoms, and, in 2017, with photons bounced off a satellite in space. Astronomers even have a plan (based on an original suggestion by Wheeler) to try it with light from distant galaxies. If that works, our choices and experiments today would determine a journey that supposedly took place billions of years in the past.

Wheeler's hunch was that the universe is like a giant feedback loop; that our observations enable the ongoing creation of not just the present and future but the past as well. He famously compared a photon observed by physicists to a "smoky dragon," with a clearly discernible head and tail (at the points of measurement), but huge clouds of uncertainty in between. Another metaphor he used was of the cosmos as "notes struck out on a piano by the observer-participants of all places and all times." It's a vision of not one big bang but billions upon billions of tiny flashes of creation, of reality as a work in progress that comes into focus, piece by piece, as we look.

Wheeler's idea of physical reality as information has led to a popular school of quantum mechanics interpretations based on studying the behavior and properties of this quantum information. But Christopher Fuchs, who was one of Wheeler's students and is now a quantum physicist at the University of Massachusetts in Boston, says he found this new direction frustrating, particularly when proponents introduced concepts such as the "unknown quantum state" that could be teleported, protected or revealed. He argues that treating information as a new kind of substance that exists in the external world, independent of our knowledge of it, was missing Wheeler's point: "If a quantum state just was information, how could it be unknown? Surely it had to be known by someone or something?"

Instead, Fuchs took inspiration from the Bayesian approach to probability (named for the eighteenth-century British mathematician Thomas Bayes). This argues that there are no objective probabilities out there to be discovered; when we determine the probability that a certain event will happen, it is always a function of our subjective knowledge and beliefs. Fuchs and his colleagues developed an analogous interpretation of quantum mechanics, which says that the probabilities calculated by quantum physicists relate not to what's happening out in the world, but the internal knowledge of the observer. The collapse from many possibilities to one upon measurement isn't an external event, but the observer updating his or her personal beliefs. They originally called their approach quantum Bayesianism, but it's now just QBism (pronounced "cubism": its founders say it's just as revolutionary).

QBists are often accused of saying that external reality doesn't exist, but Fuchs insists the approach instead implies that "reality is *more* than any third-person perspective can capture." One of Fuchs's intellectual heroes (alongside Wheeler and Niels Bohr) is William

James, who after all of his investigations of consciousness concluded that the ultimate building block of reality is "pure experience," and that "new being comes in local spots and patches." The universe is neither objective nor subjective but both, Fuchs told *Quanta* magazine in 2015: "the stuff of the world is in the character of what each of us encounters every living moment—stuff that is neither inside nor outside, but prior to the notion of a cut between the two at all."

The point is not that any of these ideas about consciousness are definitely correct, or provable, or close to becoming generally accepted. The mainstream position, as articulated by physicists such as Sean Carroll and Brian Greene, is that conventional science has proven its explanatory power over and over again, so there's no need for a wild move like introducing consciousness into the fundamental fabric of the universe. Many quantum physicists prefer the infinite, but at least objective, realities of the Many Worlds theory; most philosophers of mind lean to the view that beyond the physical activity of neurons, there is nothing "extra" about consciousness to explain. But, nonetheless, a door is opening.

We're now witnessing an explosion of ideas, with respected scientists and philosophers rejecting the idea of a purely physical universe in which awareness has no role to play. Instead, from a wide range of different perspectives, conscious experience is being revived as a keystone of our relationship with reality. It is becoming increasingly possible, even within science, to ask whether something more than the measurable, material world—as defined by Descartes four centuries ago—must exist.

For this growing minority, the efforts to keep consciousness out are simply examples of how far people will go to hang on to an old paradigm: accepting increasingly tortuous pictures of reality—zombies, infinite universes—just as astronomers once introduced ever more elaborate systems of planetary epicycles rather than contemplate the simple (and to us, blindingly obvious) notion of the Earth going around the Sun.

✡✡✡

When Chris Hadfield stepped out of the International Space Station to float free in space, he said that what surprised and amazed him most was "the power of the presence of the world, *as told to me by my ability to see it* [my emphasis]." It's a message not simply about our planet, out there in external reality, but about the central importance of experience, literally of our ability to see the world. As Hadfield put it a few years later: "We are not machines exploring the universe, we are people." That's just as true on Earth, in our lives generally, as it is in space.

It has taken us millennia to remove our experience of the cosmos from our understanding of it, and to build the mathematical grid of physics. The resulting framework of knowledge is strong, elegant and vitally important. But it is limited. No matter how sophisticated it gets, it can only ever be an abstraction, a simplified picture drawn from measurable data alone. The laws of physics might seem impregnable, but they represent a filter that, by definition, captures only one aspect of our lived experience.

This was Henri Bergson's crucial point. The existence we perceive—that bursts through our awareness every moment—is vibrant, contradictory, illogical, unbounded. In helping us to make sense of this glorious mess, science has surpassed all expectations in the specific task for which it was designed: to extract and predict the measurable behavior of the physical world. It has given us power over our surroundings; driven our technological advancement; taken us to the Moon. But that doesn't mean its objective, third-person models encompass all that reality is.

We began this book with an expanding cosmic bubble; an epic, evolving realm of galaxies, nebulae and black holes that has existed for the past fourteen billion years. It's an inspiring and powerful story, told to us by the stars themselves. But what is our universe, really? I think the people of the Paleolithic had a valid view, too, when they saw themselves in the Earth and sky. I think the Greek astronomer Ptolemy was onto something when he talked about floating side by side with Zeus. I think Kandinsky had a point when he said that art *is* reality and Malevich when he warned that reason is (or at least can be) a cage. I think we can learn from Wheeler's view of the universe not as preexisting "out there," but as billions upon billions of creative flashes in which we all play a vital role.

And I think we should listen to the astronauts when they return from space and tell us not of physical measurements and observations but of beauty, poetry and the desperate need to care for each other and our world. They're urging us to remember the power of what we all intuitively know when we stop arguing and look up at the stars: that we are in the cosmos. That we are here.

EPILOGUE

IN HIS 1941 SHORT STORY "NIGHTFALL," THE SCIENCE FICtion author Isaac Asimov imagined a planet called Lagash. This planet is much like Earth, except that it is constantly illuminated by six suns. Its inhabitants, living in permanent daytime, have no inkling of the existence of the stars. When a rare eclipse plunges the planet into darkness for the first time in two thousand years, the spectacular sight of the heavens—and with it, the realization of the vastness of the universe, and their own insignificance—drives the Lagashians insane. They start desperate fires to block out the night, and their cities burn to the ground.

I think Asimov was right about the world-changing power of the night sky. But perhaps he got it backward. In "Nightfall," the sudden appearance of the stars triggers the collapse of civilization. On Earth, this window onto the cosmos is precisely what has inspired humanity. At the very birth of our species, it was the circling heavens that allowed people to first glean order from chaos, to derive a model of existence as an ever-repeating cycle in which light alternates with darkness, life alternates with death.

We've since built on those foundations. Observing the Sun, Moon and stars informed practical abilities such as navigation, timekeeping and ultimately the scientific methods that have led to today's sophisticated technology. Meanwhile, the patterns people saw in the sky—magical beasts, divine suns, or universal forces—fed spiritual beliefs and political structures, and ideas about the nature and meaning of reality. What happens, then, when we lose that view, when we experience Lagash's eclipse in reverse?

For the first time in history, we now live in physical isolation from the wider universe. Scientists' insights come not from their own eyes but from photons hitting electronic detectors, creating data for computers to crunch. In everyday life, too, developments such as electric lights, intensive agriculture and global travel shield us from the physical cycles of the cosmos—as well as from nature more generally. We're oblivious to the messages carved out by the Moon and the planets. Our clocks and GPS tell us when and where we are.

It's becoming clear that this separation isn't good for us. Biologists are realizing the extent to which living organisms are dependent on natural cycles of light, temperature and magnetic fields to function, communicate and survive. Meanwhile, psychologists studying awe and transcendent states are warning that a healthy society needs access to a bigger picture. To foster creativity, connection and compassion takes more than analyzing data; we need to experience the universe's mystery and vastness for ourselves. It's a message echoed by astronauts, who return from their cosmic travels urging us to treat each other better, and to recognize our planet as the fragile haven that it is.

But this isn't just a practical split, the straightforward result of technology that has reduced our reliance on the stars. It is entwined with a much deeper philosophical separation. Science has been so successful in predicting the behavior of the physical world that we've started to see our mechanical explanations of the universe as more real than the experience we started with in the first place. What we personally feel or see when we contemplate the cosmos or look at the stars has become irrelevant not just to practicalities of everyday life but to our very understanding of what reality is, and where we fit within it. We've jumped from saying that science is a useful tool for investigating one particular aspect of the world to saying that the scientific view is all there is.

That matters, because the direction science takes—the questions we ask, the answers we listen to, the way in which we use the results—

depends on our underlying beliefs and assumptions. We've seen throughout this book how models of the cosmos have always seeped into every aspect of people's lives, and that is just as true now. If we deny the importance of experience, if we say that only what's countable exists, if we view our planet as an inert, raw material and ourselves as isolated machines, then those underlying principles become self-fulfilling. They guide the kind of knowledge, and society, that we build.

Perhaps it's no surprise, then, that we've ended up with schools that teach equations but not awe; and medical systems that for all their expertise, kill millions each year through overuse of physical drugs and treatments, while marginalizing psychological approaches and denying conditions that don't show up on scans. That we count money instead of valuing happiness; judge our opinions and experiences by the number of likes they collect; and increasingly hand over the running of our lives and societies to blind AI. That we obsessively build the stuff around us into clever new technologies, while driving the biosphere that sustains us to destruction. That we're so focused on screens, we barely notice when streetlights erase our view of the stars.

I think we're at a crucial turning point in our understanding of what the cosmos is. For the ancients, perception was reality. Over thousands of years, that perspective has slowly been reversed. The seeds of a new way of thinking were sown with the numerical models of the Babylonians and the celestial spheres of the Greeks. It later flowered with Descartes, Galileo and the scientists who followed, as they nurtured an increasingly powerful idea of the universe as a mathematical realm constructed of physical particles and forces. They created a system so persuasive that over the last century, it has finally spilled its bounds and all but killed off the very minds that gave rise to it. We're in danger of erasing ourselves from the universe altogether.

This is the cosmology that drives our modern world. Those who disagree with it have been dismissed as irrational, mystical or anti-science. But there's now a growing recognition, even among scientists, that physics alone cannot reveal the intrinsic nature of a photon, let alone the universe, or ourselves. And that means the role of mind in the cosmos is something we can talk about again; something we can choose. There's another way forward: not to reject science, or invoke the supernatural, but to consider conscious experience as a natural, fundamental part of the reality that science describes.

From both practical and philosophical perspectives, then, our personal connection with the cosmos is not a marginal, worthless bit of candy, worth discarding for technological convenience, but part of the essence of what makes us human. Looking back over the history of our relationship with the cosmos shows how we've banished gods, debunked myths and written our own evidence-based creation story. Stripping out subjective meaning and focusing on quantifiable observations has given us an epic power to understand and shape the world that dwarfs anything that has gone before. But unchecked, it has the potential to be a cold, narcissistic, destructive force.

This is a book about how we closed our eyes to the stars. The challenge now is to open them again.

NOTES

PROLOGUE

xi **thousands being visible:** Bob King, "9,096 Stars in the Sky—Is That All?" *Sky & Telescope*, September 17, 2014, https://www.skyandtelescope.com/astronomy-resources/how-many-stars-night-sky-09172014/.

xi **a few dozen:** Peter Christoforou, "How Many Naked Eye Stars Can Be Seen in the Night Sky?" *Astronomy Trek*, March 12, 2017, http://www.astronomytrek.com/how-many-naked-eye-stars-can-be-seen-in-the-night-sky/.

xi **can no longer see:** Fabio Falchi et al., "The New World Atlas of Artificial Night Sky Brightness," *Science Advances* 2 (2016): e1600377.

xi **"chemical scum":** David Dugan, dir., *Reality on the Rocks*, 1995; London: Windfall Films, Ltd.; produced for Channel 4 television in three episodes; quoted in Raymond Tallis, "You, Chemical Scum, You," *Philosophy Now*, March/April 2012, https://philosophynow.org/issues/89/You_Chemical_Scum_You.

CHAPTER 1: MYTH

1 **curious pattern:** Michael Rappenglück, "A Palaeolithic Planetarium Underground—the Cave of Lascaux (part 1)," *Migration & Diffusion* 5 (2004): 93–111.

2 **one of the earliest constellations:** Arkadiusz Sołtysiak, "The Bull of Heaven in Mesopotamian Sources," *Culture and Cosmos* 5 (2001): 3–21.

2 **On September 12, 1940:** Eyewitness accounts of the cave's discovery: Brigitte Delluc and Giles Delluc, "Lascaux, les dix premières années sous la plume des témoins," in *Lascaux Inconnu*, ed. Arlette Leroi-Gourhan and Jacques Allain, *XIIe Supplément à Gallia Préhistoire* (Paris: Éditions du Centre National de la Recherche Scientifique, 1979), 21–34.

4 **known elsewhere, too:** Jo Marchant, "A Journey to the Oldest Cave Paintings in the World," *Smithsonian Magazine*, January 2016, 80–95, https://www.smithsonianmag.com/history/journey-oldest-cave-paintings-world-180957685/.

4 **plant-based brushes:** Norbert Aujoulat, *Lascaux: Movement, Space and Time* (New York: Abrams, 2005); Jean-Michel Geneste, *Lascaux* (Paris: Gallimard, 2012).

4 **a rich parade:** For a useful summary, see David Lewis-Williams, *The Mind in the Cave: Consciousness and the Origins of Art* (London: Thames & Hudson, 2002).

4 **male versus female:** Robert Kelly and David Thomas, *Archaeology: Down to Earth*, 5th ed. (Belmont, CA: Wadsworth, 2013), 200.

4 **Norbert Aujoulat:** Life of Norbert Aujoulat: Judith Thurman, "First Impressions," *New Yorker*, June 23, 2008, https://www.newyorker.com/magazine/2008/06/23/first-impressions; Jacques Jaubert and Jean Clottes, "Norbert Aujoulat (1946–2011)," *Bulletin de la Société Préhistorique Française* 108 (2011): 781–91; Jean-Philippe Rigaud and Jean-Jacques Cleyet-Merle,

"Norbert Aujoulat (1946–2011)," *PALEO: Revue d'archaéologie préhistorique* 22 (2011): 9–13.

5 **the first time he saw:** Aujoulat, *Lascaux*, 9–10.

7 **"The Great Bear":** Marcel Baudouin, "La grande ourse et le phallus du ciel," *Bulletin de la Société Préhistorique de France* 18 (1921): 301–8.

7 **U.S. archaeologist Alexander Marshack:** Alexander Marshack, *The Roots of Civilization: The Cognitive Beginnings of Man's First Art, Symbol and Notation* (London: Weidenfeld & Nicolson, 1972).

8 **"I was fascinated"** [and following quotes]: Author's telephone interview with Michael Rappenglück, January 12, 2018.

8 **to test how well:** Michael Rappenglück, "The Pleiades in the 'Salle des Taureaux,' Grotte de Lascaux. Does a Rock Picture in the Cave of Lascaux Show the Open Star Cluster of the Pleiades at the Magdalenian Era (ca. 15,300 BC)?" *Astronomy & Culture* (January 1997): 217–25.

9 **farmers in the Andes:** Benjamin Orlove et al., "Ethnoclimatology in the Andes," *American Scientist* 90 (2002): 428–35.

9 **Native American peoples:** Rappenglück, "The Pleiades," 221–222.

10 **Another aurochs:** Michael Rappenglück, "Palaeolithic Timekeepers Looking at the Golden Gate of the Ecliptic; the Lunar Cycle and the Pleiades in the Cave of La-Tête-du-Lion (Ardèche, France)—21,000 BP," *Earth, Moon and Planets* 85–86 (2001): 391–404.

10 **in El Castillo cave:** Michael Rappenglück, "Ice Age People Find Their Ways by the Stars: A Rock Picture in the Cueva di El Castillo (Spain) May Represent the Circumpolar Constellation of the Northern Crown (CrB)," *Migration & Diffusion* 1 (2000): 15–28.

11 **Julien d'Huy probes the origins:** Julien d'Huy, "A Cosmic Hunt in the Berber Sky: A Phylogenetic Reconstruction of a Palaeolithic Mythology," *Les Cahiers de l'AARS*, Saint-Lizier: Association des amis de l'art rupestre saharien (2013): 93–106; Julien d'Huy, "The Evolution of Myths," *Scientific American* 315 (2016): 62–69; Julien d'Huy and Yuri Berezkin, "How Did the First Humans Perceive the Starry Night?—On the Pleiades," *The Retrospective Methods Network Newsletter* 12–13 (2016–2017): 100–122; Julien d'Huy, "Lascaux, les Pléiades et la Voie lactée: à propos d'une hypothèse en archéoastronomie," *Mythologie française* 267 (2017): 19–22.

12 **animal-tooth pendants:** Tõnno Jonuks and Eve Rannamäe, "Animals and Worldviews: A Diachronic Approach to Tooth and Bone Pendants from the Mesolithic to the Medieval Period in Estonia," in *The Bioarchaeology of Ritual and Religion*, ed. Alexandra Livarda et al. (Oxford, UK: Oxbow, 2017), 162–78.

13 **lifestyle appears to have been very similar:** Brian Hayden and Suzanne Villeneuve, "Astronomy in the Upper Palaeolithic?" *Cambridge Archaeological Journal* 21 (2011): 331–55.

13 **round grass houses:** Lynn Gamble, *The Chumash World at European Contact: Power, Trade and Feasting among Complex Hunter-gatherers* (Berkeley: University of California Press, 2008), 1–16.

14 **John Peabody Harrington:** Harrington's life and work: Jan Timbrook, "Memorial to Dee Travis Hudson (1941–1985)," *Journal of California and Great*

Basin Anthropology 7 (1985): 147–54; Catherine Callaghan, "Encounter with John P. Harrington," *Anthropological Linguistics* 33 (1991): 350–55; Lisa Krieger, "Long Gone Native Languages Emerge from the Grave," *Mercury News*, December 23, 2007, https://www.mercurynews.com/2007/12/23/long-gone-native-languages-emerge-from-the-grave/.

14 ***Crystals in the Sky:*** Travis Hudson and Ernest Underhay, *Crystals in the Sky: An Intellectual Odyssey Involving Chumash Astronomy, Cosmology and Rock Art* (Los Altos, CA: Ballena Press, 1978). See also: Travis Hudson and Thomas Blackburn, "The Integration of Myth and Ritual in South-Central California: The 'Northern Complex,'" *Journal of California Anthropology* 5 (1978): 225–50; Edwin Krupp, "Hiawatha in California," *Astronomy Quarterly* 8 (1991): 47–64.

15 **These celestial secrets:** Hayden and Villeneuve, "Astronomy."

15 **a plant called *Datura*:** Richard Applegate, "The Datura Cult among the Chumash," *Journal of California Anthropology* 2 (1975): 7–17.

16 **his seminal study:** Mircea Eliade, *Shamanism: Archaic Techniques of Ecstasy* (Princeton, NJ: Princeton University Press, 1964).

16 **really do enter:** Michael Winkelman, "Shamanism and the Alteration of Consciousness," in *Altering Consciousness* (Santa Barbara, CA: Praeger, 2011), 159–80; Michael Hove et al., "Brain Network Reconfiguration and Perceptual Decoupling during an Absorptive State of Consciousness," *Cerebral Cortex* 26 (2016): 3116–24; Pierre Flor-Henry et al., "Brain Changes During a Shamanic Trance: Altered Modes of Consciousness, Hemispheric Laterality, and Systemic Psychobiology," *Cogent Psychology* 4 (2017), DOI: 10.1080/23311908.2017.1313522.

16 **the spirit worlds they visit:** Michael Harner, *Cave and Cosmos: Shamanic Encounters with Another Reality* (Berkeley, CA: North Atlantic Books, 2013).

17 **In their 1998 book:** David Lewis-Williams and Jean Clottes, *The Shamans of Prehistory: Trance and Magic in the Painted Cave* (New York: Harry Abrams, 1998).

17 **a best-selling book:** Lewis-Williams, *The Mind in the Cave*. See also: David Lewis-Williams, *A Cosmos in Stone: Interpreting Religion and Society through Rock Art* (Lanham, MD: AltaMira Press, 2002), 321–42; David Lewis-Williams, "Rock Art and Shamanism," in *A Companion to Rock Art*, ed. Josephine McDonald and Peter Veth (Hoboken, NJ: Wiley-Blackwell, 2012), 17–33.

19 **"as a web of life":** Author's telephone interview with Sandra Ingerman, February 6, 2018.

19 **a nighttime ceremony:** Interview with Jo Bowlby, London, January 27, 2018.

20 **Marcel Ravidat and his friends:** Delluc and Delluc, *Lascaux Inconnu*.

20 **the Cosmic Hunt:** Julien d'Huy, "Un ours dans les étoiles, recherche phylogénétique sur un mythe préhistorique," *Préhistoire du Sud-Ouest, Association Préhistoire quercinoise et du Sud-Ouest* 20 (2012): 91–106.

21 **Neolithic rock painting:** Enn Ernits, "On the Cosmic Hunt in North Eurasian Rock Art," *Folklore* 44 (2010): 61–76.

21 **the birdman is a shaman:** Michael Rappenglück, "A Palaeolithic Planetarium Underground—the Cave of Lascaux (part 2)," *Migration & Diffusion* (2004): 6–47; Michael Rappenglück, "Possible Astronomical Depictions in

Franco-Cantabrian Paleolithic Rock Art," in *Handbook of Archaeoastronomy and Ethnoastronomy*, ed. Clive Ruggles (New York: Springer, 2015), 1205–12.

CHAPTER 2: LAND

23 **Just before dawn:** Michael O'Kelly's work on Newgrange and solstice discovery: "Michael J. O'Kelly," Newgrange.com, accessed November 5, 2019, http://www.newgrange.com/michael-j-okelly.htm; Simon Welfare and John Fairley, *Arthur C. Clarke's Mysterious World* (New York: A&W Publishers, 1980), 91–93; Michael O'Kelly and Claire O'Kelly, *Newgrange: Archaeology, Art and Legend* (London: Thames & Hudson, 1982); Michael O'Kelly, "The Restoration of Newgrange," *Antiquity* 53 (1979): 205–10.

26 **Klaus Schmidt was searching:** Andrew Curry, "Gobekli Tepe: The World's First Temple?" *Smithsonian*, November 2008, www.smithsonianmag.com/history/gobekli-tepe-the-worlds-first-temple-83613665/.

27 **"I had two choices":** Elif Batuman, "The Sanctuary: The World's Oldest Temple and the Dawn of Civilization," *New Yorker*, December 19/26, 2011, www.newyorker.com/magazine/2011/12/19/the-sanctuary.

27 **the hill is packed:** Discussions of recent archaeological findings: Klaus Schmidt, "Göbekli Tepe—the Stone Age Sanctuaries. New Results of Ongoing Excavations with a Special Focus on Sculptures and High Reliefs," *Documenta Praehistorica* 37 (2010): 239–56; Jens Notroff et al., "What Modern Lifestyles Owe to Neolithic Feasts. The Early Mountain Sanctuary at Göbekli Tepe and the Onset of Food-Production," *Actual Archaeology*, January 2015, 32–49; Oliver Dietrich et al., "Markers of 'Psycho-cultural' Change: The Early-Neolithic Monuments of Göbekli Tepe in Southeastern Turkey," in *Handbook of Cognitive Archaeology: Psychology in Prehistory*, ed. Tracy Henley et al. (New York: Routledge, 2020), 311–31.

28 **like "an amalgamation":** Steven Mithen, "Did Farming Arise from a Misapplication of Social Intelligence?," *Philosophical Transactions of the Royal Society B* 362 (2007): 705–18.

28 **"neither the idea nor the desire":** Jacques Cauvin, *The Birth of the Gods and the Origins of Agriculture* (Cambridge, UK: Cambridge University Press, 2000), 72.

28 **this small region:** Manfred Heun et al., "Site of Einkorn Wheat Domestication Identified by DNA Fingerprinting," *Science* 278 (1997): 1312–14; Simcha Lev-Yadun et al., "The Cradle of Agriculture," *Science* 288 (2000): 1602–3.

29 **"a by-product of the ideology":** Steven Mithen, *After the Ice: A Global Human History, 20,000–5,000 BC* (Cambridge, MA: Harvard University Press, 2006), 67.

29 **"coequal part of nature":** Notroff, "What Modern Lifestyles Owe to Neolithic Feasts."

29 **are from skulls:** Julia Gresky et al., "Modified Human Crania from Gobekli Tepe Provide Evidence for a New Form of Neolithic Skull Cult," *Science Advances* 3 (2017): e1700564.

30 **"transcendent sphere":** Schmidt, "Göbekli Tepe."

30 **"porthole stones":** Schmidt, "Göbekli Tepe."

30 **death, and particularly with skulls:** David Lewis-Williams and David Pearce, *Inside the Neolithic Mind: Consciousness, Cosmos and the Realm of the Gods* (London: Thames & Hudson, 2005), ch. 3. See also Jens Notroff et al., "Gathering of the Dead? The Early Neolithic Sanctuaries of Göbekli Tepe, Southeastern Turkey," in *Death Rituals, Social Order and the Archaeology of Immortality in the Ancient World*, ed. Colin Renfrew et al. (Cambridge, UK: Cambridge University Press, 2016), 65–80.

31 **"replete with substances":** Ian Hodder, "The Vitalities of Çatalhöyük," in *Religion at Work in a Neolithic Society: Vital Matters*, ed. Ian Hodder (Cambridge, UK: Cambridge University Press, 2014), 3.

31 **"the material expression":** David Lewis-Williams, "Constructing a Cosmos: Architecture, Power and Domestication at Catalhoyuk," *Journal of Social Archaeology* 4 (2004): 28–59.

31 **The Barasana people:** Stephen Hugh-Jones, "The Pleiades and Scorpius in Barasana Cosmology," *Journal of Skyscape Archaeology* 1 (2015): 111–24.

32 **his 2005 book:** Lewis-Williams and Pearce, *Neolithic Mind*, ch. 4.

32 **rising or setting:** Donna Sutcliff, "The Sky's the Topic," *Current Anthropology* 53 (2012): 125; Giulio Magli, "Sirius and the Project of the Megalithic Enclosures at Gobekli Tepe," *Nexus Network Journal* 18 (2016): 337; Martin Sweatman and Dimitrios Tsikritsis, "Decoding Gobekli Tepe with Archaeoastronomy: What Does the Fox Say?" *Mediterranean Archaeology and Archaeometry* 17 (2017): 233–50.

32 **Notroff isn't convinced:** Jens Notroff et al., "More Than a Vulture: A Response to Sweatman and Tsikritsis," *Mediterranean Archaeology and Archaeometry* 17 (2017): 57–74.

32 **at least partly subterranean:** Email interview with Jens Notroff, October 2019.

32 **a disk and crescent:** Schmidt, "Göbekli Tepe."

33 **"moon-deity":** Ludwig Morenz, "Media-evolution and the Generation of New Ways of Thinking: The Early Neolithic Sign System (10th/9th Millennium cal BC) and Its Consequences," *Our Place in the World*, John Templeton Foundation Newsletter, September 2014, https://www.academia.edu/8350091/Newsletter_August_2014.

33 **altered states of consciousness:** David Lewis-Williams and Thomas Dowson, "On Vision and Power in the Neolithic: Evidence from the Decorated Monuments," *Current Anthropology* 34 (1993): 55–65; David Lewis-Williams and David Pearce, "An Accidental Revolution? Early Neolithic Religion & Economic Change," *Minerva*, July/August 2006, 29–31.

33 **farming gradually spread:** Pontus Skoglund et al., "Origins and Genetic Legacy of Neolithic Farmers and Hunter-gatherers in Europe," *Science* 336 (2012): 466–69; Zuzana Hofmanová et al., "Early Farmers from across Europe Directly Descended from Neolithic Aegeans," *PNAS* 113 (2016): 6886–91; Selina Brace et al., "Ancient Genomes Indicate Population

Replacement in Early Neolithic Britain," *Nature Ecology & Evolution* 3 (2019): 765–71.

34 **around 3750 BC:** Nicki Whitehouse et al., "Neolithic Agriculture on the European Western Frontier: The Boom and Bust of Early Farming in Ireland," *Journal of Archaeological Science* 51 (2014): 181–205.

34 **"powerful transcendental network":** Robert Hensey, *First Light: The Origins of Newgrange* (Oxford, UK: Oxbow Books, Oxbow Insights in Archaeology, 2015).

35 **177 dolmen tombs:** Michael Hoskin, *Tombs, Temples and Their Orientations: A New Perspective on Mediterranean History* (Oxford, UK: Oxbow Books, 2001); Michael Hoskin, "Seven-stone Antas," in *Handbook of Archaeoastronomy and Ethnoastronomy*, ed. Clive Ruggles (New York: Springer, 2014), 1149–52.

35 **136 Irish passage tombs:** Frank Prendergast et al., "Facing the Sun," *Archaeology Ireland* 31 (2017): 10–17.

36 **purpose of these sites was shifting:** Hensey, *First Light*; Robert Bradley, *The Significance of Monuments* (New York: Routledge, 1998), ch. 7–8.

36 **"cosmological life, death and rebirth":** Lewis-Williams and Pearce, *Neolithic Mind*, ch. 9.

37 **a druid temple:** On historical theories of Stonehenge: Mike Parker Pearson, "Researching Stonehenge: Theories Past and Present," *Archaeology International* 16 (2013): 72–83.

37 **Modern excavations:** Timothy Darvill et al., "Stonehenge Remodeled," *Antiquity* 86 (2012): 1021–40; Clive Ruggles, "Stonehenge and Its Landscape," in *Handbook of Archaeoastronomy and Ethnoastronomy*, ed. Clive Ruggles (New York: Springer, 2015), 1223–37; Michael Allen et al., "Stonehenge's Avenue and 'Bluestonehenge,'" *Antiquity* 90 (2016): 991–1008; Mike Parker Pearson, "The Sarsen Stones of Stonehenge," *Proceedings of the Geologists' Association* 127 (2016): 363–69; Mike Parker Pearson, "Science and Stonehenge: Recent Investigations of the World's Most Famous Stone Circle" (Veertigste Kroonvoordracht, 2018).

38 **stone surfaces:** Parker Pearson, "Science and Stonehenge."

38 **a Malagasy colleague:** Mike Parker Pearson et al., "Materializing Stonehenge: The Stonehenge Riverside Project and New Discoveries," *Journal of Material Culture* 11 (2006): 227–61; Mike Parker Pearson, *Stonehenge: Exploring the Greatest Stone Age Mystery* (New York: Simon & Schuster, 2012).

39 **"had learned nothing":** Parker Pearson, "Researching Stonehenge."

39 **"the eternity of life":** On Durrington Walls and links to Stonehenge: Parker Pearson et al., "Materializing Stonehenge." Author's telephone interview with Mike Parker Pearson, October 15, 2019.

39 **excavated across both sites:** Parker Pearson, "Researching Stonehenge."

40 **"dark world of the dead":** Parker Pearson et al., "Materializing Stonehenge."

CHAPTER 3: FATE

43 **Hormuzd Rassam:** Hormuzd Rassam, *Asshur and the Land of Nimrod* (Cincinnati, OH: Curts & Jennings, 1897).

43 **flat-bottomed boats:** Henry Layard, *Discoveries in the Ruins of Nineveh and Babylon* (New York: Harper & Brothers, 1853).

44 **famous from the Bible:** 2 Kings 18–19; Jonah 1–3.

44 **"an experimental examination":** Rassam, *Asshur*, 24–32.

46 **"forerunners of everything":** Author's telephone interview with Jeanette Fincke, May 8, 2018.

47 **"in the month Ajaru":** *Enūma Anu Enlil*, 17.2.

47 **unearthed crateloads more:** Layard, *Discoveries*, ch. 16; David Damrosch, *The Buried Book: The Loss and Rediscovery of the Great Epic of Gilgamesh* (New York: Henry Holt & Co., 2007).

47 **as one collection:** Jeanette Fincke, "The British Museum's Ashurbanipal Library Project," *Iraq* 66 (2004): 55–60.

48 **"collect the written knowledge":** Fincke, "The British Museum's Ashurbanipal Library Project."

48 ***Epic of Gilgamesh:*** Stephen Mitchell, trans., *Gilgamesh*, (London: Profile, 2004).

48 **assistant curator George Smith:** Damrosch, *The Buried Book*, ch. 1.

49 ***Enuma Elish:*** Joshua Mark, "Enuma Elish—The Babylonian Epic of Creation—Full Text," *Ancient History Encyclopedia*, May 4, 2018, https://www.ancient.eu/article/225/enuma-elish—the-babylonian-epic-of-creation—fu/.

49 **"when on high":** Louise Pryke, "Religion and Humanity in Mesopotamian Myth and Epic," in *Religion: Oxford Research Encyclopaedias* (2016), DOI: 10.1093/acrefore/9780199340378.013.247. An alternative translation is "When Above."

49 **Priests at the temple:** Marc Linssen, *The Cults of Uruk and Babylon: The Temple Ritual Texts as Evidence for Hellenistic Cult Practice* (Leiden, the Netherlands: Brill, 2004).

50 **"gods were exalted":** Jean Bottéro, *Religion in Ancient Mesopotamia*, trans. Teresa Fagan (Chicago: University of Chicago Press, 2001), 158.

50 **"collect as many tablets":** Fincke, "The British Museum's Ashurbanipal Library Project."

50 ***namburbi*:** Hermann Hunger, "The Relation of Babylonian Astronomy to Its Culture and Society," *Proceedings of the International Astronomical Union* 5 (2009): 62–73.

50 **sign observed privately:** Author's telephone interview with Jeanette Fincke, May 8, 2018.

51 **Their wisdom was collated:** Fincke, "The British Museum's Ashurbanipal Library Project"; Jeanette Fincke, "The Oldest Mesopotamian Astronomical Treatise: Enūma Anu Enlil," in *Divination as Science: A Workshop Conducted during the 60th Rencontre Assyriologique Internationale, Warsaw, 2014*, ed. Jeanette Fincke (University Park, PA: Eisenbrauns, 2016), 107–46.

51 **"first day of Nisannu":** "Enuma Anu Enlil Text," The British Museum, accessed November 5, 2019, http://www.mesopotamia.co.uk/astronomer/explore/enuma1t.html.

51 **"country will be attacked":** Simo Parpola, "Excursus: The Substitute King Ritual," in *Letters from Assyrian Scholars to the Kings Esarhaddon and Assurbanipal*, ed. Simo Parpola (Kevelaer, Germany: Butzon & Bercker, 1983), xxii–xxvi.

51 **divided into quadrants:** Parpola, "Substitute King."

52 **built a huge palace:** Paul Tanner, "Ancient Babylon: From Gradual Demise to Archaeological Rediscovery," *Near East Archaeological Society Bulletin* 47 (2002): 11–20; Roan Fleischer, "Nebuchadnezzar II and Babylon: Building Personal Legacy through Monumentality," *Binghampton Journal of History* 18 (2017): 3–24.

52 **The tower was called Etemenanki:** Andrew George, "The Tower of Babel: Archaeology, History and Cuneiform Texts," *Archiv für Orientforschung* 51 (2005/2006): 75–95.

53 **"a structure founded":** Andrew George, *Babylonian Topographical Texts* (Leuven, Belgium: Peeters Press, 1992), 299.

53 **priest called Johann Strassmaier:** Teije de Jong, "Babylonian Astronomy 1880–1950: The Players and the Field," in *A Mathematician's Journeys*, ed. Alexander Jones et al. (New York: Springer, 2016), 265–302. See also: Gary Thompson, "The Recovery of Babylonian Astronomy" (2009–2018), Ancient Zodiacs, Star Names, and Constellations, accessed November 5, 2019, http://members.westnet.com.au/gary-david-thompson/babylon4.html.

55 **Epping was reluctant:** de Jong, "Babylonian Astronomy"; Johann Epping, *Astronomisches aus Babylon* (Vienna: Freiburg im Breisgau, 1889).

55 **a gradual progression:** James Evans, *The History and Practice of Ancient Astronomy* (New York: Oxford University Press, 1998); Mathieu Ossendrijver, "Babylonian Mathematical Astronomy," in *Handbook of Archaeoastronomy and Ethnoastronomy*, ed. Clive Ruggles (New York: Springer, 2015), 1863–70.

56 **invented the zodiac:** Evans, *Ancient Astronomy*, 39.

57 **very clever math:** Evans, *Ancient Astronomy*, 317; author's telephone interview with James Evans, May 31, 2018.

57 **using geometric techniques:** Mathieu Ossendrijver, "Ancient Babylonian astronomers calculated Jupiter's position from the area under a time-velocity graph," *Science* 351 (2016): 482–84.

58 **Quintus Curtius Rufus:** Quintus Curtius Rufus, *The History of Alexander*, trans. John Yardley (New York: Penguin Classics, 1984), 93–94.

59 **"radically new way . . . into a real theory":** Evans, *Ancient Astronomy*, 213.

60 **"in one swoop":** Evans, *Ancient Astronomy*, 23.

60 **from a landowning family:** de Jong, "Babylonian Astronomy."

61 **"their new masters":** George Bertin, "Babylonian Astronomy IV," *Nature* 40 (1889): 360.

62 **a trigonometric table:** Daniel Mansfield and Norman Wildberger, "Plimpton 322 Is Babylonian Exact Sexagisemal Trigonometry," *Historia Mathematica* 44 (2017): 395–419.

62 **must have visited:** Gerald Toomer, "Hipparchus and Babylonian Astronomy," in *A Scientific Humanist: Studies in Memory of Abraham Sachs*, ed. Erle Leichty et al., Occasional Publications of the Samuel Noah Kramer Fund, 9 (Philadelphia: The University Museum, 1988), 353–62; Alexander Jones, "The Adaptation of Babylonian Methods in Greek Numerical Astronomy," *Isis* 82 (1991): 441–53.

62 **"shocked":** Author's telephone interview with James Evans, May 31, 2018.

62 **broken fragments of ivory:** Jean-Paul Bertaux, "La découverte des tablettes: les données archéologiques," in *Les tablettes astrologiques de Grand (Vosges) et l'astrologie en Gaule romaine: actes de la table-ronde du 18 mars 1992, organisée au Centre d'études romaines et gallo-romaines de l'Université de Lyon III*, ed. Josèphe-Henriette Abry and André Buisson (Lyon: University of Lyon, 1993), 39–47.

63 **ancient rubbish dumps:** Alexander Jones, *Astronomical Papyri from Oxyrhynchus* (Philadelphia: American Philosophical Society, 1999).

63 ***Alexander Romance:*** Richard Stoneman, trans., *The Greek Alexander Romance* (New York: Penguin Classics, 1991).

64 **wealthy clients:** James Evans, "The Astrologer's Apparatus: A Picture of Professional Practice in Greco-Roman Egypt," *Journal of the History of Astronomy* 35 (2004): 1–44.

64 **dating from 410 BC:** Abraham Sachs, "Babylonian Horoscopes," *Journal of Cuneiform Studies* 6 (1952): 49–75; Francesca Rochberg, "Babylonian Horoscopy: The Texts and Their Relations," in *Ancient Astronomy and Celestial Divination*, ed. Noel Swerdlow (Cambridge, MA: MIT Press, 1999), 39–60.

65 **"which concern the reason":** Ptolemy, *Tetrabiblos* III, ch. 13.

65 **astrological predictions:** Patrick Boner, "Galileo's Astrology," *Renaissance Quarterly* 59 (2006): 222–24.

65 **strengthen and reform:** Gérard Simon, "8.3 Kepler's Astrology: The Direction of a Reform," *Vistas in Astronomy* 18 (1975): 439–48.

66 **rising in credibility:** Julie Beck, "The New Age of Astrology," *The Atlantic*, January 16, 2018, https://www.theatlantic.com/health/archive/2018/01/the-new-age-of-astrology/550034/. See also the following analysis which suggests that anywhere between 22 and 73 percent of people believe in astrology, depending on what you mean by "believe": Nicholas Campion, "How many people actually believe in astrology?" The Conversation, April 28, 2017, https://theconversation.com/how-many-people-actually-believe-in-astrology-71192.

66 **"dislocated from its cosmology":** Skype interview with Nicholas Campion, May 28, 2018.

66 **"undermining the very fabric":** Trevor Jackson, "When Balance Is Bias," *British Medical Journal* 343 (2011): d8006.

66 **"shriveling and cheapening":** Richard Dawkins, "The Real Romance in the Stars," *The Independent*, December 31, 1995, https://www.independent.co.uk/voices/the-real-romance-in-the-stars-1527970.html.

66 **approached Babylon:** Arrian, *The Campaigns of Alexander*, trans. Aubrey de Selincourt (New York: Penguin Classics, 1976), 376–78.

67 **"The king died":** Leo Depuydt, "The Time of Death of Alexander the Great: 11 June 323 BC (-322), ca. 4:00–5:00pm," *Die Welt des Orients* 28 (1997): 117–35; Jona Lendering, "Alexander's Last Days," *Livius*, accessed November, 5, 2019, https://www.livius.org/articles/person/alexander-the-great/alexander-3.6-last-days/.

68 **last cuneiform tablets:** Hermann Hunger and Teije de Jong, "Almanac W22340a from Uruk: The Latest Datable Cuneiform Tablet," *Zeitschrift für Assyriologie und Vorderasiatische Archäologie* 104 (2014): 182–94.

CHAPTER 4: FAITH

69 **What happened to Constantine:** Eusebius, *Life of Constantine*, ch. 28; Lactantius, *Liber de Mortibus Persecutorum*, ch. 44.

71 **"the favored deity":** Elizabeth Marlowe, "Framing the Sun: The Arch of Constantine and the Roman Cityscape," *Art Bulletin* 88 (2006): 223–42; Maggie Popkin, "Symbiosis and Civil War: The Audacity of the Arch of Constantine," *Journal of Late Antiquity* 9 (2016): 42–88.

71 **"shining in the sky":** Karlene Jones-Bley, "An Archaeological Reconsideration of Solar Mythology," *Word* 44 (1993): 431–43.

72 **supreme creator beings:** Lawrence Sullivan, "Supreme Beings," in *Encyclopedia of Religion*, 2nd ed., ed. Lindsay Jones and Mircea Eliade, vol. 13 (Detroit: Thomson Gale, 2005).

72 **worshipped celestial bodies:** The archaeological evidence includes a thirteenth-century basalt slab from Tel Hazor, northern Israel, that shows a pair of hands raised toward a lunar crescent, and a tenth-century cult stand from Tel Taanach that features a winged sun disk. Celestial worship is also mentioned in several biblical texts, such as 2 Kings 23:5: "He suppressed the idolatrous priests . . . those who made offerings to Baal, to the sun, moon, constellations, and all the host of heaven."

73 **"abstract and indestructible God":** David Aberbach, "Trauma and Abstract Monotheism: Jewish Exile and Recovery in the Sixth Century BCE," *Judaism* 50 (2001): 211–21.

73 **"jumping-off point":** Mircea Eliade, *Patterns in Comparative Religion*, trans. Rosemary Sheed (Lincoln: University of Nebraska Press, 1996), 95.

74 **"evil times":** Arnold Jones, *Constantine and the Conversion of Europe* (London: English Universities Press, 1948), 2.

75 **"crammed full of deities":** Marianne Bonz, "Religion in the Roman World," *PBS Frontline*, accessed November 5, 2019, https://www.pbs.org/wgbh/pages/frontline/shows/religion/portrait/religions.html.

75 **Constantine ordered his army:** Jonathan Bardill, *Constantine: Divine Emperor of the Golden Age* (Cambridge, UK: Cambridge University Press, 2011).

75 **tradition that goes back:** Bardill, *Constantine: Divine Emperor.*

76 **"dressed in silk robes":** Bradley Schaefer, "Meteors That Changed the World," *Sky & Telescope*, February 1, 2005, https://www.skyandtelescope.com/observing/celestial-objects-to-watch/meteors-that-changed-the-world/.

76 **all of Constantine's mints:** Martin Wallraff, "Constantine's Devotion to the Sun after 324," *Studia patristica* 34 (2001): 256–69.

76 **"sun-god as his guardian":** Bardill, *Constantine: Divine Emperor*, 92.

76 **a second vision:** Discussed in Bardill, *Constantine: Divine Emperor*, ch. 5.

76 **a sun dog:** Peter Weiss, "The Vision of Constantine," *Journal of Roman Archaeology* 16 (2003): 237–59.

77 **"heavenly messenger":** Eusebius, *Life of Constantine*, book 3, ch. 10.

78 **"shining like the Sun":** Discussion of Constantine's Sun worship: Wallraff, "Constantine's Devotion"; Bardill, *Constantine: Divine Emperor.*

78 **"most brilliant beams . . . a pure light":** Quoted in Bardill, *Constantine: Divine Emperor*, 330.

79 **main day of worship:** Detailed discussion of the origin of Sunday as the Christian day of worship: Samuele Bacchiocchi, "Sun-worship and the Origin of Sunday," in *From Sabbath to Sunday: A Historical Investigation of the Rise of Sunday Observance in Early Christianity* (Rome: Pontifical Gregorian University Press, 1977), 131–63.

79 **historian Marina Warner:** Marina Warner, *Alone of All Her Sex: The Myth and the Cult of the Virgin Mary* (Oxford, UK: Oxford University Press, 2016), 263.

80 **"Had Christianity not taken root":** Warner, *Alone of All Her Sex*, 266.

80 **"fought heroically":** Jacquetta Hawkes, *Man and the Sun* (London: Cresset, 1962), 199; quoted in Bacchiocchi, "Sun-worship."

80 **political "masterstroke":** Bardill, *Constantine: Divine Emperor*, 331.

81 **more imperial features:** Adam Renner, "The Nimbus in Imperial and Christian Iconography: Origin, Transformation, and Significance," accessed November 5, 2019, https://www.academia.edu/1598242/Nimbus_in_Imperial_and_Christian_Imagery.

81 **"pictured him to be":** Thomas F. Mathews, *The Clash of Gods: A Reinterpretation of Early Christian Art* (Princeton, NJ: Princeton University Press, 1999), 11.

81 **"sit on soft clouds":** Maria Shriver, *What's Heaven?* (New York: St. Martin's Griffin, 2007).

82 **According to J. Edward Wright:** J. Edward Wright, *The Early History of Heaven* (New York: Oxford University Press, 1999).

82 **what comes after death:** Diarmaid MacCulloch, *A History of Christianity: The First Three Thousand Years* (New York: Viking, 2010).

82 **"Never try to reconcile":** Homer, *Odyssey*, trans. A. T. Murray, Loeb Classical Library 104(Cambridge, MA: Harvard University Press, 1919), 435–37. Discussed in Nicholas Campion, "Was There a Ptolemaic Revolution in Ancient Egyptian Astronomy? Souls, Stars and Cosmology," *Journal of Cosmology* 13 (2011): 4174–86.

83 **have immortal souls:** "Plato's Timaeus," *Stanford Encyclopaedia of Philosophy*, December 18, 2017, https://plato.stanford.edu/entries/plato-timaeus/; Richard Poss, "Plato's Timaeus and the Inner Life of Stars," *Memorie della Societa Astronomica Italiana* 73 (2002): 287; Nicholas Campion, "Astronomy and Psyche in the Classical World: Plato, Aristotle, Zeno, Ptolemy," *Journal of Cosmology* 9 (2010): 2179–86.

83 **"fly away from earth":** Plato, *Theaeteus*, trans. Benjamin Jowett; quoted in Campion, "Astronomy and Psyche."

83 **"As we take the train":** Vincent van Gogh, letter to Theo van Gogh, c. July 9, 1888.

84 **two brothers from Berlin:** Heinrich Brugsch, "Zwei Pyramiden mit Inschriften aus den Zeiten der VI. Dynastie," *Zeitschrift für Ägyptische Sprache und Alterthumskunde* 19 (1881): 1–15; Heinrich Brugsch, *My Life and My Travels* (Berlin, 1894), ch. 7; Ronald Ridley, "The Discovery of the Pyramid Texts," *ZAS* 110 (1983): 74–80.

86 **"a last pleasure":** Brugsch, *My Life and My Travels*, ch. 7.

86 **Egyptians' view of the cosmos:** Wright, *Early History of Heaven*; Geraldine Pinch, *Egyptian Mythology: A Guide to the Gods, Goddesses and Traditions of Ancient Egypt* (New York: Oxford University Press, 2002); John Taylor, *Egyptian Mummies* (London: British Museum Press, 2010).

87 **"resting place for the corpse":** Taylor, *Egyptian Mummies*, 113.

87 **"I row in the sky":** James P. Allen, *The Ancient Pyramid Texts* (Atlanta: Society of Biblical Literature Press, 2015), 52 and 34.

87 **to point toward:** Allen, *Pyramid Texts*; Raymond Faulkner, "The King and the Star-religion in the Pyramid Texts," *Journal of Near Eastern Studies* 25 (1966): 153–61.

88 **twentieth of a degree:** Juan Antonio Belmonte, "On the Orientation of Old Kingdom Egyptian Pyramids," *Journal for the History of Astronomy* 32 (2001): S1–S20; Giulio Magli and Juan Antonio Belmonte, "Pyramids and Stars: Facts, Conjectures and Starry Tales," in *In Search of Cosmic Order: Selected Essays on Egyptian Archaeoastronomy*, ed. Juan Antonio Belmonte and Mosalam Shaltout (Cairo, Egypt: Supreme Council of Antiquities Press, 2009), ch. 10.

88 **"maniacal precision":** Giulio Magli, "A Possible Explanation of the Void in the Pyramid of Khufu on the Basis of the Pyramid Texts" (January 1, 2018), arXiv.org website, hosted by Cornell University; https://arxiv.org/abs/1711.04617.

88 **where it all started:** Campion, "Ptolemaic Revolution."

89 **Ancient biographies:** Summarized in Kitty Ferguson, *Pythagoras: His Lives and the Legacy of a Rational Universe* (London: Icon Books, 2010).

89 **"inclusion of the soul":** Campion, "Ptolemaic Revolution."

89 **"movement of the stars":** Marcus Aurelius, *Meditations*, 7.47, trans. Martin Hammond (New York: Penguin Classics, 2006).

89 **"man's spirit rises" and "wise will shine":** Ecclesiastes 3:21 and Daniel 12:3. Discussed in Wright, *Early History of Heaven*, 87.

90 **"concern with the afterlife":** MacCulloch, *History of Christianity*, 71.

90 **Constantine fell ill:** Eusebius, *Life of Constantine*, ch. 58–75; Bardill, *Constantine: Divine Emperor*, ch. 9.

91 **"a parody":** Thomas Paine, "On the Origin of Free Masonry" (London, 1818).

91 **"Out of disorder":** Plato, *Timaeus*.

91 **"heaven is my throne":** Isaiah 66:1.

91 **"liberally imparts":** Eusebius, *Life of Constantine*; quoted in Wallraff, "Constantine's Devotion."

92 **"to look with awe":** Quoted in Wallraff, "Constantine's Devotion."

92 **"The God I believe in":** Guy Consolmagno, "Astronomy and Belief," *Thinking Faith*, April 18, 2013, https://www.thinkingfaith.org/articles/20130418_1.htm.

92 **"divine and eternal animals":** Plato, *Timaeus*.

92 **"soul or . . . the whole world":** Pliny, *Natural History*, trans. H. Rackham, Loeb Classical Library 330 (Cambridge, MA: Harvard University Press, 1949), 178–79.

93 **"None come nearer to us":** Saint Augustine of Hippo, *City of God*, book 8, ch. 5.

93 **"Who cannot see":** Saint Augustine of Hippo, *City of God*, book 4, ch. 12.

93 **"rejected the notion":** Nicholas Campion, *Astrology and Cosmology in the World's Religions* (New York: New York University Press, 2012), 169.

CHAPTER 5: TIME

95 ***MS Ashmole 1796*:** John North, *God's Clockmaker: Richard of Wallingford and the Invention of Time* (London: Hambledon, 2005); Bodleian catalog entry available at http://mlgb3.bodleian.ox.ac.uk/mlgb/book/4885/.

96 **"One may observe":** John North, *Richard of Wallingford: An Edition of His Writings with Introductions, English Translation and Commentary*, vol. 2 (Oxford, UK: Oxford University Press, 1976), 366.

96 **Richard was born:** Accounts of Richard's life: North, *God's Clockmaker*; North, *Richard of Wallingford*; Thomas Walsingham, *Gesta Abbatum Monasterii Sancti Albani*, ed. Henry Riley (London: Longmans, 1867).

98 **attacked and laid siege:** Gabrielle Lambrick, "Abingdon and the Riots of 1327," *Oxoniensia* 29 (1964): 129–41.

99 **an obsession with time:** North, *God's Clockmaker*; David Landes, *Revolution in Time: Clocks and the Making of the Modern World* (Cambridge, MA: Harvard University Press, 2000); John Scattergood, "Writing the Clock: The Reconstruction of Time in the Late Middle Ages," *European Review* 11 (2003): 453–74; Lewis Mumford, *Technics and Civilization* (Chicago: University of Chicago Press, reprint ed., 2010); Jacques Le Goff, *Time, Work and Culture in the Middle Ages* (Chicago: University of Chicago Press, 1982).

99 **"temporal discipline":** Landes, *Revolution in Time*, 59.

100 **"pray and pray often":** Landes, *Revolution in Time*, 61.

101 **earliest detailed account . . . ran to the clock:** John North, "Monasticism and the First Mechanical Clocks," in *The Study of Time II*, ed. Julius Fraser and Nathaniel Lawrence (New York: Springer, 1975), 381–98; Scattergood, "Writing the Clock."

102 **"most ingenious inventions":** Landes, *Revolution in Time*, 10.

102 **rugged gearwheels:** Ibn Khalaf al-Murādī, *The Book of Secrets in the Results of Ideas. Incredible Machines from 1000 Years Ago* (Milan: Leonardo 3, 2008),

http://www.leonardo3.net/en/l3-works/publishing-house/1503-the-book-of-secrets.html.

102 **clocktower, powered by water or mercury:** Joseph Needham et al., "Chinese Astronomical Clockwork," *Nature* 177 (1956): 600–602.

102 **"perfect their work":** Robertus Anglicus, *De Sphera of Sacrobosco*, 1271; quoted in North, "Monasticism," 381–98.

102 **earliest mentions:** North, "Monasticism."

102 ***"Le roman de la rose"*:** A thirteenth-century poem by Guillaume de Lorris. More information at https://www.bl.uk/collection-items/roman-de-la-rose.

103 **two "spheres":** Cicero, *De Re Publica*, book 1, sections 21–22.

103 **mysterious bronze device:** Derek de Solla Price, "Clockwork before the Clock and Timekeepers before Timekeeping," in *The Study of Time II*, ed. Julius Fraser and Nathaniel Lawrence (New York: Springer, 1975), 368–80; Jo Marchant, *Decoding the Heavens: Solving the Mystery of the World's First Computer* (London: Windmill Books, 2009); Alexander Jones, *A Portable Cosmos: Revealing the Antikythera Mechanism, Scientific Wonder of the Ancient World* (Oxford, UK: Oxford University Press, 2017).

104 **Byzantine sundial:** Judith Field and Michael Wright, "Gears from the Byzantines: A Portable Sundial with Calendrical Gearing," *Annals of Science* 42 (1985): 87–138; Marchant, *Decoding the Heavens*.

104 **thirteenth-century astrolabe:** Marchant, *Decoding the Heavens*, 151 (and image in plate section).

105 **"wonderful wheels":** In a thirteenth-century Latin text discovered by John North and discussed in North, *God's Clockmaker*, ch. 12.

105 **impressive type of water clock:** Joseph Noble and Derek de Solla Price, "The Water Clock in the Tower of the Winds," *American Journal of Archaeology* 72 (1968): 345–55.

105 **wrote a treatise:** Roger Bacon, "Letter on Secret Works of Art and of Nature and on the Invalidity of Magic," 1248; translation by Michael Mahoney available at https://www.princeton.edu/~hos/h392/bacon.html.

105 **"greatest of secrets":** Roger Bacon, *Opus minus*, 1268, and *Opus maius*, 1267, discussed in North, *God's Clockmaker*, ch. 12.

105 **Pierre de Maricourt:** Pierre de Maricourt, "Letter on the magnet," 1269; discussed in North, *God's Clockmaker*, ch. 12.

106 **more valuable than:** Bacon, "Secret Works," discussed in North, *God's Clockmaker*, ch. 12.

106 **notes in the margin:** North, *Richard of Wallingford*, vol. 2, 309–20.

107 **"cosmic machine":** North, *God's Clockmaker*, ch. 14.

109 **"a combination of mathematical and mechanical genius" and "perhaps without equal":** North, *God's Clockmaker*, 212 and 214.

110 **built a few decades later:** North, *God's Clockmaker*, ch. 13.

110 **"great abbot":** North, *Richard of Wallingford*, preface.

110 **"original English scientist":** North, *God's Clockmaker*, xv.

110 **tradition of great clocks:** Discussed, for example, in Landes, *Revolution in Time*, ch. 4.

112 **mundane gearwork:** Nicholas Whyte, "The Astronomical Clock of Richard of Wallingford" (unpublished graduate student paper, 1990–91), http://www.nicholaswhyte.info/row.htm.

113 **"key-machine":** Mumford, *Technics*, 87.

113 **literary references to clocks:** Scattergood, "Writing the Clock."

113 **"Alas the clock":** Dafydd ap Gwilym, *Poems*, ed. and trans. Rachel Bromwich (Llandysul, UK: Gomer Press, 1982), 110–13; quoted in Scattergood, "Writing the Clock."

113 **the first thing he did:** Quoted in Landes, *Revolution in Time*, 91.

114 **counting and simple arithmetic:** Landes, *Revolution in Time*, ch. 4.

114 **"Men became powerful":** Mumford, *Technics*, 25.

115 **"glorious wheel":** Dante Alighieri, *The Divine Comedy*, vol. 3: *Paradiso*, trans. John Sinclair (New York: Oxford University Press, 1961), Canto X.

115 **complex astronomical clock:** Silvio Bedini and Francis Maddison, "Mechanical Universe: The Astrarium of Giovanni de' Dondi," *Transactions of the American Philosophical Society* 56 (1966): 1–69.

115 **"The situation is":** Nicole Oresme, *Livre du Ciel et du Monde*, 1377; quoted in Scattergood, "Writing the Clock."

115 **pushed the metaphor:** Discussed, for example, in David Wootton, *The Invention of Science: A New History of the Scientific Revolution* (New York: Harper, 2015), 436–41.

116 **"satisfied with less":** Stephen Toulmin, "From Clocks to Chaos: Humanizing the Mechanistic World-View," in *The Machine as Metaphor and Tool*, ed. Hermann Haken et al. (New York: Springer, 1993), 142.

116 **several indigenous societies:** Chris Sinha et al., "When Time Is Not Space: The Social and Linguistic Construction of Time Intervals and Temporal Event Relations in an Amazonian Culture," *Language and Cognition* 3 (2011): 137–69.

116 **The key insight:** Landes, *Revolution in Time*, ch. 7; Seth Atwood, "The Development of the Pendulum as a Device for Regulating Clocks Prior to the 18th Century," in *The Study of Time II*, ed. Julius Fraser and Nathaniel Lawrence (New York: Springer, 1975), 417–50.

118 **"true and mathematical time":** Lennart Lundmark, "The Mechanization of Time," in *The Machine as Metaphor and Tool*, ed. Hermann Haken et al. (New York: Springer, 1993), 45–65.

119 **mean time was made standard:** Lundmark, "Mechanization of Time," 57.

120 **"Our responsiveness to":** Landes, *Revolution in Time*, 2.

120 **"time famine":** Joseph Carroll, "Time Pressures, Stress, Common for Americans," Gallup, January 2, 2008, https://news.gallup.com/poll/103456/Time-Pressures-Stress-Common-Americans.aspx.

120 **In the early 1330s:** North, *God's Clockmaker*, ch. 15; North, *Richard of Wallingford*; Walsingham, *Gesta Abbatum*.

121 **"The break with Rome":** North, *God's Clockmaker*, 5.

CHAPTER 6: OCEAN

123 **sailing for two months:** Account of *Endeavour*'s approach to Tahiti: *The Journals of Captain James Cook*: vol. 1: *The Voyage of the Endeavour 1768–1771*, ed. John Beaglehole (London: Hakluyt Society, 1955), entries for March 30 to April 13, 1769.

123 **"floating coal bucket":** Joan Druett, *Tupaia: Captain Cook's Polynesian Navigator* (Santa Barbara, CA: Praeger, 2010), 58.

124 **crowd of canoes:** Accounts of *Endeavour*'s stay at Tahiti: Anne Salmond, *The Trial of the Cannibal Dog: Captain Cook in the South Seas* (Auckland, NZ: Penguin, 2004); Druett, *Tupaia*; Beaglehole, *Journals*, vol. 1; Joseph Banks, *The Endeavour Journal of Sir Joseph Banks* (Sydney: University of Sydney Library; first published 1771); William Frame and Laura Walker, *James Cook: The Voyages* (London: British Library, 2018).

125 **dung beetles that sight:** James Foster et al., "How Animals Follow the Stars," *Proceedings of the Royal Society B* 285 (2018): 20172322.

126 **Age of Discovery:** David Barrie, *Sextant: A Voyage Guided by the Stars and the Men Who Mapped the World's Oceans* (Glasgow, UK: William Collins, 2014); Ben Finney, "Nautical Cartography and Traditional Navigation in Oceania," in *The History of Cartography*, vol. 2, book 3, ed. David Woodward and Malcolm Lewis (Chicago: University of Chicago Press, 1998), 443–92.

128–129 **device he called H4:** David Landes, *Revolution in Time*, 145–57.

129 **popularized in Dava Sobel's:** Dava Sobel, *Longitude: The True Story of a Lone Genius Who Solved the Greatest Scientific Problem of His Time* (New York: Walker & Co., 1995).

132 **measure a planet's parallax:** Edmund Halley, "A New Method of Determining the Parallax of the Sun," *Philosophical Transactions* 29 (1716): 454; Michael Chauvin, "Astronomy in the Sandwich Islands: The 1874 Transit of Venus," *The Hawaiian Journal of History* 27 (1993): 185–225. History of attempts to measure solar system size from Venus transits: Donald Teets, "Transits of Venus and the Astronomical Unit," *Mathematics Magazine* 76 (2003): 335–48.

133 **"It behoves us":** Wayne Orchiston, "Cook, Green, Maskelyne and the 1769 Transit of Venus: The Legacy of the Tahitian Observations," *Journal of Astronomical History and Heritage* 20 (2017): 35–68. For more on the political background to Cook's mission, see Chauvin, "Astronomy in the Sandwich Islands."

133 **sextant and newly repaired quadrant:** On the instruments and astronomy of Cook's 1769 mission: Orchiston, "Cook, Green, Maskelyne"; Wayne Orchiston, "James Cook's 1769 Transit of Venus Expedition to Tahiti," *Proceedings of the International Astronomical Union Colloquium* 196 (2004): 52–66.

135 **"new world order":** Barrie, *Sextant*, 92.

135 **Tupaia was a high priest of the 'arioi:** Druett, *Tupaia*, 3–9.

136 **"keep him as a curiosity":** Banks, *Endeavour Journal*, July 12, 1769.

136 **"heartfelt tears":** Banks, *Endeavour Journal*, July 13, 1769.

137 **extraordinary moment in world history:** Frame and Walker, *James Cook*, 58.

137 **"the same Nation":** Beaglehole, *Journals*, vol. 1, 354.

138 **"Tupia's own hands":** Beaglehole, *Journals*, vol. 1, 294; quoted in Anne Di Piazza and Erik Pearthree, "History of an Idea about Tupaia's Chart," *Cook's Log* 35 (2012): 18. See also: Frame and Walker, *James Cook*; Finney, "Nautical Cartography," 443–92.

138 **"Sun serving them for a compass":** Beaglehole, *Journals*, vol. 1, 154.

138 **migrate by sea:** Ben Finney, "The Pacific Basin: An Introduction," in *The History of Cartography*, vol. 2, book 3, ed. David Woodward and Malcolm Lewis (Chicago: University of Chicago Press, 1998), 419–22; Ben Finney, "Colonizing an Island World," *Transactions of the American Philosophical Society* 86 (1996): 71–116.

139 **replaced by skepticism:** For a discussion of skepticism and dissenters, including the origins of the *Hokule'a* project, see Finney, "Nautical Cartography."

139 **"cultural response":** "Nainoa Thompson," Polynesian Voyaging Society, accessed November 5, 2019, http://archive.hokulea.com/index/founder_and_teachers/nainoa_thompson.html. This biography includes information from Gisela Speidel, "The Ocean Is My Classroom," *Kamehameha Journal of Education* 5 (1994): 11–23; as well as speeches given by Nainoa Thompson in 1997 and 1998. For more on *Hokule'a*, see Patrick Karjala et al., "Kilo Hōkū—Experiencing Hawaiian, Non-instrument Open Ocean Navigation through Virtual Reality," *Presence* 26 (2017): 264–80.

139 **"from being castaways":** Gary Kubota, "Ben Finney, Polynesian Voyaging Society Founder, Dies at 83," *Honolulu Star Advertiser*, May 24, 2017, https://www.staradvertiser.com/2017/05/24/breaking-news/ben-finney-a-founder-of-the-polynesian-voyaging-society-dies-at-83/.

139 **archaeological and genetic findings:** For example, Alice Storey et al., "Radiocarbon and DNA Evidence for a Pre-Columbian Introduction of Polynesian Chickens to Chile," *PNAS* 104 (2007): 10335–39; Shane Egan and David Burley, "Triangular Men on One Very Long Voyage: The Context and Implications of a Hawaiian-style Petroglyph Site in the Polynesian Kingdom of Tonga," *Journal of the Polynesian Society* 118 (2009): 209–32; Andrew McAlister et al., "The Identification of a Marquesan Adze in the Cook Islands," *Journal of the Polynesian Society* 122 (2013): 257–73.

140 **scholar David Lewis:** David Lewis, *We, the Navigators: The Ancient Art of Landfinding in the Pacific* (Honolulu: University of Hawaii Press, 1972).

140 **used a "star compass":** Finney, "Nautical Cartography"; Anne Di Piazza, "A Reconstruction of a Tahitian Star Compass Based on Tupaia's 'Chart for the Society Islands with Otaheite in the Center,'" *Journal of the Polynesian Society* 119 (2010): 377–92.

140 **"great triggerfish":** Finney, "Nautical Cartography."

140 **ancient chant:** Teuira Henry, "Birth of the Heavenly Bodies," *Journal of the Polynesian Society* 16 (1907): 101–4.

141 **teaching schedule for priests:** Stan Lusby et al., "Navigation and Discovery in the Polynesian Oceanic Empire: Part One," *Hydrographic Journal* 131/132 (2010): 17–25.

141 **Rua-Nui's pillars:** Stan Lusby et al., "Navigation and Discovery in the Polynesian Oceanic Empire: Part Two," *Hydrographic Journal* 134 (2010): 15–25.

142 **"a way of being":** Di Piazza and Pearthree, "History of an Idea."

142 **Cook's chart of Tahiti:** James Cook, cartographer, "Chart of the Island Otaheite," 1769, 24 x 41cm, available at https://collections.rmg.co.uk /collections/objects/540641.html.

142 **first people we know:** General source on the history of mapmaking: *The History of Cartography, Volume 1: Cartography in Prehistoric, Ancient, and Medieval Europe and the Mediterranean*, ed. Brian Harley and David Woodward (Chicago: University of Chicago Press, 1987).

143 ***mappae mundi* were popular:** David Woodward, "Medieval Mappaemundi," in *History of Cartography*, vol. 1, ed. Harley and Woodward, 286–370.

144 **Di Piazza and Pearthree pointed out:** Anne Di Piazza and Erik Pearthree, "A New Reading of Tupaia's Chart," *Journal of the Polynesian Society* 116 (2007): 321–40; Di Piazza and Pearthree, "History of an Idea."

144 **"Reef Hole Probing":** Finney, "Nautical Cartography."

147 **"When we have technology":** Quoted in "Navigation Part of the Brain 'Is Switched Off' as Soon as You Turn On a Sat Nav," *Daily Express*, March 21, 2017, https://www.express.co.uk/news/uk/781986/Navigation-part-brain -switched-off-sat-nav-GPS; Amir-Homayoun Javadi et al., "Hippocampal and Prefrontal Processing of Network Topology to Simulate the Future," *Nature Communications* 8 (2017): 14652.

147 **less able to find their way:** Steven Tripp, "Cognitive Navigation: Toward a Biological Basis for Instructional Design," *Educational Technology and Society* 4 (2001): 41–49; Alex Hutchinson, "Global Impositioning Systems: Is GPS Technology Actually Harming Our Sense of Direction?" *The Walrus*, November 2009, https://thewalrus.ca/global-impositioning-systems/.

147 **overreliance on technologies:** Nicholas Carr, "All Can Be Lost: The Risk of Putting Our Knowledge in the Hands of Machines," *The Atlantic*, November 2013, https://www.theatlantic.com/magazine/archive/2013/11/the-great -forgetting/309516/.

CHAPTER 7: POWER

149 **argumentative but eloquent:** For details of Thomas Paine's life: R. R. Fennessy, *Burke, Paine, and the Rights of Man* (New York: Springer, 1963), 12–47; Harvey Kaye, *Thomas Paine and the Promise of America* (New York: Hill & Wang, 2005); Peter Linebaugh, *Peter Linebaugh Presents: Thomas Paine: The Rights of Man and Common Sense* (London: Verso, 2009); Edward Larkin, *Thomas Paine and the Literature of Revolution* (Cambridge, UK: Cambridge University Press, 2005); Craig Nelson, *Thomas Paine: His Life, His Time and the Birth of Modern Nations* (London: Profile, 2007).

149 **a vibrant mix:** "Descriptions of 18th-century Philadelphia before the Revolution," National Humanities Center Resource Toolbox, 2009, http:// nationalhumanitiescenter.org/pds/becomingamer/growth/text2 /philadelphiadescriptions.pdf.

150 **"most influential writer":** Kaye, *Thomas Paine and the Promise of America*; quoted in Mariana Asis and Jason Xidias, *An Analysis of Thomas Paine's "The Rights of Man"* (New York: Routledge, 2017), 62.

150 **The notice in:** Alyce Barry, "Thomas Paine, Privateersman," *Pennsylvania Magazine of History and Biography* 101 (2014): 451–61.

150 **"a pair of globes":** Thomas Paine, *The Age of Reason* (London: Watts & Co., 1945), 42. Full text at https://archive.org/details/in.ernet.dli.2015.202369/page/n5.

152 **"Their constant lectures":** Craig Nelson, "Sample Chapter: Thomas Paine," accessed November 5, 2019, http://www.craignelson.us/books/thomas-paine/sample-chapter/.

153 **"cosmic state":** Nicholas Campion, "Astronomy and Political Theory," *Proceedings of the International Astronomical Union* 5 (2009): 595–602.

153 **"No one has ever heard":** Bruno Latour, *We Have Never Been Modern* (Cambridge, MA: Harvard University Press, 2006), 107; quoted in Campion, "Astronomy and Political Theory."

154 **"Son of Heaven":** Sun Xiaochun, "Crossing the Boundaries between Heaven and Man: Astronomy in Ancient China," in *Astronomy Across Cultures*, ed. Helaine Selin and Sun Xiaochun (New York: Springer, 2000), 423–54.

154 **the rain god Chac:** Ivan Šprajc, "Astronomy and Power in Mesoamerica," in *Astronomy and Power: How Worlds Are Structured: Proceedings of the SEAC 2010 Conference, BAR International Series* 2794 (2016): 185–92.

155 **"centre of a world empire":** Ibrahim Allawi, "Some Evolutionary and Cosmological Aspects to Early Islamic Town Planning," in *Theories and Principles of Design in the Architecture of Islamic Societies* (Aga Khan Program for Islamic Architecture, 1988), 57–72. Also discussed in Nicholas Campion, "Archaeoastronomy and Calendar Cities," *Journal of Physics: Conference Series* 685 (2016): 012005.

155 **image of a powerful king:** Keith Hutchison, "Towards a Political Iconography of the Copernican Revolution," in *Astrology, Science and Society: Historical Essays*, ed. Patrick Curry (Woodbridge, Suffolk, UK: Bydell & Brewer, 1987), 95–141.

155 **"governs the family of planets":** Nicolaus Copernicus, *Complete Works*, vol. 1, trans. Edward Rosen (Baltimore: Johns Hopkins University Press, 1992).

155 **European kings started:** Hutchison, "Copernican Revolution"; Eran Shalev, "A Republic amidst the Stars: Political Astronomy and the Intellectual Origins of the Stars and Stripes," *Journal of the Early Republic* 31 (2011): 39–73.

156 **"system of the world":** Mordechai Feingold, *The Newtonian Moment: Isaac Newton and the Making of Modern Culture* (New York: Oxford University Press, 2005), 157–67.

156 **"language of mathematics":** Galilei Galileo, *Opere* 6:232; Douglas Jesseph, "Galileo, Hobbes and the Book of Nature," *Perspectives on Science* 12 (2004): 191–211.

157 **"solitary, poor, nasty":** Thomas Hobbes, *Leviathan* (Ware, UK: Wordsworth Editions, 2014), 97. See also Jesseph, "Galileo, Hobbes."

157 **Newton replaced this aimless cosmos:** Isaac Newton, *The Mathematical Principles of Natural Philosophy*, vol. 2 (London, 1729), 388; available at http://www.newtonproject.ox.ac.uk/view/texts/normalized/NATP00056. Impact discussed in Rob Iliffe, *Newton: A Very Short Introduction* (Oxford, UK: Oxford University Press, 2007).

157 **"law of nature":** John Locke, *An Essay Concerning Human Understanding* (Ware, UK: Wordsworth Editions, 2014), 718.

158 **"never enough to be admired":** Links between Locke and Newton discussed in: Lisa Downing, "Locke's Newtonianism and Lockean Newtonianism," *Perspectives on Science* 5 (1997): 285–310; Lisa Downing, "Locke's Metaphysics and Newtonian Metaphysics," in *Newton and Empiricism*, ed. Zvi Biener and Eric Schliesser (Oxford, UK: Oxford University Press, 2014), 97–118.

158 **"emblems of a new era":** Mordechai Feingold, "Partnership in Glory: Newton and Locke through the Enlightenment and Beyond," in *Newton's Scientific and Philosophical Legacy*, ed. Paul Scheurer and Guy Debrock (Dordrecht: Kluwer, 1988), 292. For general discussion of Enlightenment philosophy/politics: Jonathan Israel, *A Revolution of the Mind: Radical Enlightenment and the Intellectual Origins of Modern Democracy* (Princeton, NJ: Princeton University Press, 2011); Jonathan Israel, *Democratic Enlightenment: Philosophy, Revolution, and Human Rights, 1750–1790* (Oxford, UK: Oxford University Press, 2013).

158 **seeped through politics:** Carl L. Becker, *The Declaration of Independence: A Study in the History of Political Ideas* (New York: Vintage Books, 1958), 59–60; Nicholas Campion, "Astronomy and Culture in the Eighteenth Century: Isaac Newton's Influence on the Enlightenment and Politics," *Mediterranean Archaeology and Archaeometry* 16 (2016): 497–502; Feingold, *Newtonian Moment.*

158 **compared social bonds to gravity:** Quoted in Feingold, *Newtonian Moment.*

158 **saw celestial mechanics in:** Henry St. John, Viscount Bolingbroke, "A Dissertation upon Parties" (1733–1734), in *The Works of Lord Bolingbroke* (Philadelphia, 1841), II, 85.

158 **most powerful political concepts:** Richard Striner, "Political Newtonianism: The Cosmic Model of Politics in Europe and America," *William and Mary Quarterly*, 3rd series, 52 (1995): 583–608.

159 **"with this kind of government":** Charles de Montesquieu, *The Spirit of the Laws*, trans. Thomas Nugent (Franklin Center, PA: Franklin Library, 1984).

161 **"When the country":** Thomas Paine, *The American Crisis*, chapter vii, November 21, 1778, http://www.ushistory.org/paine/crisis/c-07.htm, quoted in Nelson, *Thomas Paine.*

161 **"In no instance hath nature":** Thomas Paine, *Common Sense* (Girard, KS: Haldeman-Julius Co., 1920), 48, https://archive.org/details/common senseoopainrich/page/n2.

161 **"an ass for a lion":** Paine, *Common Sense*, 29.

161 **"gravitating power":** Paine, *Common Sense*, 17.

161 **"to begin the world":** Paine, *Common Sense*, 84.

162 **"a fool or a fanatic":** Bernard Bailyn, *Faces of Revolution: Personalities and Themes in the Struggle for American Independence* (New York: Knopf, 1990), 67.

162 **"Pain took an idea":** Nelson, *Thomas Paine*, 82.

162 **"Independence a year ago":** John Keane, *Tom Paine: A Political Life* (Boston: Little, Brown, 1995), 145; quoted in Nelson, *Thomas Paine*, 93.

162 **"proper sphere . . . the centre of attraction":** Quoted in Shalev, "Republic amidst the Stars."

162 **"so many balls in the air":** Shalev, "Republic amidst the Stars."

163 **"republic amidst the stars":** Shalev, "Republic amidst the Stars."

164 **"in their proper orbits":** Max Farrand, ed., *The Records of the Federal Convention of 1787*, vol. 1 (New Haven, CT: Yale University Press, 1911); quoted in Striner, "Political Newtonianism."

164 **"those attractions and repulsions":** John Adams, *A Defence of the Constitutions of the Government of the United States of America* (London: Dilly, 1787–1788); quoted in Shalev, "Republic amidst the Stars."

164 **"thirteen stars, white in a blue field":** Shalev, "Republic amidst the Stars."

164 **"It did not appear to me":** Thomas Paine, letter to George Washington, July 21, 1791; quoted in Nelson, *Thomas Paine*.

165 **hugely successful pamphlet:** Edmund Burke, *Reflections on the Revolution in France, and on the Proceedings in Certain Societies in London Relative to That Event: In a Letter Intended to Have Been Sent to a Gentleman in Paris* (London, 1790); "Contagion" quoted in Nelson, *Thomas Paine*, 192.

165 **"Never were more pains":** Thomas Paine, *The Rights of Man* (London: Watts & Co., 1906), 22, https://archive.org/details/rightsmanoopaingoog/page/n9.

166 **"a hostile mob":** Nelson, "Sample Chapter."

166 **"*Vive* Thomas Paine!":** Nelson, *Thomas Paine*, 235.

167 **compared the French Republic:** Nelson, *Thomas Paine*, 253.

167 **"I saw my life":** Thomas Paine, letter to Samuel Adams, January 1, 1803, http://www.deism.com/paine_essay_sam_adams.htm.

167 **"made myself master":** Paine, *Age of Reason*, 42.

168 **"terrify and enslave":** Paine, *Age of Reason*, 2.

168 **"exceedingly necessary":** Paine, *Age of Reason*, 1.

169 **ideas from the lectures:** David Hoffman, "'The Creation We Behold': Thomas Paine's 'The Age of Reason' and the Tradition of Physico-theology," *Proceedings of the American Philosophical Society* 157 (2013): 281–303.

169 **"millions of worlds as large":** Paine, *Age of Reason*, 44.

169 **"millions of worlds equally dependent":** Paine, *Age of Reason*, 49; arguments discussed in Hoffman, "'Creation We Behold.'"

170 **"scatters it in the mind":** Paine, *Age of Reason*, 43.

170 **"now called natural philosophy":** Paine, *Age of Reason*, 28.

170 **tradition of deism:** Fennessy, *Burke, Paine, and the Rights of Man*, 12–47.

171 **published in 1794:** Franklyn Prochaska, "Thomas Paine's The Age of Reason Revisited," *Journal of the History of Ideas* 33 (1972): 561–76.

171 **"To such a pitch":** Thomas Paine, letter to George Washington, Paris, July 30, 1796, https://www.thomaspaine.org/major-works/letter-to-george-washington.html.

171 **"I had then but little":** Paine, *Age of Reason*, 62.

171 **Paine's temperature:** Nelson, *Thomas Paine*, 282.

172 **"drunken atheist":** Keane, *Tom Paine*, 451; quoted in Nelson, *Thomas Paine*.

172 **"loathsome reptile":** Alfred Aldridge, *Man of Reason: The Life of Thomas Paine* (Philadelphia: Lippincott, 1959), 269.

172 **"demi-human archbeast":** *The Complete Writings of Thomas Paine*, vol. 1, ed. Philip Foner (New York: Citadel Press, 1945), xlii.

173 **"intelligent and powerful Being":** Isaac Newton, *Mathematical Principles of Natural Philosophy*, 388.

173 **"men like Paine":** Bertrand Russell, "The Fate of Thomas Paine," in *Why I Am Not a Christian* (New York: Routledge, 2004), 70–83.

173 **vital step:** Christopher Hitchens, *God Is Not Great: How Religion Poisons Everything* (New York: Twelve, 2007); Christopher Hitchens, *Thomas Paine's Rights of Man* (New York: Atlantic Monthly Press, 2006). Thomas Paine is also championed by Humanists UK (https://humanism.org.uk/humanism/the-humanist-tradition/enlightenment/thomas-paine/) and the National Secular Society, which says: "The development of secularism cannot be understood without reference to Thomas Paine" (https://www.secularism.org.uk/thomas-paine.html).

173 **"disenchantment":** Richard Jenkins, "Disenchantment, Enchantment and Re-Enchantment: Max Weber at the Millennium," *Max Weber Studies* 1 (2000): 11–32. Also discussed in Nicholas Campion, "Enchantment and the Awe of the Heavens," *Inspiration of Astronomical Phenomena VI* 441 (2011): 415.

173 **death of God:** Friedrich Nietzsche, *The Joyous Science*, trans. R. Kevin Hill (London: Penguin Classics, 2016).

174 **"greatest discovery":** Newton's influence on Adam Smith is discussed in Feingold, *Newtonian Moment*.

174 **hundreds of psychological "laws":** Karl Teigen, "One Hundred Years of Laws in Psychology," *American Journal of Psychology* 115 (2002): 103–18.

CHAPTER 8: LIGHT

175 **an unfamiliar light:** Simon Schaffer, "Uranus and the Establishment of Herschel's Astronomy," *Journal of the History of Astronomy* 12 (1981): 11–26.

175 **a shift from observing:** Discussed, for example, in Richard Holmes, *The Age of Wonder: How the Romantic Generation Discovered the Beauty and Terror of Science* (New York: HarperCollins, 2008).

176 **endless astronomical garden:** For example, see Erasmus Darwin, *The Botanic Garden* (London: J. Johnson, 1791).

176 **Auguste Comte:** Jonathan Turner et al., "The Sociology of Auguste Comte," in *The Emergence of Sociological Theory* (Thousand Oaks, CA: Sage, 2012), 37–54.

176 **"We can imagine the possibility":** Auguste Comte, *Cours de Philosophie Positive*, vol. 2 (Paris: Baillière, 1864), 6; quoted in Barbara Becker, "Celestial Spectroscopy: Making Reality Fit the Myth," *Science* 301 (2003): 1332–33.

177 **"no proper astronomical interest":** Edward Maunder, *The Royal Observatory Greenwich* (London: The Religious Tract Society, 1900), 266–67.

177 **"surprised them to their bones":** Biman Nath, "From Chemistry to the Stars," in *The Story of Helium and the Birth of Astrophysics* (New York: Springer, 2013), 38.

177 **"smell of burning Bunsen":** Lawson Cockcroft, "A Perilous Life," *Chemistry in Britain*, May 1999, 49–50.

178 **"Beneath our feet":** Cockcroft, "A Perilous Life."

178 **physicist Gustav Kirchhoff:** Bunsen and Kirchhoff's collaboration is discussed in Owen Gingerich, "The Nineteenth-century Birth of Astrophysics," in *Physics of Solar and Stellar Coronae*, ed. Jeffrey Linsky and Salvatore Serio (Dordrecht, the Netherlands: Kluwer, 1993), 47–58; Andrew King, *Stars: A Very Short Introduction* (Oxford, UK: Oxford University Press, 2012); Nath, "Chemistry to the Stars."

178 **in many ways the opposite:** Kirchhoff's life is discussed in Klaus Hentschel, "Biographical Introduction," in *Gustav Robert Kirchhoff's Treatise "On the Theory of Light Rays" (1882)*, ed. Klaus Hentschel and Ning Yan Zhu (World Scientific, 2016), 1–18; Robert von Helmholtz, "A Memoir of Gustav Robert Kirchhoff," trans. Joseph de Perott, *Annual Report of the Board of Regents of the Smithsonian Institution*, 1890, 527–540, trans. from *Deutsche Rundschau* 14 (1888): 232–45.

180 **"surpasses all others":** Gustav Kirchhoff and Robert Bunsen, "Chemical Analysis by Observation of Spectra," *Spectroscopy* 7 (1860): 20–25, trans. from *Annalen der Physik und der Chemie* 110 (1860): 161–89.

180 **"to designate the blue":** Mary Weeks, "The Discovery of the Elements. XIII. Some Spectroscopic Discoveries," *Journal of Chemical Education* 9 (1932): 1413–34; quoted in Nath, "Chemistry to the Stars," 43.

180 **"If we could determine":** "Some Scientific Centres: The Heidelberg Physical Laboratory," *Nature* 65 (1902): 587–90.

181 **only one survivor:** Life of Joseph von Fraunhofer: Discussed in W. F. T. Schirach, "Joseph von Fraunhofer," *Monthly Notes of the Astronomical Society of South Africa* 9 (1950): 64–67; Nath, "Chemistry to the Stars."

182 **"strong and weak vertical lines":** J. S. Ames, *Prismatic and Diffraction Spectra: Memoirs by Joseph von Fraunhofer* (New York: Harper & Brothers, 1898); quoted in Nath, "Chemistry to the Stars," 24.

183 ***Approximavit Sidera:*** Nath, "Chemistry to the Stars," 26.

184 **quickly worked out:** Gustav Kirchhoff, "On the Relation between the Radiating and Absorbing Powers of Different Bodies for Light and Heat," trans. in *Philosophical Magazine and Journal of Science* 4, no. 20 (1860): 1–21; Nath, "Chemistry to the Stars," 40.

184 **"doesn't let us sleep":** Letter from Robert Bunsen to Henry Roscoe, November 15, 1859, in *The Life and Experiences of Sir Henry Enfield Roscoe* (London and New York: Macmillan, 1906), 81; quoted in Gingerich, "Birth of

Astrophysics." See also: Gustav Kirchhoff, "Über die Fraunhofer'schen Linien," *Monatsberichte der Königlichen Preussischen Akademie der Wissenschaft zu Berlin* (1859): 662–65; translation in *Philosophical Magazine*, series 4, 19 (1860): 193–97.

185 **"fetching some gold":** Helmholtz, "Memoir of Gustav Robert Kirchhoff," 232–45.

185 **"I shall never forget":** Henry Roscoe, *Ein Leben der Arbeit Errinerungen* (Leipzig: Akademische Verlagsgesellschaft, 1919); quoted in É. V. Shpol'skii, "A Century of Spectrum Analysis," *Soviet Physics Uspekhi* 2 (1960): 967; and Nath, "Chemistry to the Stars," 42.

185 **"to go to the sun":** Becker, "Celestial Spectroscopy."

186 **a well-off family:** William and Mary Huggins's life and work: Discussed in William Huggins and Mary Huggins, *An Atlas of Representative Stellar Spectra*, vol. 1 and 2 (London: William Wesley, 1899); Sarah Whiting, "Lady Huggins," *Astrophysical Journal* 42 (1915): 1–4; Charles Mills and C. F. Brooke, *A Sketch of the Life of Sir William Huggins* (London: Times Printing Works, 1936); Barbara Becker, "Dispelling the Myth of the Able Assistant: Margaret and William Huggins at Work in the Tulse Hill Observatory," in *Creative Couples in the Sciences*, ed. Helena Pycior et al. (New Brunswick, NJ: Rutgers University Press, 1996); Barbara Becker, *Unrávelling Starlight: William and Margaret Huggins and the Rise of the New Astronomy* (Cambridge, UK: Cambridge University Press, 2011).

186 **"a little dissatisfied":** William Huggins, "The New Astronomy: A Personal Retrospect," *Nineteenth Century* 41 (1897): 907–29.

187 **"a spring of water":** Huggins, "The New Astronomy."

187 **"lift a veil":** Huggins, "The New Astronomy."

187 **Becker has questioned:** Becker, "Celestial spectroscopy"; Becker, *Unravelling Starlight*.

187 **"noxious gases":** Huggins, "The New Astronomy."

187 **"The characteristic light-rays":** Huggins and Huggins, *Atlas*, vol. 1, 8.

188 **fifty prominent stars:** William Huggins and William Miller, "On the Spectra of Some of the Fixed Stars," *Philosophical Transactions of the Royal Society* 154 (1864): 413–35.

188 **"most closely connected":** Huggins and Miller, "On the Spectra," 434.

189 **"shining fluid":** Suggested by William Herschel; quoted in Stewart Moore, "Historical Note: 150 years of Astronomical Spectroscopy," *Journal of the British Astronomical Association* (August 2014): 186–87.

189 **"island universes":** This term was introduced later, but the idea was supported by Immanuel Kant, for example; discussed in Michael Crowe, *Modern Theories of the Universe: From Herschel to Hubble* (New York: Dover, 1994), 69–70.

189 **"secret place of creation":** Huggins, "The New Astronomy."

189 **this nebula was producing:** Huggins and Miller, "On the Spectra."

189 **"substance unknown":** William Huggins, "A supplement to the paper 'On the Spectra of Some of the Fixed Stars,'" *Philosophical Transactions* 154 (1864): 443.

189 **"nebulium":** Becker "Dispelling the Myth," footnote 59.

190 **"wondrous objects":** Huggins and Miller, "On the Spectra."

190 **"to my great joy":** Huggins, "The New Astronomy."

190 **"vast convulsion":** Huggins, "The New Astronomy."

190 **"a world on fire":** Huggins, "The New Astronomy."

190 **Huggins wrote:** James Clerk Maxwell, *The Scientific Letters and Papers of James Clerk Maxwell, Vol. 2: 1862–1873*, ed. P. M. Harman (Cambridge, UK: Cambridge University Press, 1995), 306.

191 **Charles Darwin . . . published:** Charles Darwin, "Inherited Instinct," *Nature* 7 (1873): 281.

191 **Barbara Becker suggests:** Becker, "Dispelling the Myth."

192 **Henry and Anna Draper:** Joseph Tenn, "The Hugginses, the Drapers, and the Rise of Astrophysics," *Griffith Observer*, October 1986, 1–15.

192 **"universal joints":** Mills and Brooke, *Sketch*, 38–40.

192 **"No imagination could fail":** Mills and Brooke, *Sketch*, 37.

192 **"a happier man":** Mills and Brooke, *Sketch*.

193 **"always puzzled me":** Mills and Brooke, *Sketch*.

194 **"By each discovery":** Quoted in George Hale, "The Work of Sir William Huggins," *Astrophysical Journal* 37 (1915): 145–53.

194 **a grand atlas:** Huggins and Huggins, *Atlas*.

194 **"one of the greatest":** Mills and Brooke, *Sketch*.

195 **"They gazed long and sadly":** Mills and Brooke, *Sketch*.

195 **all of the heavier:** William Fowler, "Experimental and Theoretical Nuclear Astrophysics: The Quest for the Origin of the Elements," Nobel lecture, 1983; George Wallerstein et al., "Synthesis of the Elements in the Stars: Forty Years of Progress," *Review of Modern Physics* 69 (1997): 995–1084.

197 **"They didn't look as if":** Gary Fildes, *An Astronomer's Tale* (London: Century, 2016), 253.

CHAPTER 9: ART

199 **Reviewers said it made:** Rosamund Bartlett and Sarah Dadswell, eds., *Victory over the Sun: The World's First Futurist Opera* (Exeter, UK: University of Exeter Press, 2012), 88–89, 95.

199 **at the Lunar Park theater:** Details of the production discussed in: Bartlett and Dadswell, *Victory over the Sun*; Charlotte Douglas, "Victory over the Sun," *Russian History* 8 (1981): 69–89; Anna Kisselgoff, "Victory over the Sun," *New York Times*, January 27, 1981, https://www.nytimes.com/1981/01/27/arts/theater-victory-over-the-sun.html; Isobel Hunter, "Zaum and Sun: The 'First Futurist Opera' Revisited," *Central Europe Review*, July 12, 1999, https://www.pecina.cz/files/www.ce-review.org/99/3/ondisplay3_hunter.html.

200 **"It's an attempt to shock":** Douglas, "Victory over the Sun."

201 **"ludicrous waste":** Andrew Clements, "Victory over the Sun," *The Guardian*, June 22, 1999, https://www.theguardian.com/culture/1999/jun/22/artsfeatures2.

202 **"did not equate the real":** Ronald Rees, "Historical Links between Cartography and Art," *Geographical Review* 70 (1980): 66.

202 **just as cartographers adopted:** Rees, "Historical Links," 60–78.

203 **invention of photography:** For example, John Berger, *Ways of Seeing* (London: Penguin, 1972; 2008), 7–34.

204 **It was shocking:** Arthur I. Miller, *Einstein, Picasso: Space, Time, and the Beauty That Causes Havoc* (New York: Basic Books, 2001), 6; Jonathan Jones, "Pablo's Punks," *The Guardian*, January 9, 2007, https://www.theguardian.com/culture/2007/jan/09/2.

204 **a surge of scientific:** On scientific developments and how they influenced artists: Linda Henderson, "Vibratory Modernism: Boccioni, Kupka, and the Ether of Space," in *From Energy to Information: Representation in Science and Technology, Art, and Literature*, ed. Bruce Clarke and Linda Henderson (Palo Alto, CA: Stanford University Press, 2002), 126–50; Linda Dalrymple Henderson, "Editor's Introduction: I. Writing Modern Art and Science—An Overview; II. Cubism, Futurism, and Ether Physics in the Early Twentieth Century," *Science in Context* 17 (2004): 423–66; Linda Henderson, "Abstraction, the Ether, and the Fourth Dimension: Kandinsky, Mondrian, and Malevich in Context," in *The Infinite White Abyss: Kandinsky, Malevich, Mondrian*, ed. Marlon Ackermann et al. (Köln, Germany: Snoeck Verlagsgesellschaft, 2014), 233–44.

205 **"is as essential":** J. J. Thomson, "Address by the President, Sir J. J. Thomson," in *Report of the 79th Meeting of the British Association for the Advancement of Science (1909)* (London: John Murray, 1910), 15; quoted in Henderson, "Writing Modern Art and Science."

205 **the ultimate source:** Henderson, "Vibratory Modernism"; Henderson, "Writing Modern Art and Science."

205 **hidden extra dimensions:** Linda Henderson, "The Image and Imagination of the Fourth Dimension in Twentieth-Century Art and Culture," *Configurations* 17 (2009): 131–60; Linda Henderson, *The Fourth Dimension and Non-Euclidean Geometry in Modern Art*, rev. ed. (Cambridge, MA: MIT Press, 2013).

206 **invisible realms:** Henderson, "Writing Modern Art and Science," 447.

206 **"language of the modern studios":** Guillaume Apollinaire, "La Peinture nouvelle: Notes d'art," *Les Soirées de Paris* 3 (April 1912): 90; discussed in Henderson, *Fourth Dimension*, ch. 2.

206 **"atmosphere condensed":** Umberto Boccioni, *Dynamisme plastique: peinture et sculpture futuristes*, ed. Giovanni Lista, trans. Claude Minot and Giovanni Lista (Lausanne, Switzerland: L'Age d'homme, 1975), 104; discussed in Henderson, "Vibratory Modernism."

206 **The extent to which:** Henderson, "Vibratory Modernism."

207 **"Everything showed me its face":** Wassily Kandinsky, "Reminiscences/Three Pictures (1913)," in *Kandinsky: Complete Writings on Art*, ed. Kenneth Lindsay and Peter Vergo (Cambridge, MA: Da Capo Press, 1994), 361.

207 **grew up in Odessa:** On Kandinsky's life: Jerome Ashmore, "Vasily Kandinsky and His Idea of Ultimate Reality," *Ultimate Reality and Meaning* 2 (1979): 228–56; "Wassily Kandinsky and His Paintings," accessed November 5, 2019, http://wassily-kandinsky.org.

207 **"brightly-colored living pictures":** Peg Weiss, *Kandinsky and Old Russia: The Artist as Ethnographer and Shaman* (New Haven, CT: Yale University Press, 1995), 4.

207 **"That it was a haystack":** Kandinsky, "Reminiscences," 363.

208 **"look down at the universe":** Roy McMullen, "Wassily Kandinsky," *Encyclopaedia Britannica*, accessed November 5, 2019, https://www.britannica.com/biography/Wassily-Kandinsky.

208 **how he felt when he attended:** Kandinsky, "Reminiscences," 363–64.

208 **"like a wild tuba":** Kandinsky, "Reminiscences," 360.

208 **"objects harmed my pictures":** Kandinsky, "Reminiscences," 369–70.

208 **Kandinsky smashed it:** Donald Kuspit, "Spiritualism and Nihilism: The Second Decade," in *A Critical History of 20th-Century Art* (Albany: State University of New York Press, 2008), http://www.artnet.com/magazineus/features/kuspit/kuspit2-17-06.asp.

208 **made three years later:** McMullen, "Kandinsky."

209 **"an artistic leap":** Kuspit, "Spiritualism and Nihilism."

209 **"collapse of the atom":** Ashmore, "Vasily Kandinsky and His Idea of Ultimate Reality," 239.

209 **"new kind of artistic communication":** Henderson, "Vibratory Modernism."

209 **transmission of thought:** Henderson, "Vibratory Modernism."

209 **"an innumerable network":** Edward Carpenter, *The Art of Creation: Essays on the Self and Its Powers* (London: George Allen, 1904); quoted in Henderson, "Vibratory Modernism."

209 **"watch for new emotions":** Ezra Pound, "The Wisdom of Poetry," *Forum* 47 (April 1912): 500; quoted in Henderson, "Vibratory Modernism."

210 **a kind of "world soul":** Linda Henderson, "Modernism and Science," in *Modernism*, ed. Astradur Eysteinsson and Vivian Liska (Amsterdam: John Benjamins, 2007), 391.

210 **joined the Theosophical Society:** Discussions of Kandinsky's theosophical influences: Sixten Ringbom, "Art in the Epoch of the Great Spiritual: Occult Elements in the Early Theory of Abstract Painting," *Journal of the Warburg and Courtauld Institutes* 29 (1966): 386–418; Sixten Ringbom, *The Sounding Cosmos: A Study in the Spiritualism of Kandinsky and the Genesis of Abstract Painting* (Turku, Finland: Äbo Akademi, 1970); Wessel Stoker, "Kandinsky: Art as Spiritual Bread," in *Where Heaven and Earth Meet: The Spiritual in the Art of Kandinsky, Rothko, Warhol and Kiefe* (Amsterdam: Rodopi, 2012); Henderson, "Abstraction, the Ether, and the Fourth Dimension."

210 **"inner sound" . . . "a cosmic event":** Wassily Kandinsky, "Point and Line to Plane," in Kandinsky, *Complete Writings*, 619; quoted in Stoker, "Kandinsky: Art as Spiritual Bread," 68.

210 **"cosmic laws":** Ashmore, "Vasily Kandinsky and His Idea of Ultimate Reality."

210 **"nightmare of materialist ideas":** Wassily Kandinsky, *Concerning the Spiritual in Art*, updated version of the Sadleir translation (New York: Wittenborn, 1972), 24; quoted in Ashmore, "Vasily Kandinsky and His Idea of Ultimate Reality."

211 **"In this era":** Kandinsky, *Complete Writings*, 97–98; quoted in Kuspit, "Spiritualism and Nihilism."

211 **"The world sounds":** Wassily Kandinsky, *Der Blaue Reiter* (1912), in Kandinsky, *Complete Writings*, 250.

211 **"reality in its own right":** Kuspit, "Spiritualism and Nihilism." See also Michel Henry, *Seeing the Invisible: On Kandinsky*, trans. Scott Davidson (New York: Continuum, 2005).

211 **far from cultural centers:** "About the Artist," The Malevich Society, accessed November 5, 2019, http://malevichsociety.org/about-the-artist/.

212 **took part in a debate:** Aleksandra Shatskikh, *Black Square and the Origin of Suprematism*, trans. Marian Schwartz (New Haven, CT: Yale University Press, 2012), ch. 1.

212 **"Reason has shut art up":** Shatskikh, *Black Square*, ch. 1.

212 **clashing quotations:** Shatskikh, *Black Square*, ch. 1.

212 **"powerless to comprehend":** Shatskikh, *Black Square*, ch. 1.

213 **"Planes reveal themselves":** Letter dated March 5, 1914, postscript; quoted in Shatskikh, *Black Square*, ch. 1.

213 **"Marvelous geometric figures":** Shatskikh, *Black Square*, ch. 1.

213 **"partial eclipse":** Shatskikh, *Black Square*, ch. 1.

213 **"ecstatic illumination":** Shatskikh, *Black Square*, 45.

213 **"fiery lightning bolts":** Shatskikh, *Black Square*, 45.

213 **"outstripped his own brush":** Shatskikh, *Black Square*, 45.

213 **"could not eat, drink or sleep":** Shatskikh, *Black Square*, 45.

213–214 **"an extreme act":** Peter Schjeldahl, "The Prophet: Malevich's Revolution," *The New Yorker*, June 2, 2003, https://www.newyorker.com/magazine/2003/06/02/the-prophet-2.

214 **"felt the night":** Shatskikh, *Black Square*, 261.

214 **"this is the very 'icon'":** Review by Alexandre Benois; quoted in Shatskikh, *Black Square*, 109.

214 **"white abyss":** Shatskikh, *Black Square*, 252.

214 **"the ultimate illusion":** El Lissitzky, "A. and Pangeometry, 1925," in Sophie Lissitzky-Küppers, *El Lissitzky: Life, Letters, Texts* (London: Thames & Hudson, 1968), 354.

214 **"there are such bodies":** Shatskikh, *Black Square*, 140.

215 **"overwhelmed by awe":** Shatskikh, *Black Square*, 140.

215 **Malevich was responding:** Henderson, "Image and Imagination of the Fourth Dimension"; Henderson, "Abstraction, the Ether, and the Fourth Dimension."

215 **"a solo 'leap'":** Shatskikh, *Black Square*, 53.

215 **Bragdon published diagrams:** Claude Bragdon, *Primer of Higher Space: The Fourth Dimension* (Rochester, NY: Manas Press, 1913); discussed in Henderson, "Image and Imagination of the Fourth Dimension"; Henderson, "Abstraction, the Ether, and the Fourth Dimension."

215 **parable about a race of people:** Claude Bragdon, *Man the Square: A Higher Space Parable* (Rochester, NY: Manas Press, 1912).

215 **"A new horizon opens":** Charles Hinton, *The Fourth Dimension* (London: Swann Sonnenschein, 1904); quoted in Henderson, "Image and Imagination of the Fourth Dimension."

215 **Malevich's Suprematist polygons:** Henderson, "Image and Imagination of the Fourth Dimension." See also: Stephen Luecking, "A Man and His Square: Kasimir Malevich and the Visualization of the Fourth Dimension," *Journal of Mathematics and the Arts* 4 (2010): 87–100; Linda Henderson, "Malevich, the Fourth Dimension and the Ether of Space One Hundred Years Later," in *Celebrating Suprematism: New Approaches to the Art of Kazimir Malevich*, ed. Christina Lodder (Leiden, the Netherlands: Brill, 2018), 44–80.

215 **"possess the power":** Peter Ouspensky, *Tertium Organum* (New York: Routledge & Kegan Paul, 1965), 145.

216 **"I lay these bones":** Kazimir Malevich, letter to Mikhail Matyushin, November 1916; quoted in Shatskikh, *Black Square*, ch. 4.

216 **"cosmos is dissolution":** Shatskikh, *Black Square*, 253.

216 **sold for $20 million:** Christie's, New York, May 13, 2015, https://www.christies.com/lotfinder/Lot/robert-ryman-b-1930-bridge-5896026-details.aspx.

216 **didn't even mention it . . . thought it was a joke:** Shatskikh, *Black Square*, ch. 5.

217 **"Zero = All":** Shatskikh, *Black Square*, ch. 5 and epilogue.

217 **trying to communicate:** Shatskikh, *Black Square*, ch. 5.

217 **"Back in the summer":** Kazimir Malevich, letter to Mikhail Matyushin, November 1917; quoted in Shatskikh, *Black Square*, ch. 5.

217 **"meteors, suns, comets and planets":** Kazimir Malevich, "God Is Not Cast Down (1920)" in *K. S. Malevich: Essays on Art 1915–1933*, ed. Troels Andersen (Copenhagen: Borgen, 1968), 193–97; quoted in Charlotte Gill, "'An Urge to Take Off from the Earth': How Malevich Embodies the Role of 'Shamanic Artist' in His Early Career," *North Street Review: Arts and Visual Culture* 17 (2014): 53–62.

217 **"If all artists could see":** Kazimir Malevich, "The Art of the Savage and Its Principle (1915)," in *Malevich: Essays on Art*, 29; quoted in Gill, "Shamanic Artist."

217 **the mystical spirituality:** Gill, "Shamanic Artist."

217 **"I am the beginning":** Troels Andersen, ed., *Malevich IV: The Artist: Infinity, Suprematism: Unpublished Writings 1913–33* (Copenhagen: Borgen, 1978), 29; quoted in Gill, "Shamanic Artist."

218 **"strain the laws":** James Housefield, *Playing with Earth and Sky: Astronomy, Geography and the Art of Marcel Duchamp* (Lebanon, NH: Dartmouth College Press, 2016), 17.

218 **"new standard of measurement":** Jonathan Williams, "Pata or Quantum: Duchamp and the End of Determinist Physics," *Tout-fait: The Marcel Duchamp Studies Online Journal* 1 (2000), https://www.toutfait.com/issues/issue_3/Articles/williams/williams.html.

218 **shaved his hair:** Housefield, *Playing with Earth and Sky*, 137–53. See also Linda Henderson, "The 'Large Glass' Seen Anew: Reflections of Contemporary Science and Technology in Marcel Duchamp's 'Hilarious Picture,'" *Leonardo* 32 (1999): 113–26.

218 **"paces back and forth":** André Breton, *Manifesto of Surrealism* (Paris, 1924); discussed in Gavin Parkinson, *Surrealism, Art and Modern Science: Relativity, Quantum Mechanics, Epistemology* (New Haven, CT: Yale University Press, 2008), 38.

218 **"the smart set . . . buried suns":** André Breton and Philippe Soupault, *Les Champs Magnétiques* (Paris, 1919); discussed in Parkinson, *Surrealism, Art and Modern Science*, 48–50.

218 **a real-life solar eclipse:** Confirmation of the theory of general relativity: Abraham Pais, *Subtle Is the Lord: The Science and the Life of Albert Einstein* (New York: Oxford University Press, 2005); Parkinson, *Surrealism, Art and Modern Science*, ch. 1. Accounts of Eddington's expedition: Frank Dyson et al., "A Determination of the Deflection of Light by the Sun's Gravitational Field, from Observations Made at the Total Eclipse of May 29, 1919," *Philosophical Transactions of the Royal Society A* 220 (1920): 291–333; Malcolm Longair, "Bending Space-time: A Commentary on Dyson, Eddington and Davidson (1920) 'A Determination of the Deflection of Light by the Sun's Gravitational Field,'" *Philosophical Transactions of the Royal Society A* 373 (2015): 20140287; Peter Coles, "Einstein, Eddington and the 1919 Eclipse," in *Historical Development of Modern Cosmology, ASP Conference Proceedings* 252 (2001): 21.

218 **"weighing light":** Arthur Eddington, "The Total Eclipse of 1919 May 29 and the Influence of Gravitation on Light," *The Observatory* 42 (1919): 121.

219 **He presented his theory:** Longair, "Bending Space-time."

219 **The data was ambiguous:** Ben Almassi, "Trust in Expert Testimony: Eddington's 1919 Eclipse Expedition and the British Response to General Relativity," *Studies in History and Philosophy of Modern Physics* 40 (2009): 57–67; Coles, "Einstein, Eddington and the 1919 Eclipse."

220 **the headline in the *Times*:** Almassi, *Trust in Expert Testimony*; Coles, "Einstein, Eddington and the 1919 Eclipse." Two days later, the *New York Times* followed up with the headline "Lights All Askew in the Heavens."

221 **triggered fierce debates:** Juan Marin, "'Mysticism' in Quantum Mechanics: The Forgotten Controversy," *European Journal of Physics* 30 (2009): 807–22. For the influence of quantum mechanics on the surrealists, see Parkinson, *Surrealism*, especially ch. 1.

223 **city of Vitebsk:** Aleksandra Shatskikh, "Vitebsk in the Career of Kazimir Malevich," in *Vitebsk: The Life of Art* (New Haven, CT: Yale University Press, 2007), 184–97.

223 **immersed in astronomy:** Aleksandra Shatskikh, "The Cosmos and the Canvas: Malevich at Tate Modern," *Tate Etc.* 31 (Summer 2014), https://www.tate.org.uk/tate-etc/issue-31-summer-2014/cosmos-and-canvas.

223 ***sputnik:*** Shatskikh, "The Cosmos and the Canvas."

223 **"architektons":** Shatskikh, "Vitebsk in the Career of Kazimir Malevich"; Shatskikh, "The Cosmos and the Canvas."

223 **a children's picture book:** El Lissitzky, *Suprematist Tale of Two Squares* (Berlin: Verlag Skythen, 1922).

224 **moved to Petrograd:** "About the Artist," The Malevich Society; Shatskikh, "Vitebsk in the Career of Kazimir Malevich," 191.

224 **"counterrevolutionary sermonizing":** Gilles Néret, *Malevich* (Köln, Germany: Taschen, 2003), 93; Brian Dailey et al., "To Look Is to Think: A Conversation with Brian Dailey," *ASAP Journal* 2 (2017): 60.

224 **signed with a black square:** Laura Cumming, "Malevich Review: An Intensely Moving Retrospective," *The Guardian*, July 20, 2014, https://www.theguardian.com/artanddesign/2014/jul/20/malevich-tate-modern-review-intensely\-moving-retrospective.

224 **marked by a white cube:** Néret, *Malevich*, 94.

CHAPTER 10: LIFE

226 **the oysters had corrected:** Frank Brown Jr., "Persistent Activity Rhythms in the Oyster," *American Journal of Physiology* 178 (1954): 510–14; Frank Brown Jr., "The Rhythmic Nature of Animals and Plants," *American Scientist* 47 (1959): 147–68; Frank Brown Jr., "Hypothesis of Environmental Timing of the Clock," in *The Biological Clock: Two Views*, ed. Frank Brown Jr., et al. (New York: Academic Press, 1970), 13–60.

226 **a cautionary tale:** For example, see Russell Foster and Leon Kreitzman, "Oscillators, Clocks and Hourglasses," in *Rhythms of Life: The Biological Clocks That Control the Daily Lives of Every Living Thing* (New Haven, CT: Yale University Press, 2004), 46.

227 **earliest known written account:** William Schwartz and Serge Daan, "Origins: A Brief Account of the Ancestry of Circadian Biology," in *Biological Timekeeping: Clocks, Rhythms and Behaviour*, ed. V. Kumar (New York: Springer, 2017), 3–22.

228 **N-rays:** Richard Noakes, "Haunted Thoughts of the Careful Experimentalist: Psychical Research and the Troubles of Experimental Physics," *Studies in History and Philosophy of Biological and Biomedical Sciences* 48 (2014): 46–56.

228 **no convincing discoveries:** Egil Asprem, "Parapsychology: Naturalizing the Supernatural, Re-enchanting Science," in *Handbook of Religion and the Authority of Science*, ed. Jim Lewis and Olav Hammer (Leiden, the Netherlands: Brill, 2010), 633–72; Richard Noakes, "The Historiography of Psychical Research: Lessons from Histories of the Sciences," *Journal of the Society for Psychical Research* (2008), http://hdl.handle.net/10036/36372; Noakes, "Haunted Thoughts"; Andreas Sommer, "Psychical Research in the

History and Philosophy of Science: An Introduction and Review," *Studies in History and Philosophy of Biological and Biomedical Sciences* 48 (2014): 38–45.

228 **Clever Hans:** Edward Heyn, "Berlin's Wonderful Horse: He Can Do Almost Everything But Talk—How He Was Taught," *New York Times*, September 4, 1904; Fabio De Sio and Chantal Marazia, "Clever Hans and His Effects: Karl Krall and the Origins of Experimental Parapsychology in Germany," *Studies in History and Philosophy of Biological and Biomedical Sciences* 48 (2014): 94–102. Clever Hans's influence on the field of animal behavior: Michael Beran, "To Err Is (Not Only) Human: Fallibility as a Window into Primate Cognition," *Comparative and Cognition Behavior Reviews* 12 (2017): 57–82; Phillip Veldhuis, "Bees, Brains and Behaviour" (master's thesis, University of Manitoba, Winnipeg, 1999).

228 **fierce skepticism:** Discussed in James Gould, "Animal Navigation: The Evolution of Magnetic Orientation," *Current Biology* 18 (2008): R482–84.

228 **tiny seaside town:** On Brown's life: Frank Brown Jr., "Biological Clocks and Rhythms," in *Discovery Processes in Modern Biology: People and Processes in Biological Discovery*, ed. W. Klemm (Huntington, NY: Krieger, 1977), 2–24.

229 **pipefish and jacks:** On the inhabitants of Sargasso Sea: http://www.sargassoseacommission.org/about-the-sargasso-sea.

229 **filled with creatures:** Brown, "Rhythmic nature."

229 ***Anchistioides antiguensis*:** Brown, "Biological Clocks and Rhythms"; J. Wheeler and Frank Brown, "The Periodic Swarming of *Anchistioides antiguensis* (Schmitt) at Bermuda," *Zoological Journal of the Linnean Society* 39 (1936): 413–28.

229 **two monthly *A. antiguensis* swarms:** Wheeler and Brown, "Periodic Swarming."

229 **live inside sponges:** Guidomar Soledade et al., "New Records of Association between Caridean Shrimps (Decapoda) and Sponges (Porifera) in Abrolhos Archipelago, Northeastern Brazil," *Nauplius: Journal of the Brazilian Crustacean Society* 25 (2017): e2017027.

229 **summer teaching job:** Brown, "Biological Clocks and Rhythms."

229 **warned him to avoid:** Brown, "Rhythmic Nature"; "Environmental Timing."

229 **blanch and blacken each day:** Brown, "Rhythmic Nature"; "Environmental Timing"; "Biological Clocks and Rhythms."

230 **handful of other scientists:** J. Bonner, "Erwin Bünning (23 January 1906–4 October 1990)," *Proceedings of the American Philosophical Society* 138 (1994): 318–20; Serge Daan and Eberhard Gwinner, "Obituary: Jürgen Aschoff (1913–98)," *Nature* 396 (1998): 418; M. Chandrashekaran, "Biological Rhythms Research: A Personal Account," *Journal of Biosciences* 23 (1998): 545–55; Kim Kiser, "Father Time," *Minnesota Medicine*, November 2005, 26–30; Woodland Hastings, "Colin Stephenson Pittendrigh: A Memoir," *Resonance* 11 (May 2006): 81–86; Germaine Cornelissen, "Reminiscences: In Memoriam of Franz Halberg," *World Heart Journal* 5 (2013): 197–98; Germaine Cornelissen, "Franz Halberg: A Maverick Ahead of His Time," *Herald of the International Academy of Sciences. Russian Section* 1 (2018): 78–84.

230 **private, internal timers:** Schwartz and Daan, "Origins."

231 **experiments that ran for years:** Brown, "Rhythmic Nature"; "Environmental Timing."

232 **parapsychology to paranoia:** Chandrashekaran, "Biological Rhythms Research"; Kiser, "Father Time"; Cornelissen, "Reminiscences."

232 **"exogenous rhythm of the unicorn":** LaMont Cole, "Biological Clock in the Unicorn," *Science* 125 (1957): 874–76.

232 **"hit us very hard":** Brown, "Biological Clocks and Rhythms," 13.

232 **"circadian":** Foster and Kreitzman, *Rhythms of Life*, 41.

232 **a prestigious conference:** Foster and Kreitzman, *Rhythms of Life*; *A Blog Around the Clock*, "Clock Tutorial #2a: Forty-five Years of Pittendrigh's Empirical Generalizations," blog entry by Bora Zivkovic, July 3, 2006, https://blog.coturnix.org/2006/07/03/clocktutorial_3_fortyfive_year/.

233 **wasn't invited at first:** Brown, "Biological Clocks and Rhythms."

233 **came down to temperature:** Frank Brown Jr., "Response to Pervasive Geophysical Factors and the Biological Clock Problem," *Cold Spring Harbor Symposia on Quantitative Biology* 25 (1960): 57–71; Brown, "Environmental Timing"; Foster and Kreitzman, *Rhythms of Life.*

233 **capacity to germinate:** Erwin Bünning, "Endogenous Rhythms in Plants," *Annual Review of Plant Physiology* 7 (1956): 71–90.

233 **"chasing a ghost":** Colin Pittendrigh, "Circadian Rhythms and the Circadian Organization of Living Systems," *Cold Spring Harbor Symposia on Quantitative Biology* 25 (1960): 159–84. (Comments appear in the published discussion following Pittendrigh's presentation.) Also discussed in Schwartz and Daan, "Origins."

233 **increasingly rejected:** Brown, "Biological Clocks and Rhythms."

233 **Halberg eventually admitted:** Brown, "Biological Clocks and Rhythms."

233–234 **Pittendrigh ignored Brown's arguments:** Hastings, "Colin Stephenson Pittendrigh." See also Chandrashekaran, "Biological Rhythms Research: A Personal Account," which says that Brown's name was hardly ever mentioned, even in passing, in Erwin Bünning's weekly seminars held 1964–67.

234 **refused to name:** Jürgen Aschoff, "Circadian Rhythms in Man," *Science* 148 (1965): 1427–32.

234 **dedicated isolation facility:** Aschoff, "Circadian Rhythms"; Michael Globig, "A World without Day or Night," *Max Planck Research* 2 (2007): 60–61.

234 **the first volunteer:** Aschoff, "Circadian Rhythms"; Anna Wirz-Justice et al., "Rütger Wever: An Appreciation," *Journal of Biological Rhythms* 20 (2005): 554–55.

235 **daily rhythms continued:** Aschoff, "Circadian Rhythms."

235 **"desynchronization":** Aschoff, "Circadian Rhythms"; Jürgen Aschoff et al., "Desynchronization of Human Circadian Rhythms," *Japanese Journal of Physiology* 17 (1967): 450–57.

235 **three mutant fly strains:** Ronald Konopka and Seymour Benzer, "Clock Mutants of *Drosophila melanogaster*," *PNAS* 68 (1971): 2112–16.

236 **named the gene "period":** Pranhitha Reddy et al., "Molecular Analysis of the Period Locus in *Drosophila melanogaster* and Identification of a Transcript Involved in Biological Rhythms," *Cell* 38 (1984): 701–10.

236 **feedback loops . . . every type of life:** Carlos Ibáñez, "Scientific Background: Discoveries of Molecular Mechanisms Controlling the Circadian Rhythm," *The Nobel Assembly*, 2017, https://www.nobelprize.org/uploads/2018/06/advanced-medicineprize2017.pdf.

236 **one of the hottest areas:** Pietro Cortelli, "Chronomedicine: A Necessary Concept to Manage Human Diseases," *Sleep Medicine Reviews* 21 (2015): 1–2; Z. Chen, "What's Next for Chronobiology and Drug Discovery," *Expert Opinion on Drug Discovery* 12 (2017): 1181–85; Linda Geddes, *Chasing the Sun: The New Science of Sunlight and How It Shapes Our Bodies and Minds* (London: Wellcome, 2019).

237 **causing health problems:** Russell Foster and Till Roenneberg, "Human Responses to the Geophysical Daily, Annual and Lunar Cycles," *Current Biology* 18 (2008): R784–94; Kristin Uth and Roger Sleigh, "Deregulation of the Circadian Clock Constitutes a Significant Factor in Tumorigenesis: A Clockwork Cancer. Part I: Clocks and Clocking Machinery," *Biotechnology & Biotechnological Equipment* 28 (2014): 176–83; Ruth Lunn et al., "Health Consequences of Electric Lighting Practices in the Modern World: A Report on the National Toxicology Program's Workshop on Shift Work at Night, Artificial Light at Night, and Circadian Disruption," *Science of the Total Environment* 607/608 (2017): 1073–84.

237 **probable human carcinogen:** Thomas Erren et al., "Shift Work and Cancer: The Evidence and the Challenge," *Deutsches Ärzteblatt International* 107 (2010): 657–62.

237 **"human biology remains":** Foster and Roenneberg, "Human Responses."

237 **"ticking time bomb":** P. Lewis et al., "Ticking Time Bomb? High Time for Chronobiological Research," *EMBO Reports* 19 (2018): e46073.

237 **"condemn whole sectors":** Russell Foster and Katharina Wulff, "The Rhythm of Rest and Excess," *Nature Reviews Neuroscience* 6 (2005): 407–8.

237 **most medical conditions:** Michael Smolensky, "Diurnal and Twenty-four Hour Patterning of Human Diseases: Cardiac, Vascular, and Respiratory Diseases, Conditions, and Syndromes," *Sleep Medicine Reviews* 21 (2015): 3–11; Michael Smolensky, "Diurnal and Twenty-four Hour Patterning of Human Disease: Acute and Chronic Common and Uncommon Medical Conditions," *Sleep Medicine Reviews* 21 (2015): 12–22.

237 **how we'll respond:** Franz Halberg et al., "From Biological Rhythms to Chronomes Relevant for Nutrition," in *Not Eating Enough: Overcoming Underconsumption of Military Operational Rations*, ed. Bernadette Marriott (Washington, DC: National Academy Press, 1995), 361–72.

237 **even seasonal changes:** Foster and Roenneberg, "Human Responses."

238 **"slaves to the Sun":** Nicola Davis and Ian Sample, "Nobel Prize for Medicine Awarded for Insights into Internal Biological Clock," *The Guardian*, October 2, 2017, https://www.theguardian.com/science/2017/oct/02/nobel-prize-for-medicine-awarded-for-insights-into-internal-biological-clock.

238 **a Paleolithic figure:** Helen Benigni, "The Emergence of the Goddess: A Study of Venus in the Paleolithic and Neolithic Era," in *The Mythology of Venus* (Lanham, MD: University Press of America, 2013), ch. 1; Jules Cashford, *The Moon: Myth and Image* (London: Cassell, 2003), 20–21.

238 **passage of time long before:** Eliade, *Patterns in Comparative Religion*, 155; quoted in Cashford, *The Moon*, 22.

238 **The word for Moon:** Eliade, *Patterns in Comparative Religion*, 155.

239 **nine lunar months:** Walter Menaker and Abraham Menaker, "Lunar Periodicity in Human Reproduction: A Likely Unit of Biological Time," *American Journal of Obstetrics & Gynecology* 77 (1959): 905–14.

239 **"gave birth in the mind":** Joseph Campbell, "Mythologies of the Primitive Hunters and Gatherers," Part One of *The Way of the Animal Powers*, vol. 1 of *Historical Atlas of World Mythology* (New York: Perennial Library, 1988), 68; quoted in Benigni, "Emergence of the Goddess." A similar idea is discussed in Eliade, *Patterns in Comparative Religion*: "The moon measures, but it also unifies . . . The whole universe is seen as a pattern, subject to certain laws."

239 **"heavenly body above all others":** Eliade, *Patterns in Comparative Religion*, 154.

239 **"penetrating all things":** Pliny the Elder, *Natural History*, book 2, ch. 102.

239 **"brains in rabbits, woodcocks, calves":** Francis Bacon, *Sylva Sylvarum: A Natural History, in Ten Centuries* 9 (1627): 892.

239 **anthropologists have reported:** D. Kelley, "Mania and the Moon," *Psychoanalytic Review* 29 (1942): 406–26.

239 **from India to France:** Cashford, *The Moon*, foreword.

239 **hunt by the light of the Moon:** Noga Kronfeld-Schor, "Chronobiology by Moonlight," *Proceedings of the Royal Society B* 280 (2013): 20123088.

240 **the movement instead switches:** K. Last, "Moonlight Drives Ocean Scale Mass Vertical Migration of Zooplankton during the Arctic Winter," *Current Biology* 26 (2016): 244–51.

240 **caught sea urchins:** Harold Fox, "Lunar Periodicity in Reproduction," *Proceedings of the Royal Society of London B* 95 (1923): 523–50.

240 **shed their reproductive organs:** Alain Reinberg et al., "The Full Moon as a Synchronizer of Circa Monthly Biology Rhythms: Chonobiologic Perspectives Based on Multidisciplinary Naturalistic Research," *Journal of Biological and Medical Rhythm Research* 33 (2016): 465–79.

240 **sea comes alive:** Matthew Oldach et al., "Transcriptome Dynamics over a Lunar Month in a Broadcast Spawning Acroporid Coral," *Molecular Ecology* 26 (2017): 2514–26; Jackie Wolstenholme et al., "Timing of Mass Spawning in Corals: Potential Influence of the Coincidence of Lunar Factors and Associated Changes in Atmospheric Pressure from Northern and Southern Hemisphere Case Studies," *Invertebrate Reproduction & Development* 62 (2018): 98–108.

241 **Spring tides are vital:** Kronfeld-Schor, "Chronobiology by Moonlight"; Reinberg et al., "The Full Moon." A general review of marine species and the tides/moon: Martin Bulla et al., "Marine biorhythms: Bridging

Chronobiology and Ecology," *Philosophical Transactions of the Royal Society B* 372 (2017): 20160253.

241 **Japanese land crabs:** Kronfeld-Schor, "Chronobiology by Moonlight."

241 **follow the Moon:** Kronfeld-Schor, "Chronobiology by Moonlight"; Noga Kronfeld-Schor et al., "Chronobiology of Interspecific Interactions in a Changing World," *Philosophical Transactions of the Royal Society B* 372 (2017): 20160248.

241 **first example of a plant:** Catarina Rydin and Kristina Bolinder, "Moonlight Pollination in the Gymnosperm Ephedra (Gnetales)," *Biology Letters* 11 (2015): 20140993.

241 **"glitter like diamonds":** Andy Coghlan, "Werewolf Plant Waits for the Light of the Full Moon," *New Scientist*, April 1, 2015, https://www.newscientist.com/article/dn27277.

241 **complain it can be difficult:** Reinberg et al., "The Full Moon."

241 **a Tunisian biologist:** Reinberg et al., "The Full Moon."

242 **first molecular studies:** Juliane Zantke et al., "Genetic and Genomic Tools for the Marine Annelid *Platynereis dumerilii*," *Genetics* 197 (2014): 19–31; Masato Fukushiro, "Lunar Phase-dependent Expression of Cryptochrome and a Photoperiodic Mechanism for Lunar Phase Recognition in a Reef Fish, Goldlined Spinefoot," *PLoS ONE* 6 (2011): e28643; Florian Raible et al., "An Overview of Monthly Rhythms and Clocks," *Frontiers in Neurology* 8 (2017): 189.

242 ***Acropora gemmifera* coral:** Oldach et al., "Transcriptome Dynamics."

242 **sensitive to lunar light:** For example, Maxim Gorbunov and Paul Falkowski, "Photoreceptors in the Cnidarian Hosts Allow Symbiotic Corals to Sense Blue Moonlight," *Limnology and Oceanography* 47 (2002): 309–15.

242 **Arnold Lieber, who claimed:** Arnold Lieber, *The Lunar Effect: Biological Tides and Human Emotions* (New York: Doubleday, 1978).

242 **the term "lunatic":** Raible et al., "Monthly Rhythms and Clocks."

242 **furious backlash:** James Rotton and Ivan Kelly, "Much Ado about the Full Moon: A Meta-analysis of Lunar-lunacy Research," *Psychological Bulletin* 97 (1985): 286–306; Daniel Myers, "Gravitational Effects of the Period of High Tides and the New Moon on Lunacy," *Journal of Emergency Medicine* 13 (1995): 529–32; Foster and Roenneberg, "Human Responses"; Hal Arkowitz and Scott Lilienfeld, "Lunacy and the Full Moon: Does a Full Moon Really Trigger Strange Behavior?," *Scientific American*, February 1, 2009, https://www.scientificamerican.com/article/lunacy-and-the-full-moon/; "Full Moon and Lunar Effects," Skepdic, accessed November 5, 2019, http://skepdic.com/fullmoon.html.

243 **"paranormal":** Armando Simón, "No Effect of the Full Moon-Supermoon on the Aggressive Behavior of Incarcerated Convicts: Nailing the Coffin Shut on the Transylvania Effect," *Biological Rhythm Research* 49 (2018): 165–68.

243 **a blind spot:** Reinberg et al., "The Full Moon."

243 **must date from very early:** Raible et al., "Monthly Rhythms and Clocks."

243 **Trials investigating:** Raible et al., "Monthly Rhythms and Clocks."

243 **men have monthly hormone cycles:** Natalia Rakova et al., “Long-term Space Flight Simulation Reveals Infradian Rhythmicity in Human Na+ Balance,” *Cell Metabolism* 17 (2013): 125–31.

243 **disrupting fertility:** Reinberg et al., “The Full Moon.”

243 **sleep quality varies:** Christian Cajochen et al., “Evidence That the Lunar Cycle Influences Human Sleep,” *Current Biology* 23 (2013): 1485–88; Michael Smith et al., “Human Sleep and Cortical Reactivity Are Influenced by Lunar Phase,” *Current Biology* 24 (2014): R551–52; Ciro Della Monica et al., “Effects of Lunar Phase on Sleep in Men and Women in Surrey,” *Journal of Sleep Research* 24 (2015): 687–94.

244 **seizures and epileptic fits:** Stephan Rüegg et al., “Association of Environmental Factors with the Onset of Status Epilepticus,” *Epilepsy & Behavior* 12 (2008): 66–73.

244 **tracked patients with bipolar disorder:** Thomas Wehr, “Bipolar Mood Cycles and Lunar Tidal Cycles,” *Molecular Psychiatry* 23 (2018): 923–31; Thomas Wehr, “Bipolar Mood Cycles Associated with Lunar Entrainment of a Circadian Rhythm,” *Translational Psychiatry* 8 (2018): 151. See also Tânia Abreu and Miguel Bragança, “The Bipolarity of Light and Dark: A Review on Bipolar Disorder and Circadian Cycles,” *Journal of Affective Disorders* 185 (2015): 219–29; Thomas Erren and Philip Lewis, “Hypothesis: Folklore Perpetuated Expression of Moon-Associated Bipolar Disorders in Anecdotally Exaggerated Werewolf Guise,” *Medical Hypotheses* 122 (2019): 129–33.

245 **creepings of 34,000 snails:** Brown, “Pervasive Geophysical Factors”; Brown, “Environmental Timing.”

245 **24-hour ripple:** Nicholas Pedatella and Jeffrey Forbes, “Global Structure of the Lunar Tide in Ionospheric Total Electron Content,” *Geophysical Research Letters* 37 (2010): L06013; Adrian Hitchman and Ted Lilley, “The Quiet Daily Variation in the Total Magnetic Field: Global Curves,” *Geophysical Research Letters* 25 (1998): 2007–10; James Gould, “Magnetoreception,” *Current Biology* 20 (2010): R431–35.

246 **Earth’s geomagnetic field:** Hitchman and Lilley, “Global Curves.”

246 **some fish are sensitive:** Hans Lissman and Kenneth Machin, “The Mechanism of Object Location in *Gymnarchus niloticus* and Similar Fish,” *Journal of Experimental Biology* 35 (1958): 451–86.

246 **presented his results:** Brown, “Pervasive Geophysical Factors.”

246 **“We cannot yet explain”:** Brown, “Pervasive Geophysical Factors.”

247 **more than eighty volunteers:** Rütger Wever, “The Effects of Electric Fields on Circadian Rhythmicity in Men,” *Life Sciences and Space Research* 8 (1970): 177–87; Rütger Wever, “Human Circadian Rhythms under the Influence of Weak Electric Fields and the Different Aspects of These Studies,” *International Journal of Biometeorology* 17 (1973): 227–32; Rütger Wever, “ELF Effects on Human Circadian Rhythms,” in *ELF and VLF Electromagnetic Field Effects*, ed. Michael Persinger (New York: Plenum Press, 1974), 101–44.

247 **a “remarkable” result:** Wever, “Human Circadian Rhythms.”

247–248 **didn’t even mention the shielding:** For example: Aschoff et al., “Desynchronization of Human Circadian Rhythms”; Jürgen Aschoff,

"Circadian Rhythms"; Jürgen Aschoff, "Temporal Orientation: Circadian Clocks in Animals and Humans," *Animal Behaviour* 37 (1989): 881–96. Also see Aschoff's obituary in *Nature*, which says Aschoff built "an underground 'bunker'": Daan and Gwinner, "Obituary."

248 **millions of monarch butterflies:** Patrick Guerra et al., "A Magnetic Compass Aids Monarch Butterfly Migration," *Nature Communications* 5 (2014): 4164; Gregory Nordmann et al., "Magnetoreception: A Sense without a Receptor," *PLoS Biology* 15 (2017): e20032324.

248 **many species are expert:** James Foster, "How Animals Follow the Stars," *Proceedings of the Royal Society B* 285 (2018): 20172322.

248 **some animals detect patterns:** James Foster et al., "Orienting to Polarized Light at Night—Matching Lunar Skylight to Performance in a Nocturnal Beetle," *Journal of Experimental Biology* 222 (2019): jeb188532.

248 **Viking sailors are thought:** Jo Marchant, "Did Vikings Navigate by Polarized Light?" *Scientific American*, January 31, 2011, https://www.scientificamerican.com/article/did-vikings-navigate/.

249 **Wolfgang Wiltschko showed:** Wolfgang Wiltschko and Roswitha Wiltschko, "Magnetic Compass of European Robins," *Science* 176 (1972): 62–64.

249 **flood of evidence:** This "sea change" in attitudes is discussed in James Gould, "Animal Navigation: The Evolution of Magnetic Orientation," *Current Biology* 18 (2008): R482–84. See also Sönke Johnsen and Kenneth Lohmann, "Magnetoreception in Animals," *Physics Today* 61 (March 2008): 29–35.

249 **point themselves north or south:** Vlastimil Hart et al., "Dogs Are Sensitive to Small Variations of the Earth's Magnetic Field," *Frontiers in Zoology* 10 (2013): 80.

249 **an electrical solution:** General sources describing the three magnetoreception strategies: Gould, "Animal Navigation"; Gould, "Magnetoreception"; Henrik Mouritsen, "Long-distance Navigation and Magnetoreception in Migratory Animals," *Nature* 558 (2018): 50–59.

249 **bacteria that use chains:** Richard Blakemore, "Magnetotactic Bacteria," *Science* 190 (1975): 377–79.

249 **crystals could enable magnetosensing:** Veronika Lambinet, "Linking Magnetite in the Abdomen of Honey Bees to a Magnetoreceptive Function," *Proceedings of the Royal Society B* 284 (2017): 20162873; Nordmann et al., "Magnetoreception."

250 **"I thought, well, maybe":** Dan Cossins, "A Sense of Mystery," *The Scientist*, August 1, 2013, https://www.the-scientist.com/cover-story/a-sense-of-mystery-38949.

250 **discovered proteins called cryptochromes:** Ilia Solov'yov et al., "Magnetic Field Effects in *Arabidopsis thaliana* Cryptochrome-1," *Biophysical Journal* 92 (2007): 2711–26.

250 **cryptochromes are indeed involved:** Robert Gegear et al., "Animal Cryptochromes Mediate Magnetoreception by an Unconventional Photochemical Mechanism," *Nature* 463 (2010): 804–7.

250 **"see" magnetic field lines:** Peter Hore and Henrik Mouritsen, "The Radical-pair Mechanism of Magnetoreception," *Annual Review of Biophysics* 45 (2016): 299–344.

250 **human cryptochrome protein into fruit flies:** Lauren Foley et al., “Human Cryptochrome Exhibits Light-dependent Magnetosensitivity,” *Nature Communications* 2 (2011): 356.

251 **They are also crucial components:** Taishi Yoshii et al., “Cryptochrome Mediates Light-dependent Magnetosensitivity of Drosophila’s Circadian Clock,” *PLoS Biology* 7 (2009): 813–19; Thorsten Ritz et al., “Cryptochrome—a Photoreceptor with the Properties of a Magnetoreceptor?,” *Communicative & Integrative Biology* 3 (2010): 24–27.

251 **quintessential “gear”:** Gegear, “Animal Cryptochromes.”

251 **Cryptochromes are also among:** Fukushiro, “Lunar Phase-dependent Expression”; Oldach et al., “Transcriptome Dynamics.”

251 **tell the time using:** James Gould, “The Case for Magnetic Sensitivity in Birds and Bees (such as It Is),” *American Scientist* 68 (1980): 256–67; Gould, “Animal Navigation”; Gould, “Magnetoreception”; Thomas Erren et al., “What if . . . the Moon Provides Zeitgeber Signals to Humans?,” *Molecular Psychiatry* (August 2018), DOI: 10.1038/s41380-018-0216-0.

251 **a question posed by Brown:** Foster and Kreitzman, *Rhythms of Life*, 48; Christian Hong et al., “A Proposal for Robust Temperature Compensation of Circadian Rhythms,” *PNAS* 104 (2007): 1195–1200; Philip Kidd et al., “Temperature Compensation and Temperature Sensation in the Circadian Clock,” *PNAS* 112 (2015): E6284–92; Yoshiki Tsuchiya et al., “Effect of Multiple Clock Gene Ablations on Period Length and Temperature Compensation in Mammalian Cells,” *Journal of Biological Rhythms* 31 (2016): 48–56; Rajesh Narasimamurthy and David Virshup, “Molecular Mechanisms Regulating Temperature Compensation of the Circadian Clock,” *Frontiers in Neurology* 8 (2017): 161; Lili Wu et al., “Robust Network Topologies for Generating Oscillations with Temperature-independent Periods,” *PLoS ONE* 12 (2017): e0171263.

251 **the fundamental “tick”:** James Close, “The Compass within the Clock—Part 1: The Hypothesis of Magnetic Fields as Secondary Zeitgebers to the Circadian System—Logical and Scientific Objects,” *Hypothesis* 12 (2014): e1; James Close, “The Compass within the Clock—Part 2: Does Cryptochrome Radical-pair Based Signaling Contribute to the Temperature-robustness of Circadian Systems?” *Hypothesis* 12 (2014): e2; Yoshii et al., “Cryptochrome Mediates.”

252 **“existence of sympathies”:** “Magic, Witchcraft and Animal Magnetism,” *Journal of Psychological Medicine and Mental Pathology* 5 (1852): 292–322.

252 **The note “Rx”:** “Magic, Witchcraft and Animal Magnetism”; Mohammad Qayyum et al., “Medical Aspects Taken for Granted,” *McGill Journal of Medicine* 10 (2007): 47–49; Bob Zebroski, *A Brief History of Pharmacy: Humanity’s Search for Wellness* (New York: Routledge, 2016); Otto Wall, *The Prescription: Therapeutically, Pharmaceutically, Grammatically and Historically Considered* (St. Louis, MO: A. Gast, 1898), 200–26. (Although not everyone agrees: see George Griffenhagen, “Signs and Signboards of the Pharmacy,” *Pharmacy in History* 32 [1990]: 12–21.)

253 **Nighttime light pollution:** Fabio Falchi et al., “The New World Atlas of Artificial Night Sky Brightness,” *Science Advances* 2 (2016): e1600377.

253 **disrupting bird and turtle migration:** Kronfeld-Schor, "Interspecific Interactions"; Aisling Irwin, "The Dark Side of Light: How Artificial Lighting Is Harming the Natural World," *Nature* 553 (2018): 268–70.

253 **a global threat:** Thomas Davies and Tim Smyth, "Why Artificial Light at Night Should Be a Focus for Global Change Research in the 21st Century," *Global Change Biology* 24 (2018): 872–82.

253 **disrupt clocks and compasses:** Svenja Engels et al., "Anthropogenic Electromagnetic Noise Disrupts Magnetic Compass Orientation in a Migratory Bird," *Nature* 509 (2014): 353–56; Alfonso Balmori, "Anthropogenic Radiofrequency Electromagnetic Fields as an Emerging Threat to Wildlife Orientation," *Science of the Total Environment* 518/519 (2015): 58–60; Lukas Landler and David Keays, "Cryptochrome: The Magnetosensor with a Sinister Side?" *PLoS Biology* 16 (2018): e3000018; Rachel Sherrard et al., "Low-intensity Electromagnetic Fields Induce Human Cryptochrome to Modulate Intracellular Reactive Oxygen Species," *PLoS Biology* 16 (2018): e2006229; Premysl Bartos et al., "Weak Radiofrequency Fields Affect the Insect Circadian Clock," *Journal of the Royal Society Interface* 16 (2019): 20190285.

253 **"no clear boundary":** Frank Brown Jr., "The 'Clocks' Timing Biological Rhythms," *American Scientist* 60 (1972): 756–66.

253 **"The organism and its physical":** Frank Brown Jr., "Biological Clocks: Endogenous Cycles Synchronized by Subtle Geophysical Rhythms," *BioSystems* 8 (1976): 67–81.

253 **"a rambling talk . . . seriously claimed":** Chandrashekaran, "Biological Rhythms Research."

253 **died by the sea:** Marguerite Webb, "In Memoriam: Professor Frank A. Brown, Jr.," *Journal of Interdisciplinary Cycle Research* 15 (1984): 1–2.

254 **sea slugs orient:** Kenneth Lohmann and Arthur Willows, "Lunar-modulated Geomagnetic Orientation by a Marine Mollusk," *Science* 235 (1987): 331–34.

254 **Italian botanists found:** Monica Gagliano et al., "Acoustic and Magnetic Communication in Plants: Is It Possible?," *Plant Signaling and Behavior* 7 (2012): 1346–48.

254 **"One of the most obvious":** Gould, "Animal Navigation."

254 **wobbles of our solar system:** Mikhail Medvedev and Adrian Melott, "Do Extragalactic Cosmic Rays Induce Cycles in Fossil Diversity?," *Astrophysical Journal* 664 (2007): 879–89. See also the suggestion that radiation from a supernova wiped coastal ocean animals off the planet 2.6 million years ago in Adrian Melott et al., "Hypothesis: Muon Radiation Dose and Marine Megafaunal Extinction at the End-Pliocene Supernova," *Astrobiology* 19 (2019), DOI: 10.1089/ast.2018.1902.

255 **tidal effects from the planets:** Ching-Cheh Hung, "Apparent Relations between Solar Activity and Solar Tides Caused by the Planets," NASA technical report (Cleveland, OH: Glenn Research Center, July 2007), https://ntrs.nasa.gov/archive/nasa/casi.ntrs.nasa.gov/20070025111.pdf.

255 **"The universe is one . . . will be explored":** Brown, "Biological Clocks and Rhythms."

CHAPTER 11: ALIENS

257 **a dark green spot:** On Score's discovery of ALH84001: C. Meyer, "ALH84001," Martian Meteorite Compendium, https://curator.jsc.nasa.gov/antmet/mmc/alh84001.pdf; Paul Recer, "Even in '84, Geologist Thought Rock Was 'Special,'" Associated Press, August 22, 1996, https://www.apnews.com/6a50bda0d4f33fcee7f8cb84935630d4; Scott Sandford, "The 1984–1985 Antarctic Search for Meteorites (ANSMET) Field Program," *Smithsonian Contributions to the Earth Sciences* 30 (1986): 5–9; Kathy Sawyer, *The Rock from Mars: A Detective Story on Two Planets* (New York: Random House, 2006), 3–21.

258 **"fell out of the air":** Aristotle, *Meteorology*, part 7, trans. E. W. Webster, http://classics.mit.edu/Aristotle/meteorology.1.i.html.

258 **thousands of tons:** Samantha Mathewson, "How Often Do Meteorites Hit the Earth?," Space.com, August 10, 2016, https://www.space.com/33695-thousands-meteorites-litter-earth-unpredictable-collisions.html; Nancy Atkinson, "Getting a Handle on How Much Cosmic Dust Hits Earth," *Universe Today*, March 30, 2012, https://www.universetoday.com/94392/getting-a-handle-on-how-much-cosmic-dust-hits-earth/.

259 **"Yowza-yowza":** Meyer, "ALH84001," 1.

259 **The answers people give:** On the history of beliefs about extraterrestrial life: Michael Crowe, "A History of the Extraterrestrial Life Debate," *Zygon* 32 (1997): 147–62; Steven Dick, "The Twentieth Century History of the Extraterrestrial Life Debate: Major Themes and Lessons Learned," in *Astrobiology, History, and Society*, ed. Douglas Vakoch (New York: Springer, 2013).

260 **"only populated world":** Quoted in John Traphagan, "Science and the Emergence of SETI," in *Science, Culture and the Search for Life on Other Worlds* (New York: Springer, 2016), 41.

260 **earliest-known story:** Lucian, "A True Story," in *Selected Satires of Lucian*, ed. Lionel Casson (New York: Routledge, 2017).

260 **"changed our Earth":** Crowe, "A History."

261 **"innumerable" Suns and Earths:** Giordano Bruno, *On the Infinite Universe and Worlds*, http://www.faculty.umb.edu/gary_zabel/Courses/Parallel%20Universes/Texts/On%20the%20Infinite%20Universe%20and%20Worlds.htm

261 **plants and serpentlike monsters:** Johannes Kepler, *Somnium: The Dream, or Posthumous Work on Lunar Astronomy*, trans. Edward Rosen (New York: Dover, 2003).

261 **how organisms might be adapted:** Christiaan Huygens, *The Celestial Worlds Discover'd* (London, 1698).

261 **"era of the extraterrestrials":** Crowe, "A History."

261 **humanity's supposed intelligence:** Voltaire, "Micromégas," in *Romances, Tales, and Smaller Pieces of M. de Voltaire*, vol. 1 (London, 1752), 121–150, https://publicdomainreview.org/collections/micromegas-by-voltaire-1752/.

261–262 **"instance of arrogance!!":** Michael Crowe, *The Extraterrestrial Life Debate 1750–1900* (New York: Dover, 1999), 271; quoted in Iris Fry, "The Philosophy of Astrobiology: The Copernican and Darwinian Philosophical Presuppositions," in *The Impact of Discovering Life beyond Earth*, ed. Steven Dick (Cambridge, UK: Cambridge University Press, 2015), 23–37.

262 **"The supreme end":** Alfred Russel Wallace, "Man's Place in the Universe," *The Independent* 55 (1903): 473–83; quoted in Dick, "Twentieth Century History."

262 **forests and circular buildings:** Michael Crowe, "William and John Herschel's Quest for Extraterrestrial Intelligent Life," in *The Scientific Legacy of William Herschel*, ed. Clifford Cunningham (New York: Springer, 2017), 239–74.

262 **a great walled lunar city:** Richard Baum, "The Man Who Found a City in the Moon," *Journal of the British Astronomical Association* 102 (1992): 157–59.

262 **fictional yet widely believed:** Discussed in Crowe, "A History"; Dick, "Twentieth Century History."

262 **life abounds in the solar system:** Crowe, "William and John Herschel."

262 **"The Humanities of the heavens":** Camille Flammarion, *Les terres du ciel* (Paris, 1877), 594.

262 **inspired by Flammarion's books:** Alessandro Manara and A. Wolter, "Mars in the Schiaparelli-Lowell Letters," *Memorie della Società Astronomica Italiana* 82 (2011): 276–79.

263 **"The possibility of life":** Shostak, "Current Approaches to Finding Life beyond Earth," in Dick, *Life Beyond Earth*, 11.

263 **no detectable organics:** Results and implications discussed in Dick, "Twentieth Century History."

263 **"the most promising habitat":** Norman Horowitz, *To Utopia and Back: The Search for Life in the Solar System* (New York: W. H. Freeman, 1986), 46; quoted in Dick, "Twentieth Century History."

264 **"almost a miracle":** Francis Crick, *Life Itself: Its Origin and Nature* (New York: Simon & Schuster, 1982).

264 **life didn't begin on Earth:** Francis Crick and Leslie Orgel, "Directed Panspermia," *Icarus* 19 (1973): 341–46.

264 **"an extraordinary accident":** Euan Nisbet, *The Young Earth: An Introduction to Archaean Geology* (Boston: Allen & Unwin, 1987), 353.

264 **"Man at last knows":** Jacques Monod, *Chance and Necessity: An Essay on the Natural Philosophy of Modern Biology* (New York: Knopf, 1971), 180. A later dissenter, as opinions were beginning to shift, was the biologist Christian de Duve, in his book *Vital Dust: Life as a Cosmic Imperative* (New York: HarperCollins, 1995).

264 **"looking for fairies":** Paul Davies, "Searching for a Shadow Biosphere on Earth as a Test of the 'Cosmic Imperative,'" *Philosophical Transactions of the Royal Society A* 369 (2011): 624–32.

264 **never intended to join:** On Mittlefehldt's role in identifying ALH84001 as Martian: Sawyer, *Rock from Mars*, 46–58; Monica Grady and Ian Wright, "Martians Come out of the Closet," *Nature* 369 (1994): 356.

265 **Score was thrilled:** Sawyer, *Rock from Mars*, 59.

265 **formed from volcanic lava:** David McKay et al., "Search for Past Life on Mars: Possible Relic Biogenic Activity in Martian Meteorite ALH84001," *Science* 273 (1996): 924–30; Meyer, "ALH84001."

266 **Working with NASA geochemist:** Romanek and Gibson's studies of the carbonate grains, and their discovery of the worm-like shapes, is described in Sawyer, *Rock from Mars*, 61–82.

267 **"I kind of thought":** Michael Schirber, "Meteorite-based Debate over Martian Life Is Far from Over," Space.com, October 21, 2010, https://www.space.com/9366-meteorite-based-debate-martian-life.html. See also: Sawyer, *Rock from Mars*, 83–114.

267 **Didier Queloz was wondering:** On Queloz and Mayor's discovery of 51 Pegasi b: Author's telephone interview with Didier Queloz, June 12, 2019; Michel Mayor and Didier Queloz, "A Jupiter-mass Companion to a Solar-type Star," *Nature* 378 (1995): 355–59; Michel Mayor and Davide Cenadelli, "Exoplanets—the Beginning of a New Era in Astrophysics," *European Physical Journal H* 43 (2018): 1–41; Kevin Fong, "Life Changers: Didier Queloz," BBC World Service, first broadcast September 21, 2015, https://www.bbc.co.uk/programmes/p032k6jq.

268 **orbiting a pulsar:** Aleksander Wolszczan and Dale Frail, "A Planetary System around the Millisecond Pulsar PSR1257 + 12," *Nature* 355 (1992): 145–47.

269 **"a spiritual moment":** Michel Mayor and Pierre-Yves Frei, *New Worlds in the Cosmos: The Discovery of Exoplanets*, trans. Boud Roukema (Cambridge, UK: Cambridge University Press, 2003), 18.

269 **"your wonderful discovery":** Mayor and Frei, *New Worlds*, 22.

270 **"Other worlds are no longer":** John Wilford, "In a Golden Age of Discovery, Faraway Worlds Beckon," *New York Times*, February 9, 1997, https://www.nytimes.com/1997/02/09/us/in-a-golden-age-of-discovery-faraway-worlds-beckon.html.

270 **a visit from UCLA's William Schopf:** Schopf's visit and the collaboration with Richard Zare is described in Sawyer, *Rock from Mars*, 91–106.

271 **"gut-fluttering dread":** The prepublication scrutiny, including Dan Goldin's involvement, is described in Sawyer, *Rock from Mars*, 110–52.

271 **"rock 84001 speaks to us":** The White House, press release, "President Clinton Statement Regarding Mars Meteorite Discovery," August 7, 1996, https://www2.jpl.nasa.gov/snc/clinton.html.

271 **packed main auditorium:** Leonard David, "Remembering a Big Scoop about a Small Rock," *Space News*, September 12, 2016, http://www.spacenewsmag.com/feature/remembering-a-big-scoop-%E2%80%A8about-a-small-rock/ (includes a photo of the press conference); Sawyer, *Rock from Mars*, 153–68.

272 **the *Science* paper:** McKay, "Search for Past Life."

272 **"the biggest insult":** Sawyer, *Rock from Mars*, 166.

272 **"half-baked work . . . turd-like shapes":** Sawyer, *Rock from Mars*, 177–78.

273 **The debate shifted:** Low-temperature formation: Benjamin Weiss et al., "A Low Temperature Transfer of ALH84001 from Mars to Earth," *Science* 290

(2000): 791–95. For a skeptics' view, see Allan Treiman, "Traces of Ancient Martian Life in Meteorite ALH84001: An Outline of Status in Late 2003," NASA Office of Planetary Protection, https://planetaryprotection.nasa.gov/summary/alh84001.

273 **did form on Mars:** John Bridges et al., "Carbonates on Mars," in Justin Filiberto, *Volatiles in the Martian Crust* (Cambridge, MA: Elsevier, 2018), 89–118.

273 **interpretation of those results:** Everett Gibson et al., "Martian biosignatures: Tantalizing Evidence within Martian Meteorites," Biosignature Preservation and Detection in Mars Analog Environments, proceedings of a conference held May 16–18 2016, Lake Tahoe, Nevada, LPI Contribution No. 1912, id 2052, https://ui.adsabs.harvard.edu/abs/2016LPICo1912.2052G/abstract.

273 **a detailed study:** Kathie Thomas-Keprta et al., "Elongated Prismatic Magnetite Crystals in ALH84001 Carbonate Globules: Potential Martian Magnetofossils," *Geochimica et Cosmochimica Acta* 64 (2000): 4049–81; see also Kathie Thomas-Keprta et al., "Magnetofossils from Ancient Mars: A Robust Biosignature in the Martian Meteorite ALH84001," *Applied and Environmental Microbiology* 68 (2002): 3663–72.

273 **linear chains:** Imre Friedmann, "Chains of Magnetite Crystals in the Meteorite ALH84001: Evidence of Biological Origin," *PNAS* 98 (2001): 2176–81.

273 **critics proposed:** D. Golden et al., "Evidence for Exclusively Inorganic Formation of Magnetite in Martian Meteorite ALH84001," *American Mineralogist* 89 (2004): 681–95.

273 **she hit back:** Kathie Thomas-Keprta et al., "Origin of Magnetite Nanocrystals in Martian Meteorite ALH84001," *Geochimica et Cosmochimica Acta* 73 (2009): 6631–77; Michael Schirber, "Meteorite-based Debate over Martian Life Is Far from Over," Space.com, October 21, 2010, https://www.space.com/9366-meteorite-based-debate-martian-life.html.

274 **other Martian meteorites:** David McKay et al., "Life on Mars: New evidence from Martian Meteorites," *Proceedings Vol. 7441, Instruments and Methods for Astrobiology and Planetary Missions XII* (2009): 744102; Richard Kerr, "New Signs of Ancient Life in Another Martian Meteorite?," *Science* 311 (2006): 1858–59; Lauren White et al., "Putative Indigenous Carbon-bearing Alteration Features in Martian Meteorite Yamato 000593," *Astrobiology* 14 (2014): 170–81.

274 **"continue to support":** Everett Gibson et al., "Position Paper: Significance of the ALH84001 Research and 1996 Science Manuscript." This statement was emailed to me in June 2019, and is an updated version of an April 2017 statement published online: https://www.eatsleepshopplay.com/ekg-4; email exchange with Gibson, June 2019. He added that he thinks the case for ancient life in ALH84001 is "still strong" and is bolstered by more recent data from other meteorites. See also Charles Choi, "Mars Life? 20 Years Later, Debate over Meteorite Continues," Space.com, August 10, 2016, https://www.space.com/33690-allen-hills-mars-meteorite-alien-life-20-years.html.

274 **"bottom of the pecking order":** Adam Rogers, "War of the Worlds," *Newsweek*, February 9, 1997, https://www.newsweek.com/war-worlds-174656.

275 **"faster, better, cheaper":** "Daniel Saul Goldin," National Aeronautics and Space Administration, NASA History Division, https://history.nasa.gov/dan_goldin.html.

275 **"vindication of America's space program":** "President Bill Clinton's Statement Regarding the Mars Meteorite Discovery."

275 **"the allure of the extraterrestrial":** Sawyer, *Rock from Mars*, 197. See also Mark Peplow, "Do You Believe in Life on Mars?" *Nature News*, March 8, 2005, https://www.nature.com/news/2005/050307/full/050307-9.html.

275 **agency shifted resources:** On the shift in perspective triggered by ALH84001: Leonard David, "Moon and Mars Exploration Pioneer David McKay dies at 76," Space.com, February 25, 2013, https://www.space.com/19949-david-mckay-obituary-moon-mars.html; Dick, "Twentieth Century History," 140; Gibson et al., "Position Paper."

275 **"guiding idea":** Gibson et al., "Position Paper."

276 **most prolific planet hunter:** Alexandra Witze, "NASA Retires Kepler Spacecraft after Planet Hunter Runs out of Fuel," *Nature*, October 30, 2018.

276 **using spectroscopy to probe:** McGregor Campbell, "Red Sun Rising," *New Scientist* 239 (July 2018): 39–41; Ryan MacDonald, "And Now for the Exoweather . . . ," *New Scientist* 240 (November 2018): 38–41.

276 **analysis of Kepler data:** Erik Petigura et al., "Prevalence of Earth-size Planets Orbiting Sun-like Stars," *PNAS* 110 (2013): 19273–78.

276 **habitable zone of our nearest star:** Guillem Anglada-Escudé et al., "A Terrestrial Planet Candidate in a Temperate Orbit around Proxima Centauri," *Nature* 536 (2016): 437–40.

276 **seven Earth-size planets:** Michaël Gillon et al., "Seven Temperature Terrestrial Planets around the Nearby Ultracool Dwarf Stars TRAPPIST-1," *Nature* 542 (2017): 456–60.

276 **water vapor was detected:** Angelos Tsiaras et al., "Water Vapour in the Atmosphere of the Habitable-zone Eight-Earth-mass Planet K2-18 b," *Nature Astronomy* 3 (2019): 1086–91, DOI: 10.1038/s41550-019-0878-9.

276 **8.8 billion potentially habitable:** Petigura, "Prevalence of Earth-sized Planets"; Nancy Atkinson, "22% of Sun-like Stars Have Earth-sized Planets in the Habitable Zone," *Universe Today*, November 4, 2013, https://www.universetoday.com/106121/22-of-sun-like-stars-have-earth-sized-planets-in-the-habitable-zone/. See also Michael Gowanlock and Ian Morrison, "The Habitability of Our Evolving Galaxy," in *Habitability of the Universe before Earth*, part 2, ed. Richard Gordon and Alexei Sharov (Amsterdam: Elsevier, 2018).

277 **deep-sea hydrothermal vents:** Cristina Luiggi, "The Discovery of Deep-sea Hydrothermal Vents along the Galápagos Rift in 1977 Revealed a Biological Garden of Eden," *The Scientist*, September 1, 2012, https://www.the-scientist.com/foundations/life-on-the-ocean-floor-1977-40523.

277 **"insurmountable physical and chemical barriers":** Lynn Rothschild and Rocco Mancinelli, "Life in Extreme Environments," *Nature* 409 (2001): 1092–101. Other sources on extremophiles: Mosè Rossi et al., "Meeting Review: Extremophiles 2002," *Journal of Bacteriology* 185 (2003): 3683–89; Mark Lever et al., "Evidence for Microbial Carbon and Sulfur Cycling in Deeply Buried Ridge Flank Basalt," *Science* 339 (2013): 1305–8.

277 **other potential habitats:** Shostak, "Current Approaches to Finding Life beyond Earth"; Dirk Schulze-Makuch, "The Landscape of Life," in Dick, *Life Beyond Earth*, 81–94.

278 **hitched a lift:** Paul Davies, *The Fifth Miracle: The Search for the Origin and Meaning of Life* (New York: Simon & Schuster, 1998); Steven Benner and Hyo-Joong Kim, "The Case for a Martian Origin for Earth Life," *Instruments, Methods, and Missions for Astrobiology XVII* 9606 (2015): 96060C; Chandra Wickramasinghe, "Evidence to Clinch the Theory of Extraterrestrial Life," *Journal of Astrobiology & Outreach* 3 (2015): 1000e107.

278 **organic precursors of life:** Fred Goesmann et al., "Organic Compounds on Comet 67P/Churyumov-Gerasimenko Revealed by COSAC Mass Spectrometry," *Science* 349 (2015): aab0689; Queenie Chan et al., "Organic Matter in Extraterrestrial Water-bearing Salt Crystals," *Science Advances* 4 (2018): eaao3521.

278 **can survive deep space:** For example Natalia Novikova et al., "Study of the Effects of the Outer Space Environment on Dormant Forms of Microorganisms, Fungi and Plants in the 'Expose-R' Experiment," *International Journal of Astrobiology* 14 (2015): 137–42.

278 ***Mars Global Surveyor*:** Mario Acuna et al., "Magnetic Field and Plasma Observations at Mars: Initial Results of the Mars Global Surveyor Mission," *Science* 279 (1998): 1676–80. See also B. P. Weiss et al., "Records of an Ancient Martian Magnetic Field in ALH84001," *Earth and Planetary Science Letters* 201 (2001): 449–63.

278 **retain a thicker atmosphere:** Ramses Ramirez and Robert Craddock, "The Geological and Climatological Case for a Warmer and Wetter Early Mars," *Nature Geoscience* 11 (2018): 230–37.

278 **three-billion-year-old sedimentary rocks:** Jennifer Eigenbrode, "Organic Matter Preserved in 3-billion-year-old Mudstones at Gale Crater, Mars," *Science* 360 (2018): 1096–101.

278 **more life-friendly picture:** Colin Dundas et al., "Exposed Subsurface Ice Sheets in the Martian Mid-latitudes," *Science* 359 (2018): 199–201; Anja Diez, "Liquid Water on Mars," *Science* 361 (2018): 448–49; Yoshitaka Yoshimura, "The Search for life on Mars," in *Astrobiology*, ed. Akihiko Yamagishi et al. (New York: Springer, 2019), 367–81.

279 **Methane has also:** Marco Giuranna et al., "Independent Confirmation of a Methane Spike on Mars and a Source Region East of Gale Crater," *Nature Geoscience* 12 (2019): 326–32.

279 **insist Martian biology:** Gilbert Levin and Patricia Straat, "The Case for Extant Life on Mars and Its Possible Detection by the Viking Labeled Release Experient," *Astrobiology* 16 (2016): 798–810.

279 **different chemistries have been proposed:** Chris McKay, "What Is Life—and How Do We Search for It on Other Worlds?" *PLoS Biology* 2 (2004): 1260–63; Schulze-Makuch, "Landscape of Life."

279 **orbiting a neutron star:** Schulze-Makuch, "Landscape of Life."

279 **"shadow biosphere":** Carol Cleland, "Epistemological Issues in the Study of Microbial Life: Alternative Terran Biospheres?," *Studies in History and Philosophy of Science Part C: Studies in History and Philosophy of Biological and Biomedical Sciences* 38 (2007): 847–61; Paul Davies, "Searching for a Shadow Biosphere on Earth as a Test of the 'Cosmic Imperative,'" *Proceedings of the Royal Society A* 369 (2011): 624–32.

279 **"realm of fairies and elves":** David Toomey, *Weird Life: The Search for Life That Is Very, Very Different from Our Own* (New York: W. W. Norton & Co., 2013), 34.

280 **student named Jocelyn Bell:** On the story of Bell's discovery: Jocelyn Bell Burnell, "Little Green Men, White Dwarfs or Pulsars?," *Annals of the New York Academy of Science* 302 (1977): 685–89; Alan Penny, "The SETI Episode in the 1967 Discovery of Pulsars," *European Physical Journal H* 38 (2013): 535–47.

280 **"As the chart flowed":** Bell Burnell, "Little Green Men."

281 **"Were these pulsations":** Bell Burnell, "Little Green Men."

281 **"Little Green Men":** Penny, "The SETI Episode."

281 **in a *Nature* paper:** Giuseppe Cocconi and Philip Morrison, "Searching for Interstellar Communications," *Nature* 184 (1959): 844–46.

281 **Frank Drake soon undertook:** Dick, "Twentieth Century History," 139.

281 **"excitement rose":** Penny, "The SETI Episode," 4.

282 **"the less contact you have":** Penny, "The SETI Episode."

282 **"Here was I trying to get a PhD":** Bell Burnell, "Little Green Men."

282 **"By flicking switches":** Bell Burnell, "Little Green Men."

282 **"It was very unlikely":** Bell Burnell, "Little Green Men."

282 **finally wrote up their paper:** Antony Hewish et al., "Observation of a Rapidly Pulsating Radio Source," *Nature* 217 (1968): 709–13.

283 **Listening efforts have mostly:** On the history of SETI efforts: Douglas Vakoch and Matthew Dowd, eds., *The Drake Equation: Estimating the Prevalence of Extraterrestrial Life through the Ages* (Cambridge, UK: Cambridge University Press, 2015); Linda Billings, "Astrobiology in Culture: The Search for Extraterrestrial Life as 'Science,'" *Astrobiology* 12 (2012): 966–75; Shostak, "Current Approaches to Finding Life beyond Earth."

283 **"very hazardous to reveal":** Penny, "The SETI Episode."

283 **repeated and unexplained radio bursts:** University of California, Berkeley, "Distant Galaxy Sends Out 15 High-energy Radio Bursts," press release, August 30, 2017, https://news.berkeley.edu/2017/08/30/distant-galaxy-sends-out-15-high-energy-radio-bursts/.

284 **"exotic small-scale scientific enterprise":** Michael Michaud, "Searching for Extraterrestrial Intelligence," in Dick, *Life Beyond Earth*, 295.

284 **most sophisticated civilizations:** Susan Schneider, "Alien Minds," in Dick, *Life Beyond Earth*, 189–206.

284 **see intelligence as a tool:** Mark Lupisella, "Life, Intelligence, and the Pursuit of Value in Cosmic Evolution," in Dick, *Life Beyond Earth*, 159–74.

286 **"We are trying to determine":** Sawyer, *Rock from Mars*, 188.

CHAPTER 12: MIND

287 **flown over seventy different types of aircraft:** Chris Hadfield, *An Astronaut's Guide to Life on Earth* (New York: Macmillan, 2013).

287 **"completely technically prepared":** Chris Hadfield, talk at the Royal Geographic Society, London, December 7, 2014.

287 **"attacked by raw beauty":** Hadfield, *Astronaut's Guide*, 90.

287 **"stupefying . . . six billion people":** "Chris Hadfield's Incredible Description of Spacewalking," May 5, 2013, https://www.youtube.com/watch?v=cxxTGkBuo1c.

287 **"power of the presence":** Hadfield, Royal Geographic Society.

289 **"my feet no longer touch":** Quoted in Owen Gingerich, "Was Ptolemy a Fraud?," *Quarterly Journal of the Royal Astronomical Society* 21 (1980): 253–66.

289 **"grand and spacious":** Henri-Frédéric Amiel, *Amiel's Journal*, trans. Mrs. Humphrey (Mary) Ward (London: Macmillan, 1882), https://www.gutenberg.org/files/8545/8545-h/8545-h.htm; quoted in William James, *The Varieties of Religious Experience: A Study in Human Nature* (New York: Penguin Classics, 1983), 395.

290 **scientists' first working definition:** Dacher Keltner and Jonathan Haidt, "Approaching Awe, a Moral, Spiritual, and Aesthetic Emotion," *Cognition and Emotion* 17 (2003): 297–314.

291 **results have been surprising:** Awe results summarized in: Jo Marchant, "Awesome Awe," *New Scientist*, July 29, 2017, 33–35. More creative: Alice Chirico et al., "Awe Enhances Creative Thinking: An Experimental Study," *Creativity Research Journal* 30 (2018): 123–31. Memories improved: Alexander Danvers and Michelle Shiota, "Going off Script: Effects of Awe on Memory for Script-typical and -irrelevant Narrative Detail," *Emotion* 17 (2017): 938–52. Cuts levels of cytokines: Jennifer Stellar et al., "Positive Affect and Markers of Inflammation: Discrete Positive Emotions Predict Lower Levels of Inflammatory Cytokines," *Emotion* 15 (2015): 129–33. Activates parasympathetic nervous system: Michelle Shiota et al., "Feeling Good: Autonomic Nervous System Responding in Five Positive Emotions," *Emotion* 11 (2011): 1368–78. Less worried about personal concerns and goals, more connected . . . more ethical decisions: Paul Piff et al., "Awe, the Small Self, and Prosocial Behavior," *Journal of Personality and Social Psychology* 108 (2015): 883–99. Care less about money: Libin Jiang et al., "Awe Weakens the Desire for Money," *Journal of Pacific Rim Psychology* 12 (2018): e4. Care more about the environment: Huanhuan Zhao et al., "Relation between Awe and Environmentalism: The Role of Social Dominance Orientation," *Frontiers in Psychology* 9 (2018): 2367; Yan Yang et al., "From Awe to Ecological Behavior: The Mediating Role of Connectedness to Nature," *Sustainability* 10 (2018): 2477. Feel as if they have more time: Melanie Rudd et al., "Awe Expands People's Perception of Time, Alters Decision-making, and Enhances Well-being," *Psychological Science* 23 (2012): 1130–36.

292 **signed their names smaller:** Yang Bai et al., "Awe, the Diminished Self, and Collective Engagement: Universals and Cultural Variations in the Small Self," *Journal of Personality and Social Psychology* 113 (2017): 185–209. See also Michiel van Elk et al., "'Standing in Awe': The Effects of Awe on Body Perception and the Relation with Absorption," *Collabra* 2 (2016): 4.

292 **neuroscientists in the Netherlands:** Michiel van Elk et al., "The Neural Correlates of Awe Experience: Reduced Default Mode Network Activity during Feelings of Awe," *Human Brain Mapping* 40 (2019): 3561–74.

292 **"a vanishing self":** Author's telephone interview with Dacher Keltner, May 5, 2017.

292 **natural forces such as thunderstorms:** Alice Chirico and David Yaden, "Awe: A Self-transcendent and Sometimes Transformative Emotion," in *The Function of Emotions*, ed. Heather Lench (New York: Springer, 2018), 221–33.

292 **"Beautiful, so beautiful!":** Alison George, "One Minute with . . . Yuri Gagarin," *New Scientist*, April 9, 2011, 29.

292 **"People of the world":** Yuri Gagarin, "Yuri Gagarin" (Moscow: Novisti Press, 1977), 14 and 17; quoted in Richard Roney, "Beyond War: A New Way of Thinking," in *Breakthrough: Emerging New Thinking: Soviet and Western Scholars Issue a Challenge to Build a World beyond War*, ed. Anatoly Gromyko and Martin Ellman (New York: Walker & Co., and Moscow: Novisti Press, 1988), 5.

293 **"beautiful, warm, living object":** Jim Irwin, *To Rule the Night: The Discovery Voyage of Astronaut Jim Irwin* (Philadelphia: A. J. Holman, 1973), 60.

293 **"living, breathing organism":** *Overview*, Planetary Collective (2012; documentary film), http://weareplanetary.com/overview-short-film/. See also Ron Garan, *The Orbital Perspective: Lessons in Seeing the Big Picture from a Journey of 71 Million Miles* (San Francisco: Berrett-Koehler, 2015), 52–53.

293 **"compassion and concern":** David Yaden et al., "The Overview Effect: Awe and Self-transcendent Experience in Space Flight," *Psychology of Consciousness: Theory, Research, and Practice* 3 (2016): 3.

293 **sold his SUVs:** Ben Guarino, "The Overview Effect Will Save Earth One Rich Space Tourist at a Time," Inverse, December 18, 2015, https://www.inverse.com/article/6301-overview-everview-effect-space-tourism-environmentalism-spacex-richard-garriot.

293 **"From out there on the moon":** Alex Pasternak, "The Moon-walking, Alien-hunting, Psychic Astronaut Who Got Sued by NASA," Vice, May 14, 2016, https://www.vice.com/en_us/article/aek7ez/astronaut-edgar-mitchell-outer-space-inner-space-and-aliens.

293 **"you don't see the barriers":** Frank White, *The Overview Effect: Space Exploration and Human Evolution*, 3rd ed. (Reston, VA: American Institute of Aeronautics and Astronautics, 1987), 37.

293 **named in the 1980s:** White, The *Overview Effect*.

294 **powerful example of awe:** Yaden et al., "The Overview Effect," 1–11.

294 **"inner peace":** White, *The Overview Effect*, 41.

294 **"state of grace":** Geoff Hoffman, *iPM*, BBC Radio 4, May 2013, http://www.bbc.co.uk/programmes/b01sjn9l; quoted in Nick Campion, "The Importance

of Cosmology in Culture: Contexts and Consequences," in *Trends in Modern Cosmology*, ed. Abraao Capistrano (London: InTechOpen, 2017).

294 **"with the rest of the universe":** Yasmin Tayag, "Six NASA Astronauts Describe the Moment in Space When 'Everything Changed,'" Inverse, July 20, 2019, https://www.inverse.com/article/57841-nasa-astronauts-describe-overview-effect -everything-changed.

294 **"something so much bigger":** Chris Hadfield, "How Space Travel Expands Your Mind," *Big Think*, March 23, 2018, https://bigthink.com/videos/chris-hadfield-how-space-travel-expands-your-mind.

294 **"the power of God":** John Wilford, "James B. Irwin, 61, Ex-Astronaut; Founded Religious Organization," *New York Times*, August 10, 1991, https://www.nytimes.com/1991/08/10/us/james-b-irwin-61-ex-astronaut-founded-religious-organization.html.

294 **"in some way conscious":** Edgar Mitchell, *The Way of the Explorer: An Apollo Astronaut's Journey through the Material and Mystical Worlds*, rev. ed. (Newburyport, MA: New Page Books, 2008), 16 and 74–75.

295 **William James famously collected:** James, *Religious Experience*, 395.

295 **"cosmic consciousness":** Richard Bucke, *Cosmic Consciousness: A Study in the Evolution of the Human Mind* (New York: Dutton & Co., 1901).

295 **"immemorial and universal substrate":** Aldous Huxley, *The Perennial Philosophy* (New York: Harper & Brothers, 1947); quoted in Charles Grob et al., "Use of the Classic Hallucinogen Psilocybin for the Treatment of Existential Distress Associated with Cancer," in *Psychological Aspects of Cancer*, ed. Brian Carr and Jennifer Steel (New York: Springer, 2013), 291–308.

296 **mushrooms containing the hallucinogen:** Gordon Wasson, "Seeking the Magic Mushroom," *Life*, May 13, 1957, 100–102 and 109–20.

296 **led by pharmacologist Roland Griffiths:** Roland Griffiths et al., "Psilocybin Can Occasion Mystical-type Experiences Having Substantial and Sustained Personal Meaning and Spiritual Significance," *Psychopharmacology* 187 (2006): 268–83; Roland Griffiths et al., "Mystical-type Experiences Occasioned by Psilocybin Mediate the Attribution of Personal Meaning and Spiritual Significance 14 Months Later," *Journal of Psychopharmacology* 22 (2008): 621–32.

296 **"in the void":** Frederick Barrett and Roland Griffiths, "Classic Hallucinogens and Mystical Experiences: Phenomenology and Neural Correlates," in *Behavioral Neurobiology of Psychedelic Drugs*, ed. Adam L. Halberstadt et al. (New York: Springer, 2017), 393–430.

297 **A 2016 trial:** Roland Griffiths et al., "Psilocybin Procures Substantial and Sustained Decreases in Depression and Anxiety in Patients with Life-threatening Cancer: A Randomized Double-blind Trial," *Journal of Psychopharmacology* 30 (2016): 1181–97.

297 **wrote in his journal:** Grob et al., "Use of the Classic Hallucinogen Psilocybin."

297 **"nothing to be afraid of":** Grob et al., "Use of the Classic Hallucinogen Psilocybin."

297 **psychedelics reduce activity:** On how psychedelics affect the brain: Robin Carhart-Harris et al., "Neural Correlates of the Psychedelic State as Determined by fMRI Studies with Psilocybin," *PNAS* 109 (2012): 2138–43; Samuel Turton et al., "A Qualitative Report on the Subjective Experience of Intravenous Psilocybin Administered in an fMRI Environment," *Current Drug Abuse Reviews* 7 (2014): 117–27; Enzo Tagliazucchi et al., "Increased Global Functional Connectivity Correlates with LSD-induced Ego Dissolution," *Current Biology* 26 (2016): 1043–50; Barrett and Griffiths, "Classic Hallucinogens"; Matthew Nour and Robin Carhart-Harris, "Psychedelics and the Science of Self-experience," *British Journal of Psychiatry* 210 (2017): 177–79; Michael Schartner et al., "Increased Spontaneous MEG Signal Diversity for Psychoactive Doses of Ketamine, LSD and Psilocybin," *Scientific Reports* 7 (2017): 46421; Robin Carhart-Harris, "How Do Psychedelics Work?," *Current Opinion in Psychiatry* 32 (2019): 16–21.

297 **"My feeling is that":** Author's telephone interview with Robin Carhart-Harris, May 9, 2017.

298 **loosen those chains:** Robin Carhart-Harris et al., "Psychedelics and Connectedness," *Psychopharmacology* 235 (2017): 547–50; Robin Carhart-Harris and David Nutt, "Serotonin and Brain Function: A Tale of Two Receptors," *Journal of Psychopharmacology* 3 (2017): 1091–120.

299 **"We believe that awe deprivation":** Paul Piff and Dacher Keltner, "Why Do We Experience Awe?," *New York Times*, May 22, 2015, https://www.nytimes.com/2015/05/24/opinion/sunday/why-do-we-experience-awe.html.

300 **"with anguish":** James, *Religious Experience*, 386.

300 **"my own constitution":** James, *Religious Experience*, 379 and 388.

300 **"I still tend to think":** Michael Pollan, *How to Change Your Mind: The New Science of Psychedelics* (London: Allen Lane, 2018), 414.

301 **"to be is to be perceived":** Lisa Downing, "George Berkeley," *Stanford Encyclopedia of Philosophy*, ed. Edward Zalta (Spring 2013 ed.), https://plato.stanford.edu/archives/spr2013/entries/berkeley/.

301 **James was influenced:** William James, *A Pluralistic Universe: Hibbert Lectures at Manchester College on the Present Situation in Philosophy* (1908), lecture 6.

302 **"a mechanical explanation":** Henri Bergson, *Time and Free Will: An Essay on the Immediate Data of Consciousness*, trans. F. L. Pogson (New York: Dover, 2001), 100.

302 **"Physics is mathematical":** Bertrand Russell, *An Outline of Philosophy* (New York: Routledge Classics, 2009), 171; quoted in Galen Strawson, "A Hundred Years of Consciousness: 'A Long Training in Absurdity,'" *Estudios de Filosofia* 59 (2019): 9–43. See also Arthur Eddington, *The Nature of the Physical World: The Gifford Lectures 1927* (Cambridge, UK: Cambridge University Press, 1928), 259.

302 **It seems "rather silly":** Quoted in Galen Strawson, "Realistic Monism: Why Physicalism Entails Panpsychism," *Journal of Consciousness Studies* 13 (2006): 3–31.

303 **"moment of crisis":** Max Planck, *Where Is Science Going?* (New York: AMS Press, 1977), 65; quoted in Marin, "'Mysticism' in Quantum Mechanics."

303 **"philosophical prejudice":** Albert Einstein, *The Born–Einstein Letters: Friendship, Politics, and Physics in Uncertain Times: Correspondence between Albert Einstein and Max and Hedwig Born from 1916 to 1955 with Commentaries by Max Born* (New York: Macmillan, 2005), 218; quoted in Marin, "'Mysticism' in Quantum Mechanics."

303 **"the material universe":** Erwin Schrödinger, "Interviews with Great Scientists: No. 4.—Prof. Schrödinger," *The Observer*, January 11, 1931, 15–16; quoted in Strawson, "A Hundred Years of Consciousness."

303 **"blood transfusion":** Erwin Schrödinger, *What Is Life? With Mind and Matter and Autobiographical Sketches* (Cambridge, UK: Cambridge University Press, 2012), 130; quoted in Marin, "'Mysticism' in Quantum Mechanics."

303 **The 1953 discovery:** James Watson and Francis Crick, "Molecular Structure of Nucleic Acids," *Nature* 171 (1953): 737–38.

304 **"bisected with a knife":** Steven Pinker, *How the Mind Works* (New York: W. W. Norton & Co., 1997), 64.

304 **grumbling questions:** For example: Paul Davies, *The Goldilocks Enigma: Why Is the Universe Just Right for Life?* (London: Allen Lane, 2006); Ryan Gillespie, "Cosmic Meaning, Awe and Absurdity in the Secular Age: A Critique of Religious Non-theism," *Harvard Theological Review* 111 (2018): 461–87.

305 **the "hard problem":** David Chalmers, on the hard problem of consciousness (lecture, Science of Consciousness conference, Tucson, AZ, April 12–17, 1994), https://www.youtube.com/watch?v=_lWp-6hH_6g.

305 **"an illusion":** John Horgan, "Is Consciousness Real? Philosopher Daniel Dennett Tries, Once Again, to Explain Away Consciousness," *Cross Check* (blog), *Scientific American*, March 21, 2017, https://blogs.scientificamerican .com/cross-check /is-consciousness-real/.

305 **"robot vehicles":** Richard Dawkins, *The Selfish Gene*, 40th anniversary ed. (New York: Oxford Landmark Science, 2016), xxix.

305 **"pack of neurons":** Francis Crick, *The Astonishing Hypothesis: The Scientific Search for the Soul* (New York: Simon & Schuster, 1995), 3.

305 **"chemical scum":** David Dugan, dir., *Reality on the Rocks*, 3 episodes, aired on U.K. Channel 4, February 1995; quoted in Raymond Tallis, "You Chemical Scum, You," in *Reflections of a Metaphysical Flâneur and Other Essays* (Stocksfield, UK: Acumen Publishing, 2013), ch. 9, 163–68.

305 **"The reductionist worldview":** Steven Weinberg, *Dreams of a Final Theory: The Scientist's Search for the Ultimate Laws of Nature* (New York: Vintage Books, 1994), 53.

306 **"the mechanism by which":** Ariane Sherine, ed., *There's Probably No God: The Atheist's Guide to Christmas* (London: Friday Project, 2010), 83.

306 **"poetic naturalism":** Sean Carroll, *The Big Picture: On the Origins of Life, Meaning and the Universe Itself* (London: Oneworld, 2017), 3.

306 **"nobility of being":** Brian Greene, *Until the End of Time: Mind, Matter and Our Search for Meaning in an Evolving Universe* (New York: Penguin Random House, 2020), chs. 6, 7, 8, and 11.

306 **"no special mental realm":** "The Physics of Consciousness," interview with Sean Carroll, Closer to Truth, April 2017, https://www.closertotruth.com /interviews/54817.

306 **"The math *does* rule":** Brian Greene, *Until the End of Time: Mind, Matter and Our Search for Meaning in an Evolving Universe* (New York: Penguin Random House, 2020), 149–55.

307 **"incapable of providing":** Thomas Nagel, *Mind and Cosmos: Why the Materialist, Neo-Darwinian Conception of Nature Is Almost Certainly False* (New York: Oxford University Press, 2012).

307 **"shoddy reasoning":** Andrew Ferguson, "The Heretic," *Weekly Standard*, March 25, 2013, http://www.weeklystandard.com/andrew-ferguson/the-heretic.

307 **"not worth a damn":** Michael Chorost, "Where Thomas Nagel Went Wrong," *Chronicle of Higher Education*, May 13, 2013, https://www.chronicle.com/article/Where-Thomas-Nagel-Went-Wrong/139129.

307 **such as physicist Paul Davies and biologist Stuart Kauffman:** Davies, *Goldilocks Enigma*; Paul Davies, *The Demon in the Machine: How Hidden Webs of Information Are Solving the Mystery of Life* (London: Allen Lane, 2019); Stuart Kauffman, "Beyond the Stalemate," Cornell University, arXiv.org, https://arxiv.org/abs/1410.2127v2; Stuart Kauffman, *Reinventing the Sacred: A New View of Science, Reason and Religion* (New York: Basic Books, 2008).

307 **"the most certain thing . . . I was mocked":** Author interview with Galen Strawson in Chalk Farm, London, August 13, 2019. See also Strawson, "Realistic Monism"; Galen Strawson, "Mind and Being: The Primacy of Panpsychism," in *Panpsychism: Contemporary Perspectives*, ed. Godehard Brüntrup and Ludwig Jaskolla (New York: Oxford University Press, 2016); Strawson, "A Hundred Years of Consciousness."

308 **consciousness might extend:** Colin Klein and Andrew Barron, "Insects Have the Capacity for Subjective Experience," *Animal Sentience* 9 (2016): 1–19; Sy Montgomery, *The Soul of an Octopus: A Surprising Exploration into the Wonder of Consciousness* (New York: Simon & Schuster, 2016).

308 **even plants and slime molds:** Paco Calvo, "Are Plants Sentient?," *Plant, Cell and Environment* 40 (2017): 2858–69; Chris Reid and Tania Latty, "Collective Behaviour and Swarm Intelligence in Slime Moulds," *FEMS Microbiology Reviews* 40 (2016): 798–806; Jordi Vallverdú et al., "Slime Mould: The Fundamental Mechanisms of Biological Cognition," *Biosystems* 165 (2018): 57–70.

308 **in computers and aliens:** Murray Shanahan, "AI and Consciousness," in *The Technological Singularity* (Cambridge, MA: MIT Press, 2015), 117–50; Susan Schneider, "Alien Minds," in Dick, *Life beyond Earth*, 189–206.

308 **integrated information theory:** Giulio Tononi and Christof Koch, "Consciousness: Here, There and Everywhere?," *Philosophical Transactions of the Royal Society B* 370 (2015): 20140167.

308 **"a new generation":** Quote from David Chalmers's praise for Philip Goff's 2019 book *Galileo's Error* (see next note), https://www.penguinrandomhouse.com/books/599229/galileos-error-by-philip-goff/. For increasing popular credibility of panpsychism, see Adam Frank, "Minding Matter: The Closer You Look, the More the Materialist Position in Physics Appears to Rest on Shaky Metaphysical Ground," *Aeon*, March 2017, https://aeon.co/essays/materialism-alone-cannot-explain-the-riddle-of-consciousness; Philip Goff, "Panpsychism Is Crazy, but It Is Also Most Probably True," *Aeon*, March 2017,

https://aeon.co/ideas/panpsychism-is-crazy-but-its-also-most-probably-true; Olivia Goldhill, "The Idea That Everything from Spoons to Stones Is Conscious Is Gaining Academic Credibility," *Quartz*, January 2018, https://qz.com/1184574/the-idea-that-everything-from-spoons-to-stones-are-conscious-is-gaining-academic-credibility/. Academic books on panpsychism published since Strawson's provocative 2006 paper include David Skrbina, *Panpsychism in the West* (Cambridge, MA: MIT Press, 2007); Michael Blamaeur, ed., *The Mental as Fundamental: New Perspectives on Panpsychism* (Heusenstamm, Germany: Ontos Verlag, 2011); Brüntrup and Jaskolla, *Panpsychism*.

308 ***Galileo's Error***: Philip Goff, *Galileo's Error: Foundations for a New Science of Consciousness* (New York: Penguin Random House, 2019).

309 **a great ocean:** Freya Mathews, "Panpsychism as Paradigm," in Blamauer, ed., *Mental as Fundamental*, 141–56.

309 **"riding something alive":** Freya Mathews, *Reinventing Reality: Toward a Recovery of Culture* (Sydney: University of New South Wales Press, 2005), 111.

310 **"shining with light . . . centre of things which is beyond":** Author's Skype interview with Itay Shani, July 26, 2019. See also: Itay Shani, "Cosmopsychism: A Holistic Approach to the Metaphysics of Experience," *Philosophical Papers* 44 (2015): 389–437; Itay Shani and Joachim Keppler, "Beyond Combination: How Cosmic Consciousness Grounds Ordinary Experience," *Journal of the American Philosophical Association* 4 (2018): 390–410.

311 **sees panpsychism as vital:** Freya Mathews, *For the Love of Matter: A Contemporary Panpsychism* (Albany, NY: State University of New York Press, 2003); Mathews, *Reinventing Reality.*

311 **a younger contemporary of Einstein and Bohr:** Charles Misner et al., "John Wheeler, Relativity, and Quantum Information," *Physics Today*, April 2009, 40–46.

311 **"constantly emerging":** Tim Folger, "Does the Universe Exist If We're Not Looking?," *Discover*, June 1, 2002, http://discovermagazine.com/2002/jun/featuniverse. This is Folger's description, not a direct quote from Wheeler.

311 **"symbolizes the idea":** John Wheeler, "Information, Physics, Quantum: The Search for Links," in *Complexity, Entropy and the Physics of Information*, ed. Wojciech Zurek (Boston: Addison-Wesley, 1990), 309–36.

312 **confirmed in a lab . . . repeated in a series of variants:** Delayed choice experiments reviewed in: Xiao-song Ma et al., "Delayed-choice Gedanken Experiments and Their Realizations," *Reviews of Modern Physics* 88 (2016): 015005; Andrew Manning et al., "Wheeler's Delayed-choice Gedanken Experiment with a Single Atom," *Nature Physics* 11 (2015): 539–42; Francesco Vedovato, "Extending Wheeler's Delayed-choice Experiment to Space," *Science Advances* 3 (2017): e1701180.

312 **Astronomers even have a plan:** Laurance Doyle, "Quantum Astronomy: A Cosmic-scale Double-slit Experiment," Space.com, January 13, 2005, https://www.space.com/667-quantum-astronomy-cosmic-scale-double-slit-experiment.html.

312 **"smoky dragon":** John Wheeler, "Time Today," in *Physical Origins of Time Asymmetry*, ed. J. J. Halliwell et al. (Cambridge, UK: Cambridge University Press, 1994), 19–20; quoted in Anil Ananthaswamy, "Closed Loophole Confirms the Unreality of the Quantum World," *Quanta*, July 25, 2018, https://www.quantamagazine.org/closed-loophole-confirms-the-unreality-of-the-quantum-world-20180725/.

312 **"notes struck out on a piano":** Wheeler, "Information, Physics, Quantum," 24.

313 **billions upon billions:** Christopher Fuchs, "Nothwithstanding Bohr, the Reasons for QBism," *Mind and Matter* 15 (2017): 245–300n5.

313 **"If a quantum state":** Fuchs, "Nothwithstanding Bohr."

313 **it's just as revolutionary:** Fuchs, "Nothwithstanding Bohr."

313 **"reality is *more* than":** Christopher Fuchs, "On Participatory Realism," in *Information and Interaction*, ed. Ian Durham and Dean Rickles (New York: Springer, 2017), 113–34.

313 **One of Fuchs's intellectual heroes:** Fuchs, "Nothwithstanding Bohr."

314 **"pure experience":** William James, "A World of Pure Experience," *Journal of Philosophy, Psychology, and Scientific Methods* 1 (1904), http://fair-use.org/william-james/essays-in-radical-empiricism/a-world-of-pure-experience; James says here: "The universe continually grows in quantity by new experiences that graft themselves upon the older mass. . . ." Also William James, "Does 'Consciousness' Exist?," *Journal of Philosophy, Psychology, and Scientific Methods* 1 (1904): 477–91.

314 **"local spots and patches":** William James, "Lecture VIII, Pragmatism and Religion," in *Pragmatism: A New Name for Some Old Ways of Thinking* (New York: Longman, 1907).

314 **"stuff of the world":** Amanda Gefter, "A Private View of Quantum Reality," *Quanta*, June 4, 2015, https://www.quantamagazine.org/quantum-bayesianism-explained-by-its-founder-20150604/.

315 **"We are not machines":** Chris Hadfield, "What I Learned from Going Blind in Space," TED talk, March 19, 2014, https://www.youtube.com/watch?v=Zo62S0ulqhA.

EPILOGUE

317 **In "Nightfall":** Isaac Asimov, "Nightfall," *Astounding Science Fiction*, September 1941. The story was later expanded into a novel, with the planet renamed "Kalgash": Isaac Asimov and Robert Silverberg, *Nightfall* (New York: Doubleday, 1990).

319 **medical systems:** For more on how modern medicine often focuses on physical bodies and treatments while sidelining patient experiences and psychological approaches, see Jo Marchant, *Cure: A Journey into the Science of Mind over Body* (Edinburgh: Canongate, 2016).

ACKNOWLEDGMENTS

I was inspired by the passion, ideas and dedication of all the people whose work is described in this book. So I'd like to thank them—the archaeologists, astronomers, philosophers, navigators, shamans, artists—all exploring the cosmos at different times and in different ways. I'm particularly indebted to the modern-day academics who generously gave up time to discuss their ideas with me, and to check sections of the manuscript (though responsibility for any errors is of course mine). They include Michael Rappenglück, Jens Notroff, Michael Parker Pearson, Jim Evans, Nick Campion, Jeanette Fincke, Didier Queloz, Zachory Berta-Thompson, Everett Gibson, Dacher Keltner, Robin Carhart-Harris, Galen Strawson and Itay Shani. Thanks to Chris Hadfield for the plectrums, and to Jo Bowlby and Sandra Ingerman for your patience in dealing with my strange questions.

This book wouldn't exist without my agent, Will Francis, who helped me to take a vast, sprawling idea and tame it into something recognizable as a book proposal, or my editor, Simon Thorogood, for believing in the concept early on and giving me the chance to turn this book into reality. I'm so grateful to you both for your support and advice throughout the writing process. Thanks also to the book's U.S. editor, Stephen Morrow, for valuable comments and advice on the completed manuscript, and to my lovely copy editor, Eugenie Todd, for your rigorous checks and graceful edits.

I'd also like to mention the fantastic British Library, which enabled me to access sources from centuries-old historical accounts to the latest science journals, as well as providing an inspiring place to work. Thank you to Isabel Cook for chats and insightful comments on parts of the text; to Anil Ananthaswamy for valuable feedback; to Laura Donovan—it meant the world to know that when I needed to write, the kids were in such great hands; and to my wonderful friends and writing companions Gaia Vince and Emma Young—I'm so glad we're on this journey together.

Finally, thank you to Ian, for your unerring support and encouragement, even when I couldn't explain to you what on earth I was trying to write. And all my love to gorgeous Poppy and Rufus, center of my universe. This book is for you. I hope that when you grow up, you can still see the stars.

INDEX

ABOUT THE AUTHOR

Jo Marchant is an award-winning science journalist. She has a PhD in genetics and medical microbiology, and an MSc in science communication. She has worked as a senior editor at *New Scientist* and *Nature*; her articles have also appeared in *The Guardian*, *The New York Times*, *The Economist*, and *Smithsonian Magazine*. Jo Marchant's previous book, *Cure*, hit the *New York Times* best-seller hardcover nonfiction and science lists and was licensed to more than thirty countries. She is also the author of *The Shadow King* and *Decoding the Heavens*, which, like *Cure*, was short-listed for the Royal Society Prize for Science Books.